Cool Places

•

Music, drugs, clubbing and travel – this is the popular image of what it means to be young. However, young people's lives are also bound up with the nitty-gritty everyday realities of home, school, work, and are circumscribed by globalising forces of the economy and processes of marginalisation and exclusion. *Cool Places* explores these contrasting experiences of contemporary youth.

In chapters drawing on a wide range of examples – from Techno music and ecstasy in Germany, clubbing in London, global backpacking and gangs in Santa Cruz, to experiences at home of sibling rivalry, loitering on streets and seeking employment, leading contemporary writers from Geography, Cultural Studies and Sociology explore issues of representation and resistance; and geographical concepts of scale and place in young people's lives.

In the four parts which make up this book the authors consider how the media has imagined young people as a particular community with shared interests and how young people resist these stereotypes and create their own independent representations of their lives; the complex ways that youth cultures are played out across different scales; young people's experiences of everyday geographical locations from the home to the street; and the power of young people to resist adult definitions of their lives and to create new spaces and ways of living.

By drawing on a rich vein of empirical material and through employing young people's vignettes, *Cool Places* aims to place youth and youth cultures on the geographic map and to stimulate new directions for youth-oriented research.

Tracey Skelton is a Lecturer in the Department of International Studies at the Nottingham Trent University and **Gill Valentine** is a Lecturer in the Department of Geography at the University of Sheffield.

COOL PLACES

geographies of youth cultures

•

Edited by

Tracey Skelton & Gill Valentine

London and New York

First published 1998
by Routledge
11 New Fetter Lane, London EC4P 4EE

Simultaneously published in the USA and Canada
by Routledge
29 West 35th Street, New York, NY 10001

Reprinted 1998

© 1998 Edited by Tracey Skelton and Gill Valentine

The right of Tracey Skelton and Gill Valentine to be identified as the
Authors of this Work has been asserted by them in accordance with the
Copyright, Designs and Patents Act 1988

Typeset in Sabon by
J&L Composition Ltd, Filey, North Yorkshire
Printed and bound in Great Britain by
Biddles Ltd, Guildford and King's Lynn

British Library Cataloguing in Publication Data
A catalogue record for this book is available from the British Library

Library of Congress Cataloging in Publication Data
A catalog record for this book is available from the Library of Congress

ISBN 0–415–14920–7 (hbk)
ISBN 0–415–14921–5 (pbk)

CONTENTS

•

PART FOUR: SITES OF RESISTANCE

FIGURES AND PLATES

•

ACKNOWLEDGEMENTS

•

We would like to thank Alan Lodge for the cover photograph and Plates 16.1, 16.2, 19.1 and 19.2; Graham Allsopp for supplying Plates 1.1 and 1.2; Lesley Sanderson for providing Plate 4.2; Rob Strachan for Plate 6.1, Sonke Streckel for Plates 10.2, 10.3, 10.4, 10.5 and 10.6; Gwyn Kirk for Plate 18.1, and Pamela Warden for Plates 18.2, 183. Plate 4.3 is reproduced by kind permission of Yuen Fong Ling. Special thanks go to Linda Dawes for designing and redrawing all the maps which appear in the text, to Deborah Sporton for her translation services and to Kathryn Morris-Roberts for helping with the proof-reading.

We are grateful to the following for permission to reproduce their work: Linda Chui for permission to reproduce her poems which appear in Chapter 4, Angela Martin for her cartoon which appears in Chapter 5 (Figure 5.1). Figure 5.2 is reproduced by kind permission of *Disability Now*.

Every attempt has been made to obtain permission to reproduce copyright material. If any proper acknowledgement has not been made, we would invite copyright holders to inform us of the oversight.

Our thanks go to Michelle Keegan of South Nottingham College for organising the production and selection of the lino-cuts for the part-title pages which introduce each of the four sections. The students whose self-portraits feature are: Alan Tien (Part One), Tracy Smith (Part Two), Christian Poxon (Part Three) and Nicholas Wright (Part Four).

The final product owes much to the hard work of the Routledge team: Judith Ravenscroft the copy-editor, Rosalind Fergusson the proof-reader and especially Jody Ball the Desk Editor. We are grateful to Tristan Palmer for commissioning this book and Sarah Lloyd for seeing the project through.

Tracey would like to thank her Mum and Rob for such positive youth experiences and her two sisters, Tania and Janina, for sharing them with her. Gill would like to thank her Mum and Dad and brother David for their support and encouragement when she was growing up; and Gregor Russell, Liz James and Louise Say for their enduring friendships.

1

COOL PLACES

an introduction to youth and youth cultures

•

Gill Valentine, Tracey Skelton
and Deborah Chambers

Plates 1.1 and 1.2 portray very contrasting images of contemporary youth. The first captures the stereotypical image of youth as out to have a good time, carefree and perhaps rebellious. The second suggests a very different version of what it means to be young – reflecting work, a concern for others, respect for adults and perhaps a touch of 'innocence'. It is our intention in bringing this collection together to explore the diversity in young people's lives in order to place youth on the geographical map and to demonstrate youth's relevance to a range of geographical debates.

The popular imagining of youth as consumption-oriented, into sub-cultural styles based on music and drugs, and free to embark on adventur-ous travel, is considered through chapters which examine: Techno music and Ecstasy in Germany (Chapter 10), clubbing in London (Chapter 16) and backpacking round the world (Chapter 11). But as well as focusing on these 'public', and often spectacular, expressions of young people's lives, we also explore young people's mundane everyday experiences of: sibling rivalry in the parental home (Chapter 12), finding paid employment (Chap-ter 14) and hanging around on the streets (Chapter 15). While some of the chapters look at the way youth have been demonised (e.g. Chapter 9 on gangs in Santa Cruz) or stereotyped in different ways by various media (Chapters 3, 4 and 5); others explore the power of young people to resist adult definitions of their lives and to create new spaces and ways of living. For example, through public art on the streets in the USA (Chapter 18), the new age traveller movement in the UK (Chapter 19) and direct forms of political action in former East Germany (Chapter 17).

Plate 1.1 A neglected image of youth: working, concerned for others, 'innocent'.
Source: Graham Allsopp.

Before we invite you to explore some of the similarities, differences and even contradictions between these chapters, we begin in this introductory chapter by defining the category 'youth' and by considering young peoples' absence from Geography. We then go on to combine our thoughts as the editors of this book with the work of cultural theorist Deborah Chambers to consider how the discipline of cultural studies has tackled youth cultures.

DEFINING YOUTH

In contemporary Western societies the age of our physical body is used to define us and to give meaning to our identity and actions, and yet it

Plate 1.2 The popular imagining of youth: 'party animals', carefree, rebellious.
Source: Graham Allsopp.

has not always been so. The historian Aries (1962) famously observed that in the Middle Ages 'children' were missing from medieval icons. He pointed out that this was because beyond the period of infant dependency (i.e. when they had sufficient powers of strength and reason) children were treated as miniature adults, rather than as conceptually different from adults. It was not until the fifteenth century that 'children' began to be represented in icons as having a distinct nature and needs, and as separate from the adult world. This conceptualisation of the young was subsequently fostered through the development of formal education and the belief that children required long periods of schooling before they could take on adult roles and responsibilities (Prout and James, 1990). Although, initially it was only the upper classes who had the time and money to provide their offspring with a 'childhood', legislation in the late nineteenth and twentieth centuries, and more critically the introduction of mass schooling, popularised the mythical condition of 'childhood' and slowly a universal notion of what it meant to be a child developed (Hendrick, 1990); namely, that a child is temporally set apart from the adult world and that childhood is a time of innocence and freedom from the responsibilities of adulthood (although this is not necessarily the reality experienced by many children). Thus social development (e.g. in terms of a transition from irrationality to rationality; and from simplicity to complexity) is assumed to dovetail with physical development (James and Jenks, forthcoming).

In the same way that childhood has been 'discovered', so too adolescence has been invented to create a breathing space between the golden age of 'innocent' childhood and the realities of adulthood. According to Aries (1962), the emergence of what he termed this 'quarantine' period began in the early eighteenth century. Following the development of industrial capitalism the middle classes began to expand the length of their offspring's schooling in order to provide them with a better education. This period was considered a time of 'maturation' when a young person would learn the ways of the world, emerging from their 'quarantine' transformed into an adult and ready to take on adult responsibilities (Boethius, 1995). Throughout the nineteenth century this transitional stage became prolonged and young people became more separated from the adult world, as the middle classes became increasingly preoccupied with the need to control 'working class' youth as well as their own offspring (by the early twentieth century psychologists had labelled this phase 'adolescence'; see for example, Stanley Hall's book of the same title, published in 1904). Indeed, this anxiety about the undisciplinary and unruly nature of young people (especially working class youth) has been repeatedly mobilised in definitions of youth and youth cultures for over 150 years (Pearson, 1983), framed at various moments in moral panics about 'gangs', juvenile crime, violence and so on. This is a definition of 'youth as trouble' (a point we will return to later in this chapter).

In the 1950s – a period of relative affluence – the emphasis on consumption, style and leisure led to the development of a range of goods and services (magazines, record shops, clothing, dances and so on) aimed at a new market niche – the young. Thus 'the teenager' was invented (Hebdige, 1988). In contrast to the parallel construction of 'youth as trouble', this imagining of what it means to be young was a definition of 'youth-as-fun' – although as Hebdige (1988: 30) goes on to argue 'the two image clusters, the bleak portrayal of juvenile offenders and the exuberant cameos of teenage life reverberate, alternate and sometimes they get crossed'.

Thus the multiple and fluid ways that youth (embracing adolescence and the teenager) has gradually been identified and constructed over a period of several centuries serves to highlight what a complex and slippery concept it is. As Sibley argues, youth – despite all the attempts to define it – is ambiguously wedged between childhood and adulthood. He writes:

[The] child/adult illustrates a . . . contested boundary. The limits of the category 'child' vary between cultures and have changed considerably through history within Western, capitalist societies. The boundary separating child and adult is a decidedly fuzzy one. Adolescence is an ambiguous zone within which the child/adult boundary can be variously located according to who is doing the categorising. Thus, adolescents are denied access to the adult world, but they attempt to distance themselves from the world of the child. At the same time they retain some links with childhood. Adolescents may appear threatening to adults because they transgress the adult/child boundary and appear discrepant in 'adult' spaces . . . . These problems encountered by teenagers demonstrate that the act of drawing the line in the construction of discrete categories interrupts what is naturally continuous. It is by definition an arbitrary act and thus may be seen as unjust by those who suffer the consequences of the division.

(*Sibley, 1995a: 34–5*)

James (1986) has also focused on the liminial positioning of youth, pointing out that the only boundaries which define the teenage years are boundaries of exclusion which define what young people are not, cannot do or cannot be. She cites a number of legal classifications, for example the age at which young people can drink alcohol, earn money, join the armed forces or consent to sexual intercourse, to demonstrate how variable, context-specific and gendered these definitions of where childhood ends and adulthood begins are (not to mention the fact that many of these boundaries are of course also highly contested and resisted by young people). Indeed the term 'youth' is popularly used to refer to people aged 16–25 which bears no correlation with any of the diverse legal classifications of childhood or adulthood. Thus James argues that while 'The age of the physical body is used to define, control and order the actions of the social body . . . such precise accounting is relatively ineffective, representing merely unsuccessful attempts to tame time by chopping it up into manageable slices' (James, 1986: 157).

Here, notions of 'performativity' are also useful in understanding the ambiguity of the term 'youth'. Solberg (1990: 12) for example has argued that 'Conceptually children may "grow" or "shrink" in age as negotiations [with parents about what they are deemed responsible enough to do around the home] take place.' Similarly, some young people may be legally defined as adults yet may resist this definition by performing their identity in a way which is read as younger than they actually are; whereas others may actually perform their identity such that they can 'pass' as being older than the actual age of their physical body. Indeed Frankenberg (1992) argues that the whole concept of adolescence is patronising or negative because it suggests that young people do not have value in their own right (i.e. in their own 'being'), rather they are valued only to the extent to which they are in the process of 'becoming' an adult. Under this logic he claims the equivalent term for adults should be 'mortescents'. 'Consequently, for those classified as "adolescent" the very formlessness of the category which contains them is problematic: neither child nor adult the adolescent is lost in between, belonging nowhere, being no one' (James, 1986: 155).

This ambiguity of the term 'youth' is therefore mirrored in this collection of chapters. Rather than attempt to use a notional classification of biological age to construct boundaries of inclusion and exclusion around the category youth we have attempted to include a wide range of 'takes' on what it means to be young. Thus some of the authors in this book are writing about teenagers who are legally defined in a range of contexts as children (for example Sara McNamee's chapter on the home or Shane Blackman's discussion of the New Wave Girls at school); whereas others are about the lives of those who are legally – in voting terms anyway – young adults (for example, Ben Malbon's exploration of clubbing or Luke Desforges' chapter on international travel), with many chapters (such as those by Myrna Breitbart, Tim Lucas and Marion Leonard) straddling this division. Throughout the collection we also try to highlight the importance of understanding 'youth' in relation to the way age intersects with other important identities such as gender, ethnicity and disability (see for example Ruth Butler's chapter on disabled youth or David Parker's discussion of Chinese young people); rather than viewing it as a uni-dimensional category.

GEOGRAPHIES OF YOUTH

This heading, 'Geographies of youth', is rather a misnomer because although there is a significant body of work within Geography (which also straddles environmental psychology) on children's environments, including studies of children's cognition, competence, behaviour, attachment to place and access to/use of space (see for example Blaut *et al.*, 1970; Blaut, 1971; Blaut and Stea, 1971; Anderson and Tindal, 1972; Blaut and Stea, 1974; Bunge and Bordessa, 1975; G.T. Moore, 1976; Hart, 1979; Downs, 1985; R. Moore, 1986; Downs and Liben, 1987; Spencer *et al.*, 1989; Ward, 1990; Katz, 1991, 1993, 1995; Aitken and Wingate, 1993; Aitken, 1994; Sibley, 1995a, 1995b; Valentine, 1996a, 1996b, 1997a, 1997b, 1997c), geographers have been criticised for ignoring children in other areas of their work (James, 1990; Sibley, 1991; Philo,

1992) and have largely excluded the experiences of youth altogether (a notable exception being the work of Myrna Breitbart).

In a recent review of Colin Ward's book *The Child in the Country* Chris Philo (1992) argues that Ward's emphasis on young people's experiences of space has resonance with the emerging sensitivity of social-cultural geography to what he terms 'other' human groupings. Philo states that: 'social life is . . . fractured along numerous lines of difference constitutive of overlapping and multiple forms of otherness, all of which are surely deserving of careful study by geographers' (Philo, 1992: 201). Employing examples of children's agency from Ward's book, Philo uses his review to call for geographers to turn a relative neglect of children – and we might also add youth – into a positive engagement. In particular, there is a danger that geographical work on young people's lives and experiences may be corralled into an enclosure marked 'youth geographies' rather than being an integral part of all areas of main-stream geographical research and debate. While it is important to recog-nise the unique experiences of young people and the spaces they carve out for themselves, it is equally important to recognise the role which they play in all our geographies.

To date there is a small but growing body of work by Geographers (i.e. with a capital G) and academics from related disciplines who are interested in spatiality (geographies with a small g), which has high-lighted the way that public space is produced as an adult space. Studies on teenagers suggest that the space of the street is often the only autonomous space that young people are able to carve out for themselves and that hanging around, and larking about, on the streets, in parks and in shopping malls, is one form of youth resistance (conscious and unconscious) to adult power (Corrigan, 1979). However, other work has shown that teenagers on the street are considered by adults to be a polluting presence – a potential threat to public order (Baumgartner, 1988; Cahill, 1990) – and thus that they are often subject to various adult regulatory regimes including various forms of surveillance and temporal and spatial curfews (Valentine, 1996a, 1996b, 1997a). In particular, efforts to revitalise or aestheticise public space as part of initiatives to revive (symbolically and economically) North American and West Eur-opean cities have been identified as contributing to the privatisation (Berman, 1986; Fyfe and Bannister, 1996) or what Mitchell (1996) has termed the 'annihilation' of public space, by using private security forces and close circuit television to squeeze undesirable 'others' – notably youth – out of these locations. In a study of urban fringe woodland in the UK, Burgess *et al.* (1988) describe participants' anxieties about teenage delinquency, vandalism and glue-sniffing in open space, high-lighting the desires of their respondents for park keepers and wardens to regulate open areas. Such processes of exclusion are captured in Pressdee's description of the way that an Australian shopping mall is policed:

> Groups of young people are continually evicted from this opulent and warm environment, fights appear, drugs seem plentiful, alcohol is brought in, in various guises and packages. The police close in on a group of young women, their drink is tested. Satisfied that it is only Coca-Cola they are moved on and out. Not wanted. Shopkeepers and shoppers complain.
>
> *(Pressdee, 1986: 14)*

There is often a strong element of racism to such informal policing tactics. In *City of Quartz* Mike Davis (1990) documents numerous examples of curfews and policing tactics being selectively deployed against black and Chicano youth to support his argument that as a result of gang paranoia and the demonology of non-Anglo youth, large stretches of Southern California have become virtual no-go areas for young blacks or Chicanos. The racialisation of discourses about urban youth is also a theme taken up by Tim Lucas in Chapter 9 where he examines the way that moral panics about youth gangs in Santa Cruz have been racialised.

A number of geographical studies (Katz, 1991; Breitbart and Worden, 1994; Breitbart, 1995) have drawn attention to young people's ability to subvert and resist the production of public space in late capitalism. This work has particularly highlighted young people's sense of disconnection from the city, their lack of access to public space, and their attempts to resist adult oriented urban space through neighbourhood environmental activism and public art. These themes are explored in Chapters 8 and 18 of this book.

Young people's experiences of green open spaces and rural environments have received even less attention than their use of urban spaces – one exception being Burgess *et al.*'s study of urban fringe woodlands. This project involved focus group discussions, not only with adults, but also with groups of teenagers. In these discussions the young people articulated a preference for adventurous environments where they can experience what Burgess *et al.* term 'safe dangers'. In other words places where they can enjoy being stretched physically without coming to any harm (Burgess *et al.*, 1988; Countryside Commission, 1995). However, parallel with US work on young people's experiences in the city, this study also highlighted how public spending cuts are eroding young people's access to open space. For example, reductions in Local Authority budgets in the UK have led to the demise of countryside based activities and programmes for youth. The loss of these sorts of group activities have been particularly felt by girls and young women because their ability to explore open space independently is inhibited by fears for their personal safety (Countryside Commission, 1995).

The home – like public space – is another place where young people often find their use of space and time subject to surveillance and control by adults. While gender studies, particularly within Sociology, but also in Geography, have opened up the 'black box of the family' to expose how domestic resources and labour are unevenly divided (e.g. Brannen and Wilson, 1987), generational inequalities in power have been relatively neglected. David Sibley (1995b), for example, suggests that intergenerational conflict is likely to be triggered by the fact that young people have few opportunities for privacy in the home, while adults may often find their constant presence a nuisance. In particular, he suggests 'family' tensions represent a clash between adults' desire for order and young people's for disorder; and between adults' preference for firm boundaries in contrast to young people's disposition for more lax boundaries. He goes on to think about some of the ways that power is exercised in 'the family' to argue that young people's

> sense of boundary, anxieties about space and time, feelings of attachment to particular spaces, will be affected by the domestic environment, as it is shaped and manipulated by family members. Clearly, the opportunities for control, or for giving children their own spaces, will be affected by the size of the home, the way space in the home is partitioned, and the relationship between private and public space.
>
> (Sibley, 1995b: 132)

Writing elsewhere with Geoff Lowe and David Foxcroft, David Sibley (Lowe et al., 1993) has further considered the importance of family dynamics and the influence which the home environment has on adolescent drinking habits. The issue of how family members are embedded in each others' lives is a topic touched on by Sara McNamee in Chapter 12. Here she discusses power and control in the home by looking at the issue of gender in relation to young people's use of video games.

As this outline suggests, geographies of youth culture are limited both in scope and depth, focusing largely on young people's experiences of everyday spaces and their sense of spatial oppression, rather than on sub-cultural styles. But we hope that the chapters in this book will play an important part in encouraging others to include youth in the geographical imagination for, as McRobbie has argued, 'youth remains a major point of symbolic investment for society as a whole' (1993: 31).

YOUTH IN CULTURAL STUDIES

While geography has been slow to include young people's experiences within its social, cultural and disciplinary imagination, other subjects have rather built their 'reputations' upon their considerations of youth, and this is especially true of cultural studies. What follows is a summary by Deborah Chambers of cultural studies' approaches to youth. This section acts as a backdrop to the rest of the book as many of the chapters which follow draw on, or refer to, the studies outlined below.

THE DELINQUENT SOLUTION

Adolescents began to be treated as a problem for society after the Second World War, during a period in which young men, in particular, were gaining cultural and economic independence from their family of origin. Academic study of 'youth' as a distinctive social category became established during the 1950s and 1960s in the United States and Britain. The history of academic research about youth cultures reflects and reinforces the public condemnation of working class adolescents. Academic interest in teenagers was born within criminology, fuelled by moral panics concerning the nuisance value of young people on the urban streets of Western societies. Thus, the research into youth groups was marked by a preoccupation with delinquency and associated with the study of other so-called 'condemned' and 'powerless' groups in society such as the working class, migrants and the criminal.

Hence the study of youth initially began within the disciplines of criminology, psychology and sociology and was crystallised within delinquency and deviancy studies (A.K. Cohen, 1955; Cloward and Ohlin, 1960; Miller, 1958; Matza and Sykes, 1961; D.M. Downes, 1966). Criminologists, psychologists and sociologists of crime tended to focus on working class 'gangs' of youths rather than young people as a whole. The message seemed to be that all teenagers were potentially delinquent or deviant.

The British and North American literature on youth gangs clearly demonstrated a preoccupation with defining 'delinquency' and yet there was little direct research conducted on actual groups, let alone gangs, of young people themselves. Academics who theorised about delinquent youth groups based their work on limited empirical evidence. As a result, the literature was over-theorised yet empirically weak (Patrick, 1973).

Such work unconsciously equated 'youth' with young men and thus also lacked any comprehensive analysis of the lives of girls, the constitution of feminine and masculine subjectivity, or gender power relations. Since adolescence is a key phase of life in which sexual and gendered identities are explored and established, this oversight relegated the activities and meanings of girls to the margins of analyses of youth experience and neutralised the sexism invested in masculine values and practices. US gang theorists of the 1950s and 1960s such as A.K. Cohen, Cloward and Matza conducted studies of gangs which gave little understanding of the relations between the sexes or between gang members and their families, and wider institutions of society with whom they interacted. There emerged a range of explanations for US delinquency sub-cultures: they were collective resolutions to the social and psychological problems experienced by working class boys who had internalised middle class values but through a lack of education could not compete with middle class boys (A.K. Cohen, 1955; Cloward and Ohlin, 1960); delinquent behaviour was said to be a disturbing reflection or caricature of the leisure meanings of the dominant culture (Matza and Sykes, 1961; Matza, 1969).

Within British research of the same period, Downes (1966) argued that US gang research was culture-bound. He claimed that the class structure of British society encouraged working class youths to disconnect themselves from middle class aspirations and lifestyles and affirm a working class status. British researchers searched for 'proper delinquent gangs' but failed to find youth groups with a name, a leader, initiation rituals, a uniform and criminal aims (Patrick, 1973). Rather than finding stereotypical gangs most British researchers discovered small, loosely structured groups of youths.

An empirical investigation of the Teddy Boys had been conducted by Fyvel (1963) in the late 1950s. He described gangs who were organised to defend their territories and who engaged in stylised warfare with other similar gangs. Influenced by the work of Matza, Fyvel explained the styles of the Teds as a working class caricature of middle class styles of dress. This study was criticised by Downes (1968) as an example of people who reached for the nearest cliché and made it fit: 'gang warfare'. Nevertheless, Fyvel's study has acted as an important foundation to the work conducted on youth sub-cultures in the 1970s.

Stan Cohen (1967) undertook an analysis of the way the mass media manufactured an exaggerated and unverifiable picture of riots and gang warfare among Mods and Rockers at seaside resort disturbances in south-east England between 1963 and 1966. The groups of adolescents who were labelled as 'gangs' by the press and television were in fact unstructured groups with loose organisations based on territorial loyalties. The amount of damage and violence was almost negligible in contrast to the media portrayals. According to Cohen, the youths were encouraged to behave in gang fashion as a consequence of continuous media use of the word 'gang'. The over-reaction of the media and magistrates to the original events led to an amplification process whereby the expectations created by earlier events were later fulfilled, and the employment of gang stereotypes justified the measures taken by the disciplinary authorities. The complex relationship between media representations of young people and those same young people's identities and behaviour in reaction to such representations which was established throughout the 1960s is one which continues to the present day.

The fascination with male urban 'gangs' and delinquency continued. In the early 1970s, James Patrick (1973) conducted a participant observation study of a violent Scottish gang in Glasgow called 'the Young Team' which strongly resembled Yablonski's (1967) accounts of violent gangs in New York. In contrast to cities in England at the time, there were many structural conditions in Glasgow that were conducive to a gang sub-culture. Slum housing, high unemployment, a long tradition of heavy alcohol consumption, violence, and high rates of lung cancer and heart disease were evidence of interconnected and cumulative forms of inequality. Patrick's work demonstrated the ways in which poverty, poor education and lack of opportunities produce frustration, rage and a sense of powerlessness (1973).

Downes argued in 1966 that the reasons for the absence of gangs in England were the lack of 'teeming slums', the lack of high unemployment and the absence of prominent numbers of adolescent children of immigrants – all of which were conditions experienced in New York, Chicago and Glasgow where structured violent gangs existed (Patrick, 1973). The fact that many areas of poverty were 'discovered' in England and Wales after Downes's statement led to more sophisticated analyses of cultures of poverty, and of youth. As we see from the chapters by Tim Lucas, and Susan Ruddick, approaches to gangs in the US have undergone fundamental reconsiderations as gangs have moved increasingly towards self-help and community action, and have become a feature of life for young urban women as well as men.

THE CENTRE FOR CONTEMPORARY CULTURAL STUDIES

During the mid-1970s, a distinctive contribution to youth sub-cultural theory was made by members of the Centre for Contemporary Cultural Studies (CCCS), University of Birmingham, in the UK, using Marxist categories such as social class and ideology in relation to culture and consciousness. Within *Resistance Through Rituals* edited by Hall and Jefferson (1976), researchers combined empirical studies of sub-cultures such as Teds, Mods, Rockers, Skinheads and Punks with Gramsci's (1971) theory of hegemony as a way of explaining the forms of rebellion expressed by certain youth groups. The participant observation technique of ethnography was a principal methodological tool used in such studies. Gramsci's conception of hegemony is an attempt to explain the ways in which social subjects 'consent' to and negotiate their subordination.

The youth sub-cultural research emphasised the measures used by subordinate groups to resist the dominant culture by creating their own meanings. Young people were understood to either negotiate with, or oppose, the dominant ideology, or to subvert dominant meanings by actively appropriating and transforming those meanings. The creation of new subjective meanings and oppositional lifestyles were interpreted as a cultural struggle for control over their lives.

The CCCS argued that culture is a distinctive way of life embodied in beliefs and customs, social relations, institutions and material objects. All these aspects of the sub-culture were referred to as 'maps of meaning' which shape the sub-culture and make it intelligible to its members. The most mundane object could be appropriated by youth groups and take on specific meanings. Forms of adornment such as safety pins, ripped clothing, leather, chains, tattoos might be appropriated to symbolise offensive decoration. Studies of youth sub-cultural groups examined the way in which these maps of meaning were constituted by focusing on the meanings given to the objects, institutions and practices by the group.

Two forms of empirical research on youth cultural meanings and actions became established by British Cultural Studies research: ethnographic and textual. Ethnographic research drew on anthropological and sociological methods of participant observation, and textual analyses drew on techniques of critical analysis drawn from semiotics, literary theory and structuralist anthropology. Both ethnographic and textual forms of research were informed by specific theoretical frameworks. Gramsci's theoretical conceptualisations of class and his notion of hegemony were used to explain the social position of youth (P. Cohen, 1972; Hall and Jefferson, 1976; Willis, 1978) within ethnographic studies. Lévi-Strauss's concept of bricolage and Barthes's notion of myths were used in textual analyses of youth styles such as dress, personal adornment and young people's use of artefacts such as motorbikes, comics/magazines and language (Hebdige 1979).

One of the central focuses of the work of the CCCS and subsequently British cultural studies was working class youth sub-cultures. Phil Cohen began the youth sub-cultural project of the CCCS in 1972 in an historicised study entitled 'Subcultural conflict and working class community'.

Cohen gave an account of the destruction of working class communities brought about by the building of housing estates in working class areas of the East End of London in the 1950s. The new housing plans, based on the middle class concept of status and private ownership, replaced the traditional structure of the working class environment which had been based on the concept of community and collective identity. The disjunction between the middle class physical structure of the housing estates and its working class occupants led to conflict between those who clung to nostalgic class loyalties and traditional working class puritanism, and those who aspired to bourgeois upward mobility and the new hedonisms of consumption. The contradictions existing within the parent culture were articulated in youth sub-cultural styles of gangs of mods, skinheads and crombies among the adolescent groups who occupied the urban space on the housing estates.

Cohen claimed that by attempting to retrieve some of the socially binding elements destroyed in their parent culture, youth sub-cultural styles expressed and magically resolved the ideological contradictions which remained hidden or unresolved in the parent culture. Working class youth sub-cultures were seen to be the product of a conscious perception of their economic positions in society as wage labourers. The contradictions of advanced capitalism were experienced by the working class within the workplace and resolved in the realm of leisure through the adoption of oppositional lifestyles. Adolescents therefore expressed these lifestyles in an 'imagined reality'. Young people failed to overcome those class contradictions that they struggled against because the attempt to solve such material problems took place in the sphere of leisure, leaving the problems of work intact. Cohen's approach called for three dimensions of analysis in the study of youth sub-cultures: an historical analysis allowed the examination of class fractions; a structural or semiotic analysis allowed the study of the system of style, dress, music, slang and rituals; and an ethnographic analysis explained the sub-cultural group's 'lived out' daily practices (Turner, 1990).

In *Learning to Labour: How Working Class Kids Get Working Class Jobs*, Paul Willis (1977) undertook a comprehensive and methodologically more sophisticated study of a group of twelve working class boys whom he called the 'lads'. Willis studied his subjects through the use of participant observation including attending classes, informal interviews, diaries and group discussions; he made comparative studies with five other mixed background groups in the same school. He interviewed teachers, parents and careers officers, and examined the structures and ideological systems of discipline and control operated by the school. He gained a thorough knowledge of the locality and followed the 'lads' through to the first six months of their employment.

By resisting the ideologies of school and adopting masculine working class sub-cultural codes of work, Willis found that the 'lads' were well prepared for the lack of choices in employment and for entering unskilled jobs. On entering employment, the lads discovered the culture of the shop floor to be quite familiar because it echoed the structures, discourse and power relations of school. The notion of style was used in analysing the way school life was experienced and practised by the boys as lived culture, and the ways in which the boys inverted and disrupted the values of the school. Their objective to do no school work reproduced working class culture in adulthood and allowed the boys to recognise that school is a deceit for the working class since it is impossible to aspire educationally to rise above the working class. Willis therefore offered a wider political analysis of the ideologies of working class culture.

In *Subculture: The Meaning of Style*, Hebdige (1979) used semiotics, literary criticism and structural anthropology to interpret the ambiguous meanings produced by the dress, music, language, gestures, postures and behavioural styles of publicly condemned youth sub-cultural groups such as Teddy Boys, Mods, Rockers, Rastafarians, Skinheads and Punks. Unlike Phil Cohen's historical approach in which youth sub-cultural styles were related to the parent culture that produced them, Hebdige used a textual analysis, drawn from semiotics and the anthropologist Lévi-Strauss (1966), and took the notion of symbolic forms of resistance of youth sub-cultures outlined in *Resistance Through Rituals* as his starting point. He documented the appropriation of elements of British West Indian black culture such as ska and reggae music by skinheads and explained their caricature of working class dress as a challenge to the embourgeoisement of working class culture.

According to Hebdige, meanings are contested within the realm of style whereby opposing definitions are stylistically created to conflict with the dominant culture. Hebdige (1988) later claimed that he may have over-emphasised the equation of the subordinate with the resistant for certain youth groups who are not overtly political. The development of a link between youth sub-cultures and the signification of resistance was reinforced by the explicit political agenda of punk. Hebdige admits that in the desire to gain methodological objectivity within ethnography there is a tendency to adopt a strong theoretical framework influenced by the researcher's political wishes to subvert political consensus. This self-critique is echoed by McRobbie in 1993, in her criticism of earlier feminist studies of female youth.

FEMINIST CONTRIBUTIONS AND CRITIQUES

The preoccupation with youth as deviant, spectacular and male was critiqued by feminist researchers in the late 1970s and early 1980s. The youth cultural approach of the 1970s failed to account for the cultural activities, forms and meanings of girls. In *Resistance Through Rituals*, which dealt mainly with 'spectacular male youth', McRobbie and Garber (1976) criticised the way in which such studies focused exclusively on boy's sub-cultural styles and their entrance into waged work. Some of these shortcomings were partially overcome in the 1980s through feminist contributions to Cultural Studies. Researchers such as McRobbie and Garber (1976) and Griffin (1985) focused not only on spectacular youth actions on the street and the playground, but also examined girls in ordinary and everyday contexts, in the domestic sphere, relations between the sexes within daily leisure practice, and forms of regulation of young people (McRobbie and Nava, 1984).

The Women's Studies Group at the CCCS, University of Birmingham, edited a collection of essays entitled *Women Take Issue* in 1978. Various aspects of women's, including young women's, experiences were analysed from a feminist perspective in what is considered to constitute the foundational work on feminist analyses of popular culture. This early feminist work criticised the tendency for male academics to celebrate male sub-cultural practices as 'oppositional' or 'resistant' to dominant social relations even when such practices were sexist or racist, and to generalise about youth from male samples and practices. Moreover, the emphasis on masculine public discourse obscured the context of the family and the domestic arena in which male youth lived and negotiated with parents and siblings (Roman and Christian-Smith, 1988). McRobbie (1980) pinpointed the weaknesses and shortcomings of the sub-cultural 'classics' about male youth, in which the sub-cultural theorists failed to transcend the male connotations of 'youth' by avoiding analyses of the relationship between youth and the family and thereby marginalising the whole question of women and sexual divisions among youth. It was fashionable to study deviance in the 1970s, but it was not considered fashionable to study the family during the same period, when the functionalist model of the family was being heavily criticised.

Studies by male researchers failed to examine how young women were located within the contradictions of domesticity and waged work. Labelling the existing literature as irrevocably male-biased, McRobbie and others called for a shift of attention to girls' cultures and to the construction of ideologies about girlhood and the way femininity is articulated in a range of institutions and cultural forms such as school, the family, law and the popular media.

In terms of the methodology of feminist work McRobbie recognised the close links between women's personal experience and the areas selected for study by feminist academics, and argued that women's auto-biographies inform their research and their commitment to their chosen subject. She claimed that, by contrast, male academics often fail to admit how their own experience influences their choice of subject-matter.

Within an exploration and analysis of the nature of women's and girls' leisure, feminist scholars addressed questions such as 'what role do fantasy escapes, hedonism and imaginary solutions play in their lives?' (McRobbie, 1980). A range of textual and ethnographic studies were conducted on topics such as girls' comics (McRobbie, 1982) and magazines, school (Griffin, 1985), women's magazines (Winship, 1987), soap opera such as *Dallas* (Ang, 1985) and dance (McRobbie and Nava, 1984). Feminine bonding, cultures of romance and consumption, and use of clothes, make-up and pop music (see for example Lewis, 1990; Nava, 1992; Gannetz, 1989 and 1995) were examined as distinctive elements of female sub-cultures. What there was little space for in such studies were the girls who did not do these things, who were what might be termed 'tomboys', who rejected hyper-femininity. Apart from the occasional autobiographical essay by lesbians, girls who did not conform to the above 'distinctive elements' of female youth culture, lesbian-identified or not, were not considered at all.

What this early 1980s research concluded was that, at that time and in those case studies: girls' leisure was more restricted than that of boys; they were often unable to engage in spectacular leisure activities which were dirty, dangerous or hedonistic, such as motorcycle riding or hanging around the urban streets; girls spent more time in the home, supervised by parents; unlike boys, girls' leisure was not structured first by the move from school to work but by their relationship to men. The studies also argued that for the majority of adolescent women the main objective was to attract a boyfriend and that femininity was constructed to secure a future married life (Griffin, 1985).

Chapters in this book by Claire Dwyer, Shane Blackman, Sarah McNamee and Marion Leonard demonstrate that there is now a much broader consideration of what young women do and what constitutes the 'distinctive elements' of their cultures.

'DISCURSIVE CODES OF FEMININITY': NORTH AMERICAN FEMINIST CULTURAL POLITICS OF THE 1980s

American feminist scholars of popular culture Roman and Christian-Smith (1988) claim that by studying the girls' sub-cultures in isolation from their relations with boys, and by exaggerating female youth meanings of opposition, resistance and dissent, the early feminist cultural studies repeated the problems existing within the research of their male counterparts.

The early CCCS feminist projects tended to romanticise young women's 'cultures of femininity' in the attempt to authenticate girls' cultures of romance and domesticity. Family life and domesticity were privileged as the principal origins of women's subordination. The strategies of research employed by such investigations of all-female groups to counter the existing masculine discourses led to 'equally naive representations of the young women's practices as exemplars of autonomous feminine cultures of "resistance"' (Roman and Christian-Smith, 1988: 17).

According to Roman and Christian-Smith, feminist ethnographic studies of cultural texts consumed typically by women (such as romantic fiction, magazines and soap opera) tended to privilege consumption over production. The interaction between readers and texts remained under-examined, leading to notions of a homogeneous and fixed feminine readership. Popular cultural texts are 'polysemic', that is, capable of generating many diverse signs and meanings (see Fiske, 1987). Thus, there was a call for combining textual and ethnographic approaches. In this way, both commercially produced 'popular cultural forms' (such as magazines, clothes, popular music, comics, television, films) and the 'lived cultures' (such as youth sub-cultural groups) of those who use them were to be inter-related within analyses of the texts and artefacts young people use in exploring and expressing their identities.

Within the call for approaches that allow for an understanding of the inter-connection between representational and lived social relations, the contributors to the North American book, *Becoming Feminine: The Politics of Popular Culture* (1988), address the question of whether culture is an adequate concept for explaining both the material conditions of women's subordination and women's subjective understanding of them (Smith, 1988; McCarthy, 1988). The concept of culture has been important in emphasising the active agency of youth by overcoming the assumption that such groups are 'passive bearers of dominant ideology' (Roman and Christian-Smith, 1988: 25). Femininity is not simply an effect of patriarchal oppression. Yet attempts to understand differences of race, class, age, sexual orientation and location require new approaches in analysing subjectivity and meaning.

Dorothy Smith (1988) argued that the concept of culture neglects the actual knowledge that young women make use of in becoming feminine. She recommended the use of a Foucauldian concept of 'textually mediated discourse' as a new form of analysis of social relations. This concept is capable of explaining how social relations are mediated by the codes of femininity that originate in most public discourses and popular texts which have no single local source or historical agent. Smith contended that through expanded transnational communication, social relations have been globally reorganised by textually mediated discourse as a material condition of contemporary capitalism. For example, the large number of women who are discontented with their body image arises in the relation between texts and women who find in texts images reflecting upon the imperfections of their bodies (Smith, 1988). Thus, the discourses within public and popular cultural texts, such as women's magazines, are organisers of local relations.

Leslie Roman's (1988) examination of the punk slam dance centred on the way middle and working class punk girls form their gender identities and class relations around the ritual of slam dancing. She analysed ethnographic data semiotically, drawing upon the girls' discourses about their modes of participation in the dance as well as their personal histories of sexual abuse, family violence, waged work and independence from their families. Roman demonstrated the way class differences and conflict constrain the gender-specific alliances built by the girls in their effort to challenge masculine forms of control over the dance. By uncovering class-specific forms of feminine subjectivity concerning meanings of intimacy, sexual pleasure and danger, Roman showed that combining ethnography with semiotics was effective in overcoming problems of authorship associated with the narrative realism of naturalistic ethnography.

GLOBAL CULTURE AND THE POLITICS OF DIFFERENCE

As mentioned earlier, the notion of ethnic cultures does not feature centrally in the work of the CCCS despite the fact that black styles of dress, music, dance and fashion have had a profound impact on the urban youth cultures and leisure institutions of 'Western' societies (Hebdige, 1979 and 1988). As Connell (1983) points out, the Birmingham research did not tackle centrally questions of race and of geographical region. Stuart Hall reminds us that 'Western Europe did not have, until recently, any ethnicity at all. Or didn't recognize any' (1992: 22).

In an ethnographic account of audio-visual culture among South Asian families in west London, Marie Gillespie (1993) distinguishes between the VCR viewing habits of second generation South Asians and those of their parents and grandparents. She points out that the dominant assumptions about the archaic marriage and family customs of

South Asian cultures, such as arranged marriages, have remained unchallenged by the exclusion and marginalisation of young Asian voices from debates about their lives. Gillespie attempts to re-present 'their voices' within their interpretation of popular Indian films. She found that young people use the films to 'negotiate, argue, and agree about a wider range of customs, traditions, values, and beliefs . . . . The films function as tools for eliciting attitudes and views on salient themes; family affairs and problems, romance, courtship and marriage were often discussed' (1993: 15). Gillespie stresses that a sense of ethnic, national and cultural identity is intersected by identities based on age, gender, peer group and neighbourhood. In a similiar way, Clare Dwyer in Chapter 3 explores the way young Muslim women talk about the media and the contextual processes by which it is used in their own identities, and David Parker in Chapter 4 demonstrates the importance of cultural production in Hong Kong for Chinese youth identities in the UK.

In *'There Ain't No Black in the Union Jack'*, Paul Gilroy states that because black people, including black youth, have been located in the structures of British society in a variety of ways, they are 'not reducible to the disabling effect of racial subordination' (1987: 155). Black expressive cultures draw on a multiplicity of black histories and politics. Elements of new black cultures are derived from colonisation in Africa, the Caribbean and the Indian sub-continent.

Stuart Hall points to Cornel West's (1990) essay, 'The new cultural politics of difference', in defining the contemporary moment of black popular culture as centred on the emergence of a global cultural production and circulation which is displacing and hegemonically shifting the definition of culture from high culture to (American) mainstream popular culture, effected by the impact of civil rights and black struggles on the decolonisation (in Frantz Fanon's [1967] sense) of the cultures of the black diaspora. New technologies of communication have generated a global perspective from the history of slavery of the African diaspora and allowed the international export of new world black cultures to white and then 'Third World' markets (Wallis and Malm, 1984). Black musicians have, for example, addressed international audiences through the forms of blues, gospel, soul, reggae, hip-hop and rap. As Gilroy states, new definitions of 'race' have emerged from the international flow of cultural commodities such as books and music so that 'a new structure of cultural exchange has been built up across the imperial networks which once played host to the triangular trade of sugar, slaves and capital' (1987: 157).

Thus, the internationalisation of what Gilroy calls the 'expressive culture of black communities', linking the Caribbean, America, Europe and Africa, has demonstrated not only the need to undertake a global analysis of the political dimensions of black youth sub-cultures (and increasingly of many other youth sub-cultures) but the need to transcend the narrow research focus on national cultures. The 'race' and 'nation state' nexus is no longer adequate. Gilroy argues that:

> culture does not develop along ethnically absolute lines but in complex, dynamic patterns of syncretism in which new definitions of what it means to be black emerge from raw materials provided by the black populations elsewhere in the diaspora.
>
> *(Gilroy, 1987: 13)*

In addressing the absence of a history of the musical cultures of black Britain, Gilroy analyses the popular traditions of music and dance, taking into account lived and formed relations based on gender, age, class and locality and the experiences of racism and expressions of resistance by first and subsequent generation blacks. In doing so, he makes a major contribution to the origins and characteristics of British popular culture in which white youth have been centrally involved in the consumption of black cultures (Gilroy, 1987). Recent studies of black popular youth musics such as rap have moved beyond the exclusive study of the body and sexuality and addressed the politics of content and form, and expressions of dissent (Rose, 1989 and 1990; Wheeler, 1991; Tate, 1992; Baker, 1993).

This book provides several chapters where 'race' and youth are discussed: Claire Dwyer considers the identities of young Asian women (Chapter 3); Paul Watt and Kevin Stenson demonstrate the ways in which young white, Asian and black people react to and use certain spaces within a single town (Chapter 15); and Sophie Bowlby, Sally Lloyd Evans and Robina Mohammad discuss Muslim Asian women's experiences and attitudes towards work and employment (Chapter 14). Part Two, 'Matters of Scale', also debates the relationship between youth cultures from the global to the local. In particular Doreen Massey (Chapter 7) sets out some of the theoretical implications of talking about scale in relation to youth cultures.

CONCLUSION: MAKING A DIFFERENCE?

Although youth cultural studies since the period of the 1950s and 1960s have been marked by a shift away from the treatment of youth as inevitably 'deviant' and male, youth centred definitions of their lives remain largely absent. Young people have not been enfranchised by the research conducted on their lives. The history of youth cultural studies of the last four decades tells us more about the politics of academic research than it does about young people. However, this is not an inevitable outcome for potential future youth cultural studies. A reassessment of the definition of cultural studies has been prompted by the problems of inter-relating theory and methodology, by the unequal power relations between academic researchers and their youthful subjects, by the social changes that have taken place (such as Thatcherism in 1980s Britain), and by the recognition of the intersection of local cultural practices with the trends of globalisation of popular culture.

Within the discipline of Geography a debate has also taken place about the politics and ethics of research, about the nature of power relations between researchers and research participants; questions of insider/outsider research; and the role of research in allowing marginalised groups a voice. These debates have been framed within the broader context of the discipline, rather than purely in relation to work on youth (see the special issue of *Environment and Planning D: Society and Space*, 1992). In large part these methodological concerns have been stimulated by work within feminist Geography, and latterly within critical Geography too (see *Professional Geographer*, 1994; Madge *et al.*, 1997).

The debate about ethics and power in the research process is especially pertinent in relation to working with young people. This has been the subject of particular concern within Sociology. Specifically, researchers have faced accusations that young participants do not tell the truth or make things up to satisfy the interviewer and that young people do not have enough experience to comment knowledgeably on their own lives (Mayall, 1994). These criticisms have been strongly rebutted. Numerous studies have demonstrated the importance of recognising young people as competent agents in their own lives (Goode, 1986; Qvortrup, 1994). All research accounts, whether provided by a 17 year old gang member or a middle age businessman, are just that – 'accounts', which are mediated by the tellers' experiences, by their perceptions of the researcher and of the research context, and by their own agendas. Thus all research accounts are equally likely to be a cocktail of the 'experienced', the 'perceived' and the 'imagined'.

However, while young people are important actors in their own right, it is also true that many young people's lives (especially those under 18, or those who, for whatever reason, do not, or cannot, live independently of their 'families') are strongly mediated by others – at home, at school and so on. Young people's relatively marginalised position can place them in situations of obligation to researchers, especially if the researcher is positioned as an authority figure (e.g. on the basis of their age, social or financial position), although not all young people may feel themselves to be marginalised or disempowered within society – this is especially true of privileged young people – and not all researchers are in positions of 'authority', personally or financially, as many postgraduate students can testify. This is particularly relevant in situations where young people are approached in contexts such as school, or youth clubs, where they may perceive implicit pressure from the institution or organisation to participate in the study. In a ground-breaking document on the ethics of researching young people's lives Priscilla Alderson (1995) stresses the importance of: allowing young people to opt into academic research, rather than putting them into a situation where they have to opt out; providing young people with comfortable ways of saying 'no' to particular questions or situations; enabling participants to withdraw at any point in the research, allowing those involved to ask researchers questions about their own lives; and being accessible to respondents beyond the actual times that they are participating in the project. Alderson also advocates allowing young participants to have a voice in the research design, and adapting methods to suit the abilities of different individuals (e.g. young people with speech difficulties may not wish to take part in a verbal interview).

Finally, Alderson (1995) highlights the importance of respecting and maintaining young people's confidence, and the sensitivity of some of the issues involved in doing so when the participants are legally minors. She argues that if young people disclose information about situations where they are at risk (in relation to drug taking, violence and so on) – which may indeed be unrelated to the research project – then the researcher should encourage them to talk to someone who can help, or to gain agreement that the researcher could do this for them.

Given that young people are invariably marginalised within the wider society and have little, if any, input in public policy debates which directly impact on their lives, empirical research with young people provides an opportunity to increase our understanding of their lives and in some situations to contribute to academic or public debates which play a part in social construction of youth. It is crucial therefore that young people should be able to play a part (if they so wish) in the discussion of the research findings and where possible the dissemination of this material.

In the following chapters (particularly Chapters 5, 14, 15, 18 and 20) there are a whole range of examples of where the authors have tried to represent young people's accounts of their own experiences by including interview material in their work. In some cases whole conversations are cited so the dynamics of the group are made explicit (Chapters 3, 4, 12 and 13). In other chapters authors employ extended pieces of transcribed material (which are boxed) in order to try to articulate multiple accounts from young people of their lives.

We hope that the shared concern within Cultural Studies, Geography and Sociology with the politics and ethics of child and youth oriented research will open up possibilities for more synergy between the disciplines in order to develop youth centred research methodologies. In particular, we hope that this book will play a part in inspiring young geographers to tell their own stories.

Finally, in this introduction we want to outline a number of areas which from our discussion of the geographical and cultural studies literature we consider remain un- or under-researched. One key area which needs further attention is the role played by youth sub-cultures in production and marketing for consumption. McRobbie (1993) suggests that the intersection of sub-cultures, the hidden economy and institutions such as fashion, image and music form a complex process whereby youth styles are commercially appropriated and become part of transnational popular cultural youth forms. However, young people then reinterpret those forms, invent new forms from their own productive creativity and conspire to render the commercial forms obsolete, and so the process begins again. The dynamism of such perpetual processes are in need of further research, research which actually recognises the energy young people put into creating their cultures and does not represent them as passive consumers and victims of commercialism.

Another area which research on youth has neglected is that of privileged youth cultures. While working class youth sub-cultures have been extensively documented, privileged youth from elite social classes have not been the focus of academic research (although they have been the subject of the occasional documentary or journalistic exposé). Here researchers need to consider the consequences of how power and privilege are (re)produced and played out in space.

While much youth cultural research has centred on music, fashion and drugs there has been very little consistent research on questions of sex, sexuality and gender. In particular, most of the early cultural studies work focused on constructions of heterosexual identity, the ways in which young lesbian and gay people live their lives is an unresearched area. Indeed, we were unable to find a contribution on this theme for this collection.

The emphasis on resistance and spectacular forms of youth cultures has led to a neglect of the young people who conform in many ways to social expectations. We have an inadequate understanding of young people who perform well at school, have good and positive relationships with their parents and other adults, who participate in a range of activities which do not cause harm or annoyance – who basically get on with their lives as young people, but who at the same time have to face an enormous range of social, cultural, educational and financial pressures.

While this collection does not address all of the gaps we have identified we hope that it will go some way towards stimulating new directions for youth oriented research. To ease navigation through the diverse material included in this volume we have divided the chapters which follow into four parts. Through the chapters which make up Part One 'Representations', this book considers how the media have defined and redefined youth – imaging young people as a particular community with shared interests; how young people talk about the media and the contextual processes by which they are used in their own identities; the way the media and other cultural forms stereotype different groups of youth; and how young people resist the images created of them and indeed produce their own independent representations of their lives. Part Two, 'Matters of scale', explores relationships across geographical boundaries. Different chapters adopt different takes on the concept, making connections across scales in multiple ways. Some consider local inflections of national issues; others look at national interpretations of global trends and problems. In Part Three 'Place', each chapter addresses young people's experiences of a different everyday location (home, school, workplace, the street and the club). Finally, in Part Four 'Sites of resistance', the chapters consider how young people find their lives defined by others (the state, planning authorities and so on) and describe some of the ways that they have sought to resist these definitions and renegotiate both urban and rural spaces.

Although we have chosen to adopt this framework for the collection, this is not to imply that each part is discrete and bounded. Rather, our four structuring themes may also be referred to in chapters which have been placed under a different heading. To give two examples, while Marion Leonard's chapter (6) on zines is placed in the section on representation, her discussion also makes reference to the themes of both scale and resistance; and while Tim Lucas's chapter (9) on gangs in Santa Cruz is located in the section on scale, his discussion is bound up with issues of representation – particularly the demonisation of young people. Thus we hope that you will enjoy mapping your own path through this collection, and making your own connections between the diverse range of chapters which follow.

REFERENCES

Aitken, S. (1994) *Children's Geographies*, Washington, D.C.: Association of American Geographers.
—— and Wingate, J. (1993) 'A preliminary study of the self-directed photo-

graphy of middle-class, homeless and mobility impaired children', *Professional Geographer* 45: 66–72.

Alderson, P. (1995) *Listening to Children: Children, Ethics and Social Research*, Ilford: Barnardo's.

Anderson, J. and Tindal, M. (1972) 'The concept of home range: new data for the study of territorial behaviour', in W. Mitchell (ed.) *Environmental Design: Research and Practice*, Los Angeles: University of California Press.

Ang, I. (1985) *Watching Dallas: Soap Opera and the Melodramatic Imagination*, London: Methuen.

Anyon, J. (1981) 'Social class and school knowledge', *Curriculum Inquiry* 11: 13–42.

Aries, P. (1962) *Centuries of Childhood*, New York: Vintage Press.

Baker, H.A., Jr (1993) *Black Studies, Rap and the Academy*, Chicago: University of Chicago Press.

Barthes, R. (1973) *Mythologies*, London: Paladin.

Baumgartner, M. (1988) *The Moral Order of the Suburbs*, New York: Oxford University Press.

Bennett, T. (1991) 'Youth and the arts', *Culture and Policy* 2, 2 and 3, 1, Special Double Issue on Youth and the Arts: 189–200.

Berger, B.N. (1969) 'Sociologist on a bad trip', *Transactions*, February 1969: 54–6.

Berman, M. (1986) 'Take it to the streets: conflict and community in public space', *Dissent*, Fall: 470–94.

Bhabha, H. (1986) 'Signs taken for wonders: questions of ambivalence and authority under a tree outside Delhi', in H.L. Gates (ed.), *'Race', Writing and Difference*, Chicago: University of Chicago Press, pp. 163–84.

Blaut, J. (1971) 'Space, structure and maps', *Tijdschrift voor Economische en Sociale Geografie* 62: 1–11.

—— and Stea, D. (1971) 'Studies of geographic learning', *Annals of the Association of American Geographers* 61: 387–93.

—— and Stea, D. (1974) 'Mapping at the age of three', *Journal of Geography* 73: 5–9.

——, McCleary, G. and Blaut, A. (1970) 'Environmental mapping in young children', *Environment and Behaviour* 2: 335–49.

Boethius, U. (1995) 'Youth, the media and moral panics', in J. Fornas and Goran Bolin (eds) *Youth Culture in Late Modernity*, London: Sage.

Brannen, J. and Wilson, G. (1987) (eds) *Give and Take in Families: Studies in Resource Distribution*, London: Allen and Unwin.

Breitbart, M. (1995) 'Banners for the street: reclaiming space and designing change with urban youth', *Journal of Education and Planning Research* 15: 101–14.

—— and Worden, P. (1994) 'Creating a sense of purpose: public art and Boston's orange line', *Places* 9: 80–6.

Brown, M.P. (1994) 'Funk music as genre: black aesthetics, apocalyptic thinking and urban protest in post-1965 African American pop', *Cultural Studies* 8, 3 (October): 484–508.

Buchner, P. (1990) 'Growing up in the eighties: changes in the social biography of childhood in the FRG', in L. Chisholm, P. Buchner, H. Kruger and P. Brown (eds) *Childhood, Youth and Social Change: A Comparative Perspective,* London: Falmer Press.

Burgess, J., Harrison, C.M. and Limb, M. (1988) 'People, parks and the urban green: a study of popular meanings and values for open spaces in the city', *Urban Studies* 25: 455–73.

Bunge, W. and Bordessa, R. (1975) *The Canadian Alternative: Survival, Expeditions and Urban Change, Geographical Monographs No. 2,* Toronto: York University Press.

Cahill, S. (1990) 'Childhood and public life: reaffirming biographical divisions', *Social Problems* 37: 390–402.

Campbell, A. (1981) *Girl Delinquents,* Oxford: Blackwell.

Clarke, J., Hall, S., Jefferson, T. and Roberts, B. (1976) 'Subcultures, cultures and class', in S. Hall and T. Jefferson (eds) *Resistance Through Rituals: Youth Subcultures In Post-war Britain,* London: Hutchinson.

Cloward, R. and Ohlin, L. (1960) *Delinquency and Opportunity: A Theory of Delinquent Gangs,* Glencoe, Illinois: Free Press.

Cohen, A.K. (1955) *Delinquent Boys: The Culture of the Gang,* Glencoe, Illinois: Free Press.

Cohen, P. (1972) 'Subcultural conflict and working class community', *Working Papers in Cultural Studies* 2, University of Birmingham, Centre for Contemporary Cultural Studies.

Cohen, S. (1967) *Folk Devils and Moral Panics,* London: Paladin.

Connell, R.W. (1983) *Which Way is Up? Essays on Class, Sex and Culture,* Sydney: Allen and Unwin.

Corrigan, P. (1979) *Schooling the Smash Street Kids,* London: Macmillan.

Countryside Commission (1995) *Growing in Confidence: Understanding People's Perceptions of Urban Fringe Woodland,* Northampton: Countryside Commission.

Davis, M. (1990) *City of Quartz,* London: Verso.

Downes, D.M. (1966) *The Delinquent Solution,* London: Routledge and Kegan Paul.

Downs, R. (1985) 'The representation of space: its development in children and in cartography', in R. Cohen (ed.) *The Development of Spatial Cognition,* New Jersey: Hillsdale, pp. 323–46.

—— and Liben, L. (1987) 'Children's understanding of maps', in P. Ellen and C. Thinus-Blanc (eds) *Cognitive Processes and Spatial Orientation in Animal and Man,* Dordrecht: Martinus Nijhoff, pp. 202–19.

Ellsworth, E. (1988) 'Illicit pleasures: feminist spectators and Personal Best', in

L.G. Roman and L.K. Christian-Smith (eds) *Becoming Feminine: The Politics of Popular Culture,* London: Falmer Press.

Environment and Planning D: Society and Space (1992) special issue, 10.

Fanon, F. (1967) *Black Skin, White Masks,* New York: Grove Press.

Fiske, J. (1987) *Television Culture,* London: Routledge.

Frankenberg, R. (1992) Contribution to Conference on the Consent of Disturbed and Disturbing Young People, London: Institute of Education, University of London.

Fyfe, N. and Bannister, J. (1996) 'City watching: closed circuit television surveillance in public spaces', *Area* 28, 1: 37–46.

Fyvel, T.R. (1963) *The Insecure Offenders,* Harmondsworth: Penguin.

Gannetz, H. (1989) 'Tjejer och stil', *Tvarsnitt* 2.

—— (1995) 'The shop, the home and femininity as a masquerade', in J. Fornas and G. Bolin (eds) *Youth Culture in Late Modernity,* London: Sage.

Gates, H.L. (ed.) (1986) *'Race', Writing and Difference,* Chicago: University of Chicago Press.

Gillespie, M. (1993) 'Technology and tradition – audio-visual culture among South Asian families in west London', in A. Gray and J. McGuigan (eds) *Studying Culture: An Introductory Reader,* London: Edward Arnold.

Gilroy, P. (1987) *'There Ain't No Black in the Union Jack': The Cultural Politics of Race and Nation,* London: Hutchinson.

—— (1993) *'The Black Atlantic: Modernity and Double Consciousness,* Cambridge, Mass.: Harvard University Press.

Goldthorpe, J.H., Lockwood, D., Bechhofer, F. and Platt, J. (1968) *The Affluent Worker in the Class Structure,* Cambridge: Cambridge University Press.

Goode, D. (1986) 'Kids, culture and innocents', *Human Studies* 9: 83–103.

Gramsci, A. (1971) *Selections from the Prison Notebooks,* London: Lawrence and Wishart.

Gray, A. and Mcguigan, J. (eds) (1993) *Studying Culture: An Introductory Reader,* London: Edward Arnold.

Griffin, C. (1985) *Typical Girls? Young Women from School to Job Market,* London: Routledge and Kegan Paul.

Hall, S. (1992) 'What is this "black" in black popular culture?', in M. Wallace and G. Dent (eds) *Black Popular Culture,* Seattle: Bay Press.

—— and Jefferson, T. (eds) (1976) *Resistance Through Rituals: Youth Subcultures in Post-war Britain,* London: Hutchinson.

Hart, R. (1979) *Children's Experience of Place,* New York: Irvington.

Hebdige, D. (1979) *Subculture: The Meaning of Style,* London: Methuen.

—— (1988) *Hiding in the Light: On Images and Things,* London: Routledge.

Hendrick, H. (1990) 'Constructions and reconstructions of British childhood: an interpretive survey, 1800 to present', in A. Prout and A. James (eds) *Constructing and Reconstructing Childhood,* Basingstoke: Falmer Press.

James, A. (1986) 'Learning to belong: the boundaries of adolescence', in A.P.

Cohen (ed.) *Symbolising Boundaries: Identity and Diversity in British Cultures*, Manchester: Manchester University Press.

—— and Jenks, C. (forthcoming) 'Public perceptions of chldhood criminality', *British Journal of Sociology.*

James, S. (1990) 'Is there a "place" for children in geography?' *Area* 22, 3: 278–83.

Katz, C. (1991) 'Sow what you know: the struggle for social reproduction in rural Sudan', *Annals of the Association of American Geographers* 8: 488–514.

—— (1993) 'Growing girls/closing circles: limits on the spaces of knowing in rural Sudan and US cities', in C. Katz and J. Monk (eds) *Full Circles: Geographies of women over the Life Course*, London: Routledge, pp 88–106.

—— (1994) 'Textures of global change: eroding ecologies of childhood in New York and Sudan', *Childhood* 2: 103–10.

—— (1995) 'Power, space and terror: social reproduction and the public environment', paper presented at Conference on Landscape Architecture, Social Ideology and the Politics of Place, Cambridge, Mass.: Harvard University. Available from the author.

Lesko, N. (1988) 'The curriculum of the body: lessons from a Catholic high school', in L.G. Roman and L.K. Christian-Smith (eds) (1988) *Becoming Feminine: The Politics of Popular Culture*, London: Falmer Press.

Lévi-Strauss, C. (1966) *The Savage Mind*, London: Weidenfeld and Nicolson.

Lowe, G., Foxcroft, D.R. and Sibley, D. (1993) *Adolescent, Drinking and Family Life*, Reading: Harwood Academic Publishers.

Madge, C., Raghuram, P., Skelton, T., Willis, K. and Williams, J. (1997) 'Methods and methodologies in feminist geographies: politics, practice and power', in Women and Geography Study Group, *Feminist Geographies: Explorations in Diversity and Difference*, London: Longman.

Matthews, M.H. (1992) *Making Sense of Place: Children's Understandings of Large-Scale Environments*, Hemel Hempstead: Harvester Wheatsheaf.

Matza, D. (1969) *Becoming Deviant*, New York: Prentice Hall.

—— and Sykes, G.M. (1961) 'Juvenile delinquency and subterranean values', *American Sociological Review* 26 (Summer 1961): 330–2.

Mayall, B. (1994) 'Introduction', in B. Mayall (ed.) *Children's Childhoods Observed and Experienced*, London: Falmer Press.

McCarney, J. (1980) *The Real World of Ideology*, Brighton: Harvester.

McCarthy, C. (1988) 'Marxist theories of education and the challenge of a cultural politics of non-synchrony', in L.G. Roman and L.K. Christian-Smith (eds) (1988) *Becoming Feminine: The Politics of Popular Culture*, London: Falmer Press.

McRobbie, A. (1980) 'Settling accounts with subcultures: a feminist critique', *Screen Education* 34: 37–49.

—— (1982) 'Jackie: an ideology of adolescent femininity', in B. Waites, T. Bennett and G. Martin (eds), *Popular Culture: Past and Present*, London: Croom Helm in association with The Open University Press.

—— (1993) 'Shut up and dance: youth culture and changing modes of femininity', *Cultural Studies* 7, 3: 406–26.

—— and Garber, J. (1976) 'Girls and subcultures', in S. Hall and T. Jefferson (eds) *Resistance Through Rituals: Youth Subcultures in Post-war Britain*, London: Hutchinson.

—— and Nava, M. (1995) 'Living on the edge: children as outsiders', *Yijscrift Voor Economischeet Socials Geografie* 86, 5: 456–66.

—— and Nava, M. (eds) (1984) *Gender and Generation*, London: Methuen.

Miller, W.B. (1958) 'Lower class as a generating milieu of gang delinquency', *Journal of Social Issues* 14: 111–21.

Mitchell, D. (1996) 'Aesthetics, anti-homelessness laws and the making of a brutal public sphere', paper presented at the Association of American Geographers Annual Conference, Charlotte, North Carolina, 13 April. Available from the author: Dept. of Geography, University of Colorado, Boulder, Colorado.

Moore, G.T. (1976) 'Theory and research on the development of environmental knowing', in G.T. Moore and R.G. Golledge (eds) *Environmental Knowing*, Stroudsberg: Dowden, Hutchinson and Ross.

Moore, R. (1986) *Childhood's Domain: Play and Place Development*, London: Croom Helm.

Nava, M. (1992) *Changing Cultures: Feminism, Youth and Consumerism*, London: Sage.

Patrick, J. (1973) *A Glasgow Gang Observed*, London: Eyre Methuen.

Pearson, G. (1983) *Hooligan: A History of Respectable Fears*, London: Macmillan.

Philo, C. (1992) 'Neglected rural geographies: a review', *Journal of Rural Studies* 8: 193–207.

Pressdee, M. (1986) 'Agony or ecstasy: broken transitions and the new social state of working class youth in Australia', occasional paper, South Australian Centre for Youth Studies, S.A. College of Adult Education, Magill, South Australia.

Professional Geographer (1994) 46, 1.

Professional Geographer (1995) 47, 1.

Prout, A. and James, A. (1990) 'A new paradigm for the sociology of childhood? Provenance, promise and problems', in A. James and A. Prout (eds) *Constructing and Reconstructing Childhood: Contemporary Issues in the Sociological Study of Childhood*, London: Falmer Press.

Qvortrup, J. (1994) 'Childhood matters: an introduction', in J. Qvortrup, M. Bardy, G. Sgritta and H. Wintersberger (eds) *Childhood Matters: Social Theory, Practics and Politics*, Aldershot: Avebury Press.

Roman, L.G. (1988) 'Intimacy, labour, and class: ideologies of feminine sexuality in the punk slam dance' in L.G. Roman and L.K. Christian-Smith (eds) *Becoming Feminine: The Politics of Popular Culture*, London: Falmer Press.

—— and Christian-Smith, L.K. (eds) (1988) *Becoming Feminine: The Politics of Popular Culture*, London: Falmer Press.

Rose, T. (1989) 'Orality and technology: rap music and Afro-American cultural resistance', *Popular Music and Society* 13, 4: 35–44.

—— (1990) 'Never trust a big butt and a smile', *Camera Obscura* 23: 109–31.

Schwarz, B. (1994) 'Where is Cultural Studies?' *Cultural Studies* 8, 3: 377–93.

Sibley, D. (1991) 'Children's geographies: some problems of representation', *Area* 3: 269–70.

—— (1995a) *Geographies of Exclusion*, London: Routledge.

—— (1995b) 'Families and domestic routines: constructing the boundaries of childhood', in S. Pile and N. Thrift (eds) *Mapping the Subject*, London: Routledge.

Smith, D. (1995) *Geographies of Exclusion*, London: Routledge.

Smith, D.E. (1988) 'Femininity as discourse', in L.G. Roman and L.K. Christian-Smith (eds) *Becoming Feminine: The Politics of Popular Culture*, London: Falmer Press.

Solberg, A. (1990) 'Negotiating childhood: changing constructions of age for Norwegian children', in A. James and A. Prout (eds) *Constructing and Reconstructing Childhood: Contemporary Issues in the Sociological Study of Childhood*, London: Falmer Press.

Spencer, C., Blades, M. and Morsley, K. (1989) *The Child in the Physical Environment: The Development of Spatial Knowledge and Cognition*, Chichester: John Wiley.

Storey, J. (1993) *An Introductory Guide to Cultural Theory and Popular Culture*, London: Harvester Wheatsheaf.

Tate, G. (1992) *Flyboy in the Buttermilk: Essays on Contemporary America*, New York: Simon and Schuster.

Thomas, C. (1980) 'Girls and counter-school culture', *Melbourne Working Papers*, University of Melbourne.

Turner, G. (1990) *British Cultural Studies: An Introduction*, Sydney: Unwin Hyman.

Valentine, G. (1996a) 'Children should be seen and not heard?: the production and transgression of adults' public space', *Urban Geography* 17, 2: 205–20.

—— (1996b) 'Angels and devils: moral landscapes of childhood', *Environment and Planning D: Society and Space* 14: 581–99.

—— (1997a) '"Oh yes I can". "Oh no you can't ": Children's and parents' understandings of kids' competence to negotiate public space safely', *Antipode* 29, 1: 65–89.

—— (1997b) '"My son's a bit dizzy." "My wife's a bit soft." Gender, children and cultures of parenting', *Gender, Place and Culture: A Journal of Feminist Geography* 4, 1: 37–62.

—— (1997c) 'A safe place to grow up? Parenting, perceptions of children's safety and the rural idyll', *Journal of Rural Studies* (forthcoming).

Waites, B., Bennett, T. and Martin, G. (eds) (1982) *Popular Culture: Past and Present*, London: Croom Helm in association with The Open University Press.

Walker, J.C. (1986) 'Romanticising resistance, romanticising culture: problems in Willis's theory of cultural production', *British Journal of Sociology of Education* 7, 1: 59–80.

—— (1988) *Louts and Legends*, Sydney: Allen and Unwin.

Wallis, R. and Malm, K. (1984) *Big Sounds from Small Peoples*, London: Constable.

Ward, C. (1977) *The Child in the City*, Harmondsworth: Penguin.

—— (1990) *The Child in the Country*, London: Bedford Square Press.

Watson, I. (1993) 'Education, class and culture: the Birmingham ethnographic tradition and the problem of the new middle class', *British Journal of Sociology of Education* 14, 2: 179–97.

West, C. (1990) 'The new cultural politics of difference' in R. Ferguson, M. Gever, T.T. Minh-ha and C. West (eds) *Out There: Marginalisation and Contemporary Cultures*, Cambridge: MIT Press.

Wheeler, E.A. (1991) '"Most of my heros don't appear on no stamps": the dialogics of rap', *Black Music Research Journal*, 11, 2: 193–216.

Willis, P. (1977) *Learning to Labour: How Working Class Kids Get Working Class Jobs*, Westmead: Saxon House.

—— (1978) *Profane Culture*, London: Routledge and Kegan Paul.

—— (1990) *Common Culture: Report of the Gulbenkien Inquiry*, Milton Keynes: Open University Press.

Winship, J. (1987) *Inside Women's Magazines*, London: Pandora.

Women's Studies Group, Centre for Contemporary Cultural Studies (ed.) (1978) *Women Take Issue*, London: Hutchinson.

Yablonski, L. (1967) *The Violent Gang*, Harmondsworth: Penguin.

one

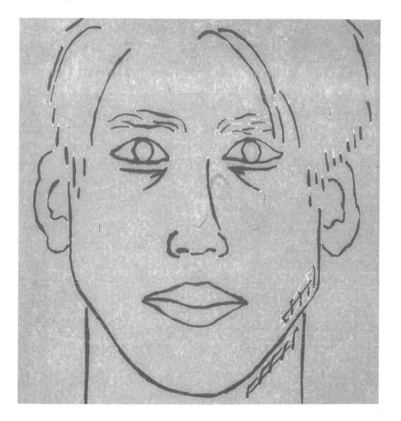

representations

Issues of representation have been at the heart of much work on youth. The young are popularly imagined to be deviant and subversive (youth-as-trouble) or exotic, stylish and laid-back (youth-as-fun). Both representations are evident in David Oswell's chapter which opens this section. He explores the debate about television as a bad influence on the young, paying particular attention to the relationship between television and 'moral panics' about youth violence; and he considers the way television has struggled to imagine young people as a 'community' of viewers.

The following three chapters all contest the ways in which young people and their cultures are represented while at the same time debating the very ways in which representation as a concept is researched and theorised. Rather than viewing 'youth' as a uni-dimensional category, the chapters by Clare Dwyer, David Parker and Ruth Butler each explore how 'youth' intersects with other identities. Clare Dwyer looks at how young Muslim women talk about the media and the contextual processes by which it is used in the negotiation of their own identities. David Parker considers the ways in which young British Chinese people, residing in the UK and Hong Kong, are rethinking their identities. He clearly demonstrates the importance of Hong Kong, both in a historical and in a contemporary context, in the formulation and re-production of Chinese youth cultures. Ruth Butler's chapter exposes some of the stereotypical and negative ways disabled young people are represented in different contexts and goes on to show how the meaning of 'normal' is socially and culturally contested by disabled youth.

While these three chapters show some of the ways in which young people have to continually challenge the representations that society (and indeed academia) places upon them, the final chapter in this section, by Marion Leonard, examines the way some young women are using new forms of communications (including print and electronic media) to create their own representations of their lives. She considers how zines promote ideas about riot grrrls, enabling young women, either as producers or as audience, to create an 'underground community' and 'sub-culture' which is disparately geographically located. The chapter contests the meaning of boundaries and scale, both themes which are debated further in Part Two.

2

A QUESTION OF BELONGING

television, youth and the domestic

•

David Oswell

While I have the greatest admiration for The Children's Programme and –
knowing the difficulties under which T.V. producers work – also very
much praise for the 'adult' programmes, I DO feel that the 'in-between'
age has been left out in the cold.
                              (Petula Clark quoted in Heiress, *October 1953)[1]*

'Youth', of course, is an elastic state. The 36 delegates from seven countries . . . apparently found themselves unanimous enough on when 'youth'
ends, but not when it begins, so that the United Kingdom delegation
arrived prepared to discuss the 14–20 age group and discovered others
concerned about the impact of television on five-year-olds.
                              (Glasgow Herald, 25 February 1960)[2]

Despite the significant amount of research on youth cultures within
cultural studies, sociology and social psychology since the 1950s, there
has been surprisingly little academic work on 'youth' and television.
This is perhaps due, on the one hand, to the domestic, private nature of
television viewing and, on the other, to the presentation of 'youth' as a
public problem. Whereas youth is displayed through style, music, ritual
and resistance, television is less spectacular and urban, altogether more
ordinary and suburban (Oswell, 1994; Silverstone, 1994). Moreover, the
relationship between youth and television is also entangled within
debates concerning the distinction between 'adult' and 'child'. It is as
if the attempt to imagine young people's relation to television is always

troubled, as if youth is forever figured as in between, never to find a happy resting place. In order to unpack this we need to see how television has been construed as a particularly domestic medium and how the notion of youth television has always been something of an impossible object.

In this chapter, then, I want to look at some contemporary questions about the relationship between young people and television.[3] How is young people's television viewing constructed as a problem? How are young people addressed by broadcast television? And who identifies with 'youth television'? These questions form, if you like, particular surfaces upon which the relationship between youth and television is inscribed. However, instead of simply analysing these surfaces of the present, I want to consider their conditions of emergence. In this sense, 'how' becomes a genealogical question: a history of the present (Foucault, 1977).

YOUNG PEOPLE, TELEVISION AND MORAL PANIC

One such surface upon which 'youth' is cited and performed is the well-worn area of young people and moral panics. The articulation of young people's television viewing as a moral panic is a regular occurrence in the press and on radio and television. Similarly in public debates, the concept of the 'moral panic' has become, as it were, the calling-card of the cultural critic. The concept itself emerged in the 1970s, in the work of Stanley Cohen (1972) and Stuart Hall and others (1978), as a way of describing the construction of and reaction to certain popular cultural forms: Cohen looks at the concern about the 'Mods' and 'Rockers' in the 1960s; and Hall *et al.* look at the fabrication of the 'mugger' (as a racialised Other) in the 1970s and its effectivity in establishing a consensus which made possible the Conservative Party victory in 1979. In both these accounts the media, particularly the press, is of considerable importance in terms of its ability to mobilise a consensus and constitute young people as scapegoats or 'folk devils'. The media display youth as a public problem: as a problem shared and publicly discussed and as a problem concerning the maintenance of order and civility within public spaces. (This section is a reworking of some ideas developed in Oswell 1994b.)

Martin Barker and David Buckingham, in their different ways, draw upon these arguments to discuss young people's media consumption (Barker, 1984a; Buckingham, 1993). However, for Barker and Buckingham it is television, not youth, which is constructed as the scapegoat. Television, they argue, is conceived, primarily by the press, as all-powerful, whereas young people are constructed as 'children': innocent, manipulable and needing protection. Behavioural psychology, the reactionary press and conservative political interests converge '[t]o define young people as merely vulnerable and credulous [which] thus represents a forceful legitimation of adult power and control' (Buckingham, 1993: 4).[4]

A relatively recent British example concerns the murder of James Bulger, aged 2, in February 1993 by Jon Venables and Robert Thompson, both aged 10, and the panic that ensued following Judge Morland's statement, that the boys' watching of *Child's Play 3* was an important factor influencing their actions.[5] The following year, in March 1994, Elizabeth Newson, the distinguished Professor of Developmental Psychology at the Child Development Research Unit, University of Nottingham, presented a paper, titled 'Video violence and the protection of children', to both Houses of Parliament. As an attempt to explain the murder, she held liberal academics responsible in their failure to clearly warn of the dangers of televisual violence to young people.[6] As an attempt to sway popular opinion and public policy, the paper was clearly part of a wider campaign orchestrated around Liberal Democrat MP David Alton's proposed amendment to the Criminal Justice Bill, which was due to be debated on 12 April 1994. Alton had gained cross-party support of around 220 MPs, including former Labour leader Neil Kinnock, Liberal Democrat Sir David Steel and Conservative MP Sir Ivan Lawrence (of the Home Affairs Select Committee), to restrict the availability of 'violent' videos in the home.

There were numerous proposals for regulatory controls, such as: making it illegal for parents, or other adults, to show adult videos to young people or to leave these videos in the home where teenagers or young children might find them; improving local government's ability to enforce existing legislation; introducing identity cards for young people to prove their age when buying or renting videos; and reviving the 'Restricted 18' category, which would limit these videos to registered sex shops which young people are forbidden to enter (The *Independent*, 12 April 1994). The *Daily Express* argued that there should be more legislation to tackle the selling of 'pirate' videos at market stalls and car boot sales (13 April 1994). Lynda Lee-Potter appealed to *Daily Mail* readers to take upon themselves the 'responsibilities towards the most vulnerable members of our society' and she suggested that 'bad parents' should be fined or 'if necessary' given a custodial sentence (*Daily Mail*, 13 April 1994).[7] The *Evening Standard* similarly focused debate on the responsibility of the parent: '[t]he onus will always be on parents to consider the psychological health of their offspring. A symptom is treated, but the disease remains' (*Evening Standard*, 13 April 1994). Michael Howard, Home Secretary, as a move to head off a growing backbench revolt, announced on 12 April 1994 that he would table a new clause in the Criminal Justice Bill in the Lords which would accommodate David Alton's demands. These would include the British Board of Film Classification (BBFC) taking account of videos which are likely to cause 'psychological harm to a child' or which present 'an inappropriate model for children', the tightening of the BBFC's own classifications, and a tightening of existing penalties for video store owners who rent inappropriate films to teenagers and children.

It is clear that the Newson paper and the amendments to the Criminal Justice Bill, regarding video violence, fed into a wider set of moral panics. In an article about schooling, Beatrix Campbell referred to 'the great moral crusade against children' (The *Independent*, 13 April 1994).[8] This crusade included the campaigns waged in the press since the summer of 1993 about 11 year old boys taking heroin and other drugs, absconding from youth detention centres and sleeping with prostitutes, and joyriding in city centres and housing estates. These panics were undoubtedly linked to concerns about 'the break-up of the family' (including the continued stigmatisation of single mothers and the recent debates about the responsibilities of fathers) and to the powerlessness of the police and the judiciary to properly bring these young offenders to task.[9]

There is, as Beatrix Campbell and Judith Dawson argue, a general problem about the *ungovernability* of young people: 'The decade of the discovery of the dangers of childhood has become the decade of dangerous children' (Campbell and Dawson, 1994). The construction of the young as both *in danger* and *dangerous* has, though, a longer genealogy which we can trace back to the nineteenth century (Rose, 1985). These discourses constitute lines of consent and mark out new boundaries between 'us' and 'them' (McRobbie, 1994). The *othering* of young people is, we might argue, constituted within an ambivalence which leads both to the desire to expel these dangerous youths from the realms of decent society (to exclude them from the boundaries of citizenship) and also to the desire to protect them from further harm (to lead them out of the wilderness back into the fold). Moreover, whereas those in danger are constituted as 'children', those deemed dangerous are quite clearly constituted as 'youth'. For example, at the time of the murder of James Bulger, both Jon Venables and Robert Thompson were only 10 years old and yet they were presented and referred to as 'youths' (cf. Kember, 1995: 121). The category of 'youth' is deployed when young people are seen to no longer perform the *proper* modes of conduct of 'childhood'. 'Youth' defines a moment of disturbance: a space *in between*.

The history of these disturbances has been written as a 'history of respectable fears'. Geoffrey Pearson (1983) has discussed the emergence of moral panics of youth-as-ungovernable-mob. He refers to the campaigns against the penny dreadfuls and music hall in the mid- to late nineteenth century and against cinema in the early twentieth century. But what is made clear, when one considers the wider concerns, is that it is not simply television that is constructed as the scapegoat, but young people as well: the two feed into each other in an expansive cycle. Moreover, the visibility of 'youth' in research, debates and public discussion about young people and television does not centre simply upon the *publicity of the crowd*, but also upon the *psychology of the withdrawn and introverted self*. And although the Bulger case clearly raises questions about public order and surveillance (Kember, 1995), these issues are symptomatic of something which is altogether more homely and familial, albeit in an uncanny (*Unheimlich*) sort of way.[10] In this discourse 'youth' is imagined as the solitary viewer in a darkened room with the steely reflection of television images upon his/her face. This is the image of 'youth' as addict, caught in fantasy and disavowing reality (Oswell, 1994). This is a history, not of armed gangs on the street (of the 'hooligan'), although it is still a problem of masculinity, but of youth shying away from the public gaze.

TELEVISION FOR YOUNG PEOPLE: PROGRAMMES AND PROGRAMME POLICY

Although the display of youth-as-pathology is not one which television corporations regularly employ in their programmes aimed at young people, a similar set of dynamics, concerning the entanglement of youth within the binaries of public/private and adult/child, do come to bear on the debates and institutional practices concerning 'youth television'. On the one hand there is an attempt to display the spectacle of youth as style, music, culture and resistance; on the other, an attempt to address young people within the home. Initially this attempt to domesticate youth took form as an attempt to address all young people as 'children'.

It is enlightening to look at some of the debates within the BBC in the early 1950s. Until this time, the BBC, as it had done since the 1920s, categorised all young people as 'children'. There was, in policy terms, no difference between an 8-year-old and a 16-year-old. However, in the 1950s (at a time when young people were critical of the notion of 'children's programmes') the making of television programmes for young people became increasingly problematic. For example, Freda Lingstrom, Head of Children's Programmes (1951–6), talked about the diversity of tastes between the youngest and the oldest members of this audience.

in infancy they will cheerfully eat coal or drink their bath water; at ten their likes and dislikes are violent but changeable. At fifteen, however inarticulate they may be in expressing opinions to adults, they have formed them – indeed, by that time a large proportion will be wage-earners compelled to rely upon their own judgement.

(Lingstrom, 1953: 101)

Likewise Mary Adams, who as Head of Television Talks was responsible for setting up children's television at the BBC, stated that: 'At present, children from two years keep company with teenagers. How shall they be separated, each provided with adequate satisfaction?' (Adams, 1950: 86). Her solution was to have 'specialist' programmes for these distinct groupings *within* Children's Programmes. The distinct programmes were to be properly signposted and announcements made so as to separate the different age groupings. She also considered showing programmes for different age groupings on different days of the week.

At the same time, Mary Adams argued that young people, from pre-school child to teenager, were more homogenous than they had ever been. Adams argued that this was the result of changes in schooling, economic factors, the widespread distribution of *Children's Hour* on sound radio and the fact that 'the great majority of families with television sets are in the lower income levels' (Adams, 1950: 85). Thus, despite the recognition that there was a difference between the different age groups and that there was a problem of using a visual medium to translate the 'widest range of emotional, aesthetic and intellectual experience' (Adams, 1950: 86) to such a population divided by age, the difference was conceived as being contained within the category of the 'child audience'. Given these constraints, what should the teenager or the adolescent be offered?

Plays of action or detection? Documentaries like 'London Town', or 'How to be a Doctor'? Travel films? Advice on collecting? Hobbies, such as carpentry or metal work? Coaching for sports and athletics? Dancing lessons? Crafts? Competitions which test his [sic] knowledge and abilities? Campanology?

(Adams, 1950: 87)

For Adams this section of the child audience could be given 'realistic' settings and environments. Instead of playing with models and toys, teenagers could be shown the 'real thing' in the outside world. Television could be used to transport this audience to other cultures and other worlds. What is significant, though, is that at no time did she consider that the teenage audience might constitute a distinct and separate category of audience in its own right. The teenage audience was always seen as a particular sub-category of the child television audience. The 'teenager' defined a particular stage, along with other stages of childhood development, and programmes were scheduled according to the different age groups in the form of a 'ladder'. The division within children's programmes both divided the viewing habits of young people and provided a means through which they could progress from the programmes appropriate for one age group to another.

However, by the late 1950s, after the setting up of commercial television in the UK and the new found affluence of young people, staff at the BBC were resigned to the fact that teenagers constituted a distinct television audience. Owen Reed, Head of Children's Programmes (1956–63), argued that:

> The tastes of older children (i.e., the 12+ age group) are so different from those of younger children that there is nothing to be gained by trying to include them under the title of 'children' at all. Teen-agers are more likely to resent the appellation than to be drawn to it.
>
> (Reed, 2, T16/45/2)

Internal discussion within the BBC reflected structural and cultural changes concerning the position of the teenager in postwar Britain (Hall and Jefferson, 1976). The problem for the BBC, though, was what programmes to make for them, when they would be shown and whether they would be watched.

Freda Lingstrom had argued that the 6 p.m. to 7 p.m. slot, which was called the 'toddlers' truce' (as it was a time when mothers were expected to put their children to bed), should be used for teenage programmes: 'It occurs to me that the period 6 to 7 might well be used for handicraft demonstrations or for a series on careers and trades likely to be of interest to teenagers' (Lingstrom, 30 December 1955, T16/45/2). By Saturday 16 February 1957, after the ITV companies had petitioned Charles Hill, the Postmaster General, arguing that 'it was the responsibility of parents, not the state, to put their children to bed at the right time', the toddlers' truce was formally ended and *Six-Five Special* started (Hill, 1991: 90). It was pop music rather than handicrafts which finally won the day.

Six-Five Special had an audience of about six million and was designed to appeal, not just to the tastes and pleasures of young people, but to 'the young in spirit of all ages' (*Radio Times*, 22 February 1957: 4, quoted in Hill, 1991: 92). It was a mix between a variety show and a magazine programme: rock'n'roll with a Reithian twist. Studio bands and light entertainment were used to address teenagers within a familial viewing context, and its two presenters, originally Pete Murray and Josephine Douglas, held the show together. In competition ITV scheduled *Oh Boy!*, named after The Crickets' song, in June 1958. Unlike *Six-Five Special*, *Oh Boy!* featured non-stop music and was broadcast live from the Hackney Empire theatre in London. In 1959 the BBC launched *Juke Box Jury* as a successor to *Six-Five Special*. The show addressed both teenagers and adults alike. David Jacobs introduced excerpts from current record releases. A panel of guests were then invited to offer their opinions. There was invariably an age difference between the panel and the studio audience of young people. The latter visibly identified with the music more than their elders. And whereas the panel of critics commented on the music in a parental manner, the pop performers were able to enact a mild-mannered rebelliousness. Comments such as 'terrible' and 'what a noise' served to confirm teenage opinion that adults just did not understand them (Hill, 1991: 102). The combination of 'adult' critics and 'teenage' music allowed all members of the family audience watching at home to confirm their prejudices. Although ITV's *Oh Boy!* (in an attempt to shift away from the joint address to youth and parent) had used the format of a succession of pieces of music and very little commentary, it was not until 1963 that the BBC more concertedly addressed a youth audience with *Ready Steady Go!* Less familial than *Juke Box Jury* and more like *Six-Five Special*, *Ready Steady Go!* initiated a format which is recognisably deployed in more recent youth television: music, dancing, competitions within, as John Hill states, 'an informal party atmosphere' (Hill, 1991: 103).

IMAGINING THE YOUTH AUDIENCE
ACROSS TIME AND SPACE

Teenage programmes, scheduled during the early evening slot, attempted to produce youth as a repeated ritualisation, a form of performance which attempted to connect young people with a community of youth (cf. Butler, 1993: 2). The television schedule made possible the complex co-ordination of broadcasting practices and individual viewers. It organised different sectors of the population within specific cultural zones: times of the day and familial spaces. It was, to use Anthony Giddens' phrase, a 'time–space ordering device' (Giddens, 1991: 20). In this sense the imagining, or invention, of a youth community (cf. Anderson, 1983) was not simply a question of representation, but of a regular and repeated process.

This ritualisation was seen, in the late 1950s, to displace more traditional communities. Television, and other forms of modern communication (radio, postal service, telecommunications and transport systems), were seen to fracture old loyalties and authorities. The Crowther Committee Report on education stated that:

> Not so long ago a man accepted as natural a loyalty to his home town and county, to the church of his birth and to his father's political party. He followed their lead unquestioningly in their respective spheres: they in turn gave him significance. To change sides, to abandon an old loyalty, was something only to be done for grave cause and sometimes with serious misgiving.
>
> (Ministry of Education, 1958: 12–3)

The Committee argued that the 'mass media' were responsible for the breakdown of these earlier communities and for the creation of new forms of public opinion: 'Teenage opinion is often badly informed, fickle and superficial' (Ministry of Education, 1958: 43). Youth programmes, such as *Six-Five Special*, were exemplary in this respect. They presented new forms of television community based upon style and music. And they did so in such a manner as to displace 'youth' from its cultural and geographical place and to reconstitute it within an electronic community (cf. Morley and Robins, 1995). In this sense, my argument about the temporal and spatial relocation of youth concurs with Giddens' analysis of the temporal and spatial dynamics of modernity. Television disembeds the social relations of youth from their locality and re-embeds them across large tracts of time and space (Giddens, 1991).

It is clear, though, that with the emergence of a multi-channel and increasingly globalised television environment the temporal and spatial dynamics of youth programming have significantly shifted. Satellite and cable television no longer define youth as a point in the schedule, but as niche markets with their own particular channels. In doing so 'youth television' becomes more like radio, it becomes a day-in and day-out production, and viewing becomes very much a secondary activity. The importance of the schedule within MTV, for example, is in defining communities according to taste, rather than age. As E. Ann Kaplan has argued:

> the very fact that MTV addresses itself to a broad, generally youthful section of the American public that ranges from 12 to 34 on up, distinguishes it from earlier rock cultures, which addressed much more homogeneous groups, clearly defined in terms of values, age, and social status.
>
> (Kaplan, 1987: 8–9)

Similarly, one aspect of the new televisual forms is their address to youth as an imagined global community: not as a homogeneous style, but a pick'n'mix of taste cultures.

Nevertheless, a question that concerned television institutions in the 1950s and, to a certain extent, concerns such institutions in the 1990s (especially terrestrial television broadcasters) is whether 'youth television' is actually watched by young people. Namely, if youth television constructs imagined communities across time and space, who actually identifies with these constructions and who stays at home to consume the images? In the late 1950s the BBC conducted audience research and set up a Teenage Advisory Committee to look into this question: how successful was the BBC in appealing to young people and how successful were they at constituting a 'teenage' community? A Report of the Teenage Advisory Committee stated in 1959 that 'the 14 to 19 group is a section of the television audience for which television is less important than for almost any other group' (Teenage Advisory Committee, 1959, 2, R9/13/180). It stated that young people 'are more inclined to go out and do things', that they are starting to make girl- and boy-friends, that they might be in 'some sort of friction with their parents' and that they 'want to be with groups of people of their own age' (ibid.).[11] It seems that television was tainted too much with the familial and the domestic to keep teenagers at home.[12]

The Report also argued that research had indicated that 'there was no such thing as a "teenagers' hour" between 6pm and 7pm': 'We could find little to suggest that, as an audience, they constituted an intermediate third group between children and adults, let alone an extension of the child audience' (ibid.). Moreover, it would seem, the programme tastes of teenagers (when they stayed at home) were the same as adults. Hence young people did not seem to identify with television nor did they seem to identify with their construction on television as 'teenagers'.

Teenagers constituted an audience that did not identify with themselves as an audience. The Advisory Committee stated that the reason for this lay in the fact that it was 'not so much that their specialist interests aren't catered for, but that their *attitude* isn't acknowledged or reflected' (ibid., 7). Television producers picked up on this and clearly refused to acknowledge that 'youth television' was not desirable. On this basis broadcasters changed tactics and attempted both to represent youth as streetwise and also to reflect its attitudes. Hence the abiding connotations of youth television as ironic, critical and sassy. More recent British examples would undoubtedly include *TFI Friday* (Channel Four), *The Girlie Show* (Channel Four) and *God's Gift* (ITV). Chris Evans, the ex-Radio 1 DJ and presenter of *TFI Friday*, exemplifies the ironic laddishness of a *Loaded* boys' own culture: his football shirt hung low over his swinging butt, his Adidas trainers peeking out of his slightly flared trousers and his Elvis Costello National Health glasses bobbing in tune to Oasis.[13] He talks 'babes' and hedonism to his adoring mix-gendered studio audience. Or, on *The Girlie Show*, Rachel Williams drapes a 1940s retro raincoat over her bra-ed and briefed body and invites her predominantly male studio audience to pick out the star celebrity 'Wanker of the Week'. The former model, with pierced ears, lip and tongue, parades a sassy but inviting post-feminist stilettoed aggressiveness.[14] Nevertheless, despite the obvious effort, television broadcasters have continually failed to capture a significant mass audience of young people.[15] It is as if television always arrives on the scene too late. It always names a trend that has already happened

The attempt to constitute young people as a television audience defined by age was, and still is, to a certain extent, forlorn. Recent research indicates that the 18–30 age range watch television least of all the population. And yet youth programmes – 'television with attitude' – are watched. They appeal not just to young people, but of necessity (in order to gather a large enough audience) to a diversity of ages. A recent report in the *Evening Standard* stated that a large proportion of the audience of *God's Gift* (a programme supposedly aimed at young 'laddettes') was made up of elderly men (*Evening Standard*, 13 May 1996).[16] Youth, in this sense, is an impossible object. It is dispersed across different ages and it names an audience that refuses to recognise itself as such.

The relationship between youth and television is thus an unhappy one. Teenagers have been caught in a double bind. On the one hand the construction of television as domestic and familial has meant that young people have had little desire to stay at home and watch. Attempts by television producers to escape this problem have always fallen short of the 'real thing'. And on the other hand, those teenagers who have stayed at home watching television have been constructed as addicts and pathological. Psychologists and other experts have talked about young 'heavy viewers' as introverted and about the way these viewers have been drawn into the fantasy scenarios of television rather than the public duties of 'normal' life. Either through their 'bad press' in expert circles or through familial circumstance, teenagers have been ousted from the home.[17]

At stake in the discursive constructions I have outlined above are two dynamics. The relationship of youth and television, as exemplified in both the panics surrounding the Bulger case and the difficulties of youth television programming, is intersected by problems concerning the nature of youth *between* the categories of 'adult' and 'child' and *between* the spaces of public and private (Holland, 1992: 104). The relationship is typified by a peculiar incommensurability. When Petula Clark asserts that youth is an 'in-between age', she identifies this in terms of a geography of exclusion. The teenage television audience 'has been left out in the cold'. It has been excluded from the warmth of the home. She defines youth, quite literally, as homeless.

ACKNOWLEDGEMENTS

I would like to thank the BBC Written Archives at Caversham for letting me make use of their documents. When referring to material from the Archive, I use their file index notation (e.g. T16/45/1). I have not referred to specific sub-sections of files.

NOTES

1 Petula Clark was speaking as the President of the newly formed Teenage Televiewers' Society.
2 The quote comes from a report on the Western Union conference, in Rome 1960, on the impact of television on youth.
3 I use the term 'television' to refer to video, satellite and cable as well as terrestrial broadcast television (cf. Murdock, 1994).
4 The debate about 'video nasties' in the early to mid-1980s provides an example (Barker, 1984a). In a quick-witted strategic response to the Video Recordings Bill, Barker collected a group of academics and journalists to write a critique of what they saw as the ideological production of a panic around young people's viewing of certain horror videos. In Barker's opening essay he attacked the social scientists, psychiatrists, paediatricians and others who, as the Parliamentary Group Video Enquiry, had been constructed in certain sections of the press as 'folk heroes'. Steven Barnett refers to this in terms of 'the language of political pressure groups and tabloid editorials' (The *Independent*, 5 April 1994). It would be incorrect, though, to argue that all moral panics are predicated upon conservative political interests. Barker's fascinating work on the campaign against the 'horror comics' in the 1950s uncovers an unholy alliance between the political interests mentioned above and the Communist Party of Great Britain (Barker, 1984b).
5 The judge was later shown to be incorrect in his statement (see Petley, 1994). The film, in question, is about a young child's attempt to deal with an obnoxious little monster-doll called Chucky.
6 In her closing paragraph she stated that: 'Many of us hold our liberal ideals of freedom of expression dear, but now begin to feel that we were naive in our failure to predict the extent of the damaging material and its all too free

availability to children' (Newson, 1994: 7). This statement was widely reported in the press. The *Evening Standard* paraded the headline: 'U-Turn Over Video Nasties'. It referred to the naiveté of the experts and referred to the report as a 'confession' (*Evening Standard*, 31 April 1994). The *Guardian* carried the headline 'Video-Crime Link Stronger Than Thought, Say Child Experts' and the *Daily Telegraph* carried '"Naive" Experts Admit Threat of Violent Videos' (1 April 1994). The professional experts included psychologists, psychiatrists, criminologists, doctors and others 'professionally concerned with children' (Newson, 1994: 3).

7 She stated that:

> There are thousands of children in this country with fathers they never see and mothers who are lazy sluts.
>
> They are allowed to do what they want, when they want. They sniff glue on building sites, scavenge for food and until now, they were free to watch increasingly horrific videos. By 16, they are disturbed and dangerous . . . . It's no longer enough to look after our own sons and daughters. The time has come when we need to feel a commitment towards the sad, disturbed, neglected and abused youngsters around us. It's too simplistic and destructive to opt out of involvement.
>
> (Daily Mail, 13 April 1994)

8 The distinction between 'children' and 'youth' will become evident later in this section.

9 John Pitts argues that the panic came to a head a year earlier in 1992. He argues that this was the year when, in many people's minds, there was 'a rising sense of panic about youth crime' (Pitts, 1995: 5).

10 I discuss the psychology of addiction and the pathologisation of certain families and domestic environments (predominantly working class) at greater length elsewhere (Oswell, 1995).

11 The Crowther Report similarly reported that:

> Less than 8 per cent of the boys and girls who had left modern schools two years before were not still living at home at the time of our Social Survey (when they were, presumably, about 17) but 36 per cent of the boys and 32 per cent of the girls had only spent one or two evenings at home out of the previous seven. A further 28 per cent of the boys and 6 per cent of the girls had been out every evening.
>
> (Ministry of Education, 1959: 36)

12 This is wonderfully represented in the British New Wave films *Saturday Night and Sunday Morning* (Karel Reisz, 1960) and *The Loneliness of the Long Distance Runner* (Tony Richardson, 1962) in which the young male protagonists are clearly alienated from the cosy domesticity and femininity exemplified by watching television (Barr, 1986).

13 *Loaded* is a particularly laddish men's magazine: a mix between *GQ* and the *Sun*.

14 From these examples at least, it would seem that the gender lines of current youth programming are clearly drawn.

15 This is not to say, however, that the audience that youth programmes do attract are not seen as profit-making.

16 It reported that almost three-quarters of viewers were men and one-sixth were over 65 years old.

17 And of course, once out of the home they constituted a different kind of problem. Each problem feeds into the other.

REFERENCES

Adams, M. (1950) 'Programmes for the young viewer', *BBC Quarterly* 5: 81–9.

Anderson, B. (1983) *Imagined Communities: Reflections on the Origin and Spread of Nationalism*, London: Verso.

Barker, M. (1984a) 'Nasty politics or video nasties?', in M. Barker (ed.) *The Video Nasties: Freedom and Censorship in the Media*, London: Pluto Press.

—— (1984b) *A Haunt of Fears: The Range History of the British Horror Comics Campaign*, London: Pluto Press.

—— (1993) 'Sex, violence and videotape', *Sight and Sound* 3, 5.

Barr, C. (1986) 'Broadcasting and cinema: screens within screens', in C. Barr (ed.) *All Our Yesterdays: 90 Years of British Cinema*, London: British Film Institute.

Buckingham, D. (1993) 'Introduction: young people and the media', in D. Buckingham (ed.) *Reading Audiences: Young People and the Media*, Manchester: Manchester University Press.

Campbell, B. and Dawson, J. (1994) 'Censorship: inside stories', *Sight and Sound* 4, 9: 30–1.

Cohen, S. (1972) *Folk Devils and Moral Panics: The Creation of the Mods and Rockers*, London: MacGibbon and Kee.

Foucault, M. (1977) 'Nietzsche, Genealogy, History', in Donald F. Bouchard (ed.) *Language, Counter-Memory, Practice*, Ithaca: Cornell University Press.

Giddens, A. (1991) *The Consequences of Modernity*, Cambridge: Polity Press.

Hall, S., and Jefferson, T. (eds) (1976) *Resistance Through Rituals*, London: Hutchinson.

—— (1978) *Policing the Crisis: Mugging, the State, and Law and Order*, London: Macmillan.

Hill, J. (1991) 'Television and pop: the case of the 1950s', in J. Corner (ed.) *Popular Television in Britain: Studies in Cultural History*, London: British Film Institute.

Holland, P. (1992) *What is a Child? Popular Images of Childhood*, London: Virago.

Kaplan, E.A. (1987) *Rocking Around the Clock: Music Television, Postmodernism, and Popular Culture*, Methuen: London.

Kember, S. (1995) 'Surveillance, technology and crime: the James Bulger case', in M. Lister (ed.) *The Photographic Image in Digital Culture*, London: Routledge.

Lingstrom, F. (1953) 'Children and television', *BBC Quarterly* 8: 96–102.

McRobbie, A. (1994) 'Folk devils fight back', *New Left Review*, 203: 107–116.

Ministry of Education (1959) *Report of the Central Advisory Council for Education: England, Vol. 1· 15 to 18*, London: HMSO.

Morley, D., and Robins, K. (1995) *Spaces of Identity: Global Media, Electronic Landscapes and Cultural Boundaries*, London: Routledge.

Murdock, G. (1994) 'Money talks: broadcasting, finance and public culture', in Stuart Hood (ed.) *Behind the Screens: The Structure of British Television in the Nineties*, London: Lawrence and Wishart.

Newson, E. (1994) 'Video violence and the protection of children', unpublished paper presented to the Houses of Parliament.

Oswell, D. (1994) 'All in the family: television in the postwar period', *Media Education Journal* 17: 4–8.

—— (1994b) 'This is not a moral panic: creating child viewers as citizens,' paper presented at the Citizenship and Cultural Frontiers Conference, University of Staffordshire.

—— (1995) 'Watching with Mother: a genealogy of the child television audience', unpublished Ph.D. thesis, Open University.

Pearson, G. (1983) *Hooligan: A History of Respectable Fears*, London: Macmillan.

Petley, J. (1994) Unpublished paper presented at the 'Effects Tradition' conference, Brunel University.

Pitts, J. (1995) 'Youth crime', *Community Care*, London: Reed Business Publishing.

Rose, N. (1985) *The Psychological Complex*, London: Routledge and Kegan Paul.

Silverstone, R. (1994) *Television and Everyday Life*, London: Routledge.

Valentine, G. (1996a) 'Angels and devils: moral landscapes of childhood', *Environment and Planning D: Society and Space* 14: 581–99.

—— (1996b) 'Children should be seen and not heard: the production and transgression of adults' public space', *Urban Geography* 15: 205–20.

•••••••••••••••••••••••••••

3

CONTESTED IDENTITIES

challenging dominant representations of young British Muslim women

•

Claire Dwyer

'Other people think Muslim women are not allowed to go out they are not allowed to do this, they have to cover themselves, they're chained to the kitchen sink, but we're not like that.'

(Robina)[1]

INTRODUCTION: 'NEW' ETHNICITIES AND YOUTH CULTURE

Robina, speaking during a group discussion of young Muslim women,[2] highlights the extent to which the construction and negotiation of her cultural identity depends upon a contestation of dominant representations. This chapter examines the ways in which young British Muslim women are involved in the construction and contestation of their own identities. The analysis focuses on how young British Muslim women talk about specific aspects of youth culture – dress, music and television – as important cultural spaces through which identities are negotiated. The chapter suggests that there are many different understandings of what it means to be a young British Muslim woman, and young women themselves are actively engaged in challenging dominant representations and producing new meanings. This process of identity contestation and (re)construction is always contextual and negotiated. Identities are negotiated differently ,in different places and are constructed and contested within particular spaces.

An exploration of this process of identity construction and contestation illustrates how young people are involved in what has been termed the 'reinvention of ethnicity' (Gilroy, 1993a) and the making of 'new ethnicities' (Hall, 1988). Reflecting on the politics of anti-racism and the experiences of black people in Britain both Hall and Gilroy have opened up new ways of theorising ethnicity and identity. Stuart Hall has argued for a conceptualisation of ethnic identities which are not primordial, essential and fixed but instead can be recognised as constructed, multiple and changing. This theorisation of ethnicity depends upon an understanding of how all identities are articulated out of particular places, histories and experiences: 'The term ethnicity acknowledges the place of history, language and culture in the construction of subjectivity and identity, as well as the fact that all discourse is placed, positioned, situated and all knowledge is contextual' (Hall, 1988: 257). Both Hall and Gilroy (1993) use the notion of a 'diasporic culture' to examine the complex interconnections which encompass intercultural exchanges and transnational linkages through which cultures are both transmitted and translated to produce new ethnicities. The cultural construction of new ethnic identities thus becomes a process through which differences are engaged with and new forms of representation are produced.

Hall (1988) finds such representations in the work of black artists and film makers like *Looking for Langston* (Isaac Julien, 1988) or *Passion of Remembrance* (Maureen Blackwood and Isaac Julien, 1986). Other examples might be the work of Hanif Kureishi such as *My Beautiful Launderette* (1985) or the film *Wild West* made in Southall in 1992 (David Attwood/Initial Films). Paul Gilroy highlights youth culture as offering new spaces of representation which 'play a special role in mediating both the racial identities that are freely chosen and the oppressive effects of racism' (1993b: 61). In particular Gilroy highlights musical styles like bhangra which fuses Punjabi and Bengali folk music with hip-hop, soul and House as suggesting: '[t]he opening up of a self-consciously post-colonial space in which the affirmation of difference points forward to a more pluralistic conception of nationality' (Gilroy, 1993b: 62). Through the intercultural dialogue which the music of such artists as Apache Indian[3] produce it is suggested that new cultural spaces of identification are created where 'new ethnicities' are made and remade (Back, 1996).

This chapter explores this process of the construction of 'new ethnicities' at the level of the everyday lives of some young Muslim women. While cultural theorists have suggested important illustrations of how the representation of 'new ethnicities' might be produced through youth cultures, there is still relatively little work on how such youth cultures are actively produced, consumed and transformed by young people themselves.[4] By focusing on the 'everyday' processes by which identities are constructed and contested through cultural practices such as dress styles, listening to music and watching television this chapter suggests that the contextual and spatial dimensions of such processes must be taken into account. Looking specifically at the spaces of the school and the neighbourhood it is clear that young women negotiate their identities differently within different places and in relation to their parents and other family members, their Muslim and non-Muslim peers and teachers.[5] They also construct their identities within particularly discursive frameworks and in relation to dominant discourses.

THE LOCAL CONTEXT

The research was conducted in a suburban town north-west of London. The 1991 Census suggested that of a population of 85,000 some 10 per cent were recorded as 'non-white' and about half of these were defined as Asians, the majority of whom were of Pakistani heritage. While Census data does not record religious identities it would appear that about 5 per cent of the population are Muslims – a figure which would include small numbers of black-Caribbean and black-African Muslims. The South Asian Muslim population had its origins in migration in the 1960s from rural north-west Pakistan, particularly from areas of Azad Kashmir such as Mirpur. These male migrants often settled first in other parts of the UK, particularly in the north-east in towns like Preston or Blackburn, before moving further south to find work and eventually reuniting their families in the late 1970s and early 1980s. Most of these families retain close ties with relatives in Pakistan and some family reunification is still taking place.

Although previously prosperous with flourishing local engineering and manufacturing industries, the town has undergone industrial decline since the early 1980s and the base of work for those who are unskilled has shrunk. While local unemployment levels are approximately 7 per cent it is estimated that unemployment levels among South Asian men, including recent school leavers, may be as high as 50 per cent. The South Asian Muslim population in the town is concentrated within a distinct area and a number of community surveys have identified issues of social and economic marginalisation, particularly the isolation of women. While there are a number of social clubs organised for young people, by both the Local Authority and the Muslim Community Centre, for many young women school remains an important social space and it was for this reason that the research was concentrated in schools. The research was carried out in two girls' schools, with significantly different intakes and profiles, allowing a comparison between young women from different socio-economic backgrounds.

REPRESENTATIONS OF MUSLIM WOMEN

Central to the argument presented in this chapter is that for young Muslim women the construction of their own identities is produced through a challenge to dominant representations of 'Muslim Women'. This category is produced through the intersection of racialised discourses about British Asian populations and a reworking of Orientalist discourses about the nature of Islam. Discourses are understood here as a set of widely held shared beliefs, or 'commonsense' understandings, which are repeatedly reproduced through different media and institutional practices. Racialised discourses about 'Asian culture' construct Asian women as passive victims of oppressive cultures (Brah and Minhas, 1985; Parmar, 1984). Such discourses intersect with Orientalist discourses which construct Islam as antithetical to Western culture and 'Muslim Women' as the embodiment of a repressive and 'fundamentalist' religion (Said, 1978; Kabbani, 1986).

Through such discourses young Muslim women are defined as 'caught between two cultures' of home and school, torn by a 'culture clash' between the 'secular/modern' world of the school and the 'traditional/ fundamentalist' world of the home (Knott and Khokher, 1993). As Sara, one of the research participants, explains:

> 'That is just one of the stereotypes. That you have your father, and your grandfather who says you shouldn't do this, and you shouldn't go out, and the timid mother, and the timid daughter who is really confused and doesn't know what to do.'
>
> *(Sara)*

Avtar Brah has argued that the intersection of racialised discourses results in 'culturalist explanations' of the lives of young Muslim women. Such explanations deny the agency of women themselves as 'concrete historical subjects with varying social and personal biographies and social orientations' (Brah, 1993: 443). They also rely upon a static understanding of culture rather than seeing culture as a process through which social meanings are produced and contested. Instead Brah offers a framework to understand the position of South Asian young Muslim women in the labour market, which seeks to show how 'structure, culture and agency are conceptualised as inextricably linked, mutually inscribing formations' (Brah, 1993: 442). Thus an examination of how young Muslim women articulate their own identities requires a recognition of the multiplicity of subject positions which they occupy as well as the extent to which 'their everyday lives are constituted in and through matrices of power embedded in intersecting discourses and material practices' (Brah, 1993: 449). Brah does not deny the impact of patriarchal discourses and practices in the lives of South Asian young Muslim women (Afshar, 1994; Ali, 1992) but locates these within a broader social formation. Thus the analysis of the narratives of young Muslim women presented here seeks to show how they negotiate their own identities in relation both to particular discourses and in and through particular places and moments.

RESISTING THE DRESS CODE

'One girl was asking me "Are you a Muslim? . . . It's just that you don't look it."'

(Humaira)

Much of the early writing on youth cultures emphasised the significance of fashion and style in the creation of sub-cultural identities and affiliations and, in particular, the role of dress as a contested boundary marker between different group identities (Hebdige, 1991; Hewitt, 1986; McRobbie, 1989). In the construction of dominant representations of 'Muslim Women' different styles of dress have been used as highly significant markers of difference and their bodies have become contested sites of cultural representation.[6] A pertinent recent example is the media coverage of the marriage of British socialite Jemima Goldsmith to former Pakistani cricket captain Imran Khan which was dominated by discussion of how she would find herself forced to change her appearance and dress.[7] Brah and Minhas (1985: 16) illustrate how 'changing from school uniform to Shalwar Kameez'[8] is the charged metaphor through which the Western/traditional, school/home dichotomy structuring many accounts of the lives and experiences of South Asian women is reinforced. Through the wearing of 'ethnic/traditional' or 'Western' clothes oppositional identities are constructed. Such representations offer an either/or identity option and suggest that identities can be straightforwardly read from the body and its adornment.

For the young women interviewed dress and style were important topics for discussion. They were well aware not only of how different meanings were attached to individuals depending on what they were wearing but also how these meanings shifted in different spaces. In their reflections on the meanings of different dress styles the participants illustrated how they seek to define their own identities by subverting or redefining the codes associated with different styles of dress. This process of negotiation is complex since individuals are positioned within a variety of discourses which produce a constellation of different meanings around dress. On the one hand they seek to challenge the 'traditional'/ 'Western' dichotomy which structures dominant representations. At the same time they also negotiate the expectations of a local Asian community which places a high premium on female sexual purity and morality. Thus it is through a monitoring of the dress of young women, particularly in the streets of the neighbourhood, that the cultural integrity of the community is upheld. While some young women respond to this by highlighting the inconsistencies underlying assumptions about particular dress styles, others have challenged them through the creation of 'new' Muslim identities which are expressed through particular dress styles.

Although in many discussions participants would distinguish between 'Western clothes' and 'Asian clothes' they often sought to challenge the meanings associated with this dichotomy. Thus while many agreed that they might wear different clothes at school from those that they wore at home they denied that wearing different clothes in different places had any special significance. Several respondents pointed out that they alternated daily between wearing a *shalwar kameez* or wearing 'English clothes' and their choices were simply a reflection of how they felt that day. As Robina explains:

> 'I wore them [*shalwar kameez*] today because I couldn't be bothered to iron my other clothes . . . I sometimes wear the scarf on my head if it's cold.'
> *(Robina)*

Other group members also sought to challenge the meanings which others might attach to their wearing 'Western clothes', as the following example illustrates:

> *Wendy*: I suppose it's hard because in your religion it says cover your ankles and then you see everybody with short skirts and fashion and stuff and I suppose you just want to copy everybody else.
> *Sarah*: No, no . . . with English clothes, they're sensible for school and stuff like that, but personally I prefer Asian clothes. Because we've got more, I don't want to be horrible about English clothes . . . but we've got much more variety.

Sarah contradicts Wendy's suggestion that how she dresses is constrained by her religion and also that she would choose to wear 'English clothes' outside school if she was permitted to. She also challenges the assumption that wearing 'English clothes' is superior or more fashionable than wearing 'Asian clothes'. Thus while Sarah agrees that she might wear different clothes in different spaces she resists Wendy's interpretation of what this means. In the same way Ghazala resists the assumptions made about her behaviour and attitudes because she wears 'Western clothes' to school:

> 'It's like this girl she was saying you're modern Westernised people why are you hanging round with . . . like, people with scarves on their heads and that. She didn't expect it. But they're my friends.'
>
> *(Ghazala)*

The simple opposition between 'Asian' and 'Western' was also undermined when respondents considered the ways in which *hybrid* styles were being created where fashions were being blended together. As two participants reflect:

> *Rozina*: Ours is really like the English fashion, the flares now and short tops . . . and really long shalwar kameez tops as well, with long slits at the side . . . they vary, and now they've gone to short kameez tops with flares, so it's really just like trousers now isn't it?
> *Yasmin*: Yeah, you could just wear a pair of trousers and a long shirt and it's just like that, and just put the scarf on the side, that's all we do.

Such connections parallel the commodification of Asian styles by the fashion industry (Bhachu, 1993: 111). For some of the young women, it became difficult to define an opposition between Asian and Western clothes – was the wearing of Indian appliquéd waistcoats with jeans, or long shirts over flared trousers, or long silk skirts and embroidered shirts wearing Asian or Western clothes? And did it matter? Yet while there was evidence, particularly for some more middle class girls, that school could become a space for the experimentation with new styles and hence new cultural identities a recurrent theme was the extent to which you were always judged, and defined, by what you wore.

While participants recognised the contradictions of attaching meanings to different styles of dress they were well aware of how they were always caught within a negotiation of these cultural meanings – even as such negotiations shifted in different places. Thus Sameera explains that at school she is always positioned by her dress as being 'typical' or representative of a group:

> 'I'm constantly thinking about what people will think of me, they must think that I'm really typical. Even when I haven't got a scarf on my head,

but I'm like in Asian clothes, I'm so paranoid. Oh people must think typical . . . you know that I'm from the dark ages and that.'

(Sameera)

Like Ghazala she resists this interpretation with its assumption that 'we are all the same' by challenging the fixed meanings of 'Asian' and 'Western' clothes and attempting to produce new meanings which challenge the 'traditional'/'modern' dichotomy. If participants resist one set of assumptions from their white non-Muslim peers within the social spaces of the school, within other spaces such as the street they negotiate other assumptions. Many respondents talked about the degree of surveillance which was exercised on them, by parents and other adults, when they were out in the streets.

Rozina: If you just walk down the streets and you've got trousers on and one lady says 'I saw her' and that's all they do they gossip.
Shamin: They say 'I saw so-and-so's daughter and she's started going out with boys' . . . just because you're wearing English clothes.

In expressing their resistance to this equation between 'rebelliousness' and 'English clothes' various participants not only suggested how false assumptions about your behaviour were assumed from your dress but also recognised the ways in which dress could function as a 'cover-up'. This was particularly true of the most contested item of 'appropriate' Muslim dress, the headscarf:

'If there is a girl with a scarf on her head, right, and she's been out with all these guys, she'll get away with it because she's got that cover, it doesn't matter how bad she is, she'll get away with it.'

(Ghazala)

These comments suggest the possibility that dress styles can be used 'strategically' within certain places in order to escape parental approbation or to safely negotiate particular spaces such as the 'public space' of the streets.[9]

These possibilities were also recognised by a small number of the young women who were interested in expressing a more explicitly Islamic identity which was also reflected in their adoption of different styles of dress. Such young women sought greater independence through a recourse to Islam which challenged the ways in which culture and religion were often seen as synonymous by their parents. As Husbana explains:

'They mix up religion and culture as well, like it doesn't say in the religion
or anything, it just says that you've got to be covered . . . but the women
don't see it like that, it's like you've got to wear Asian clothes.'

(Husbana)

Thus by adopting headscarf – or *hijab*[10] – which were often more Middle
Eastern in style than the loosely tied scarves or *dupattas*[11] worn by their
mothers, and by wearing long, ankle length skirts, some of the young
women were producing new styles of orthodox Muslim dress. Such styles
offered a challenge to existing dress codes while also providing young
women with new cultural spaces within which to assert their independence and own identities. As Sameera explains:

'It's like the other day I came in and the skirt I was wearing was right
down to my ankles. My uncle was sitting there and my mother said "Go
and change your clothes" and I said "no". And I was right to take a stand
because I knew I was right, she has to realise that I was doing nothing
wrong.'

(Sameera)

If the self-conscious adoption of new Muslim dress styles is open to
contestation by parents it has also provoked opposition by other young
women who do not wish to adopt such an explicitly religious identity but
feel themselves judged in relation to those who do.

'You can't just think oh well, I'll wear a scarf on my head that will make
me a believer . . . . Although we're Muslims we're not going to stick a
scarf on our head and everything, I'm not prepared to do that yet.'

(Sarah)

Hence for young Muslim women dress style is a highly significant
marker of cultural identity which is constantly negotiated. As this
analysis has suggested young women themselves are actively engaged
in the contestation of different meanings attached to dress style. By
adopting new styles and challenging meanings which are attached to
particular dress styles – such as 'Western', 'Asian' and 'Muslim' – young
women also construct and contest their own identities. This process of
contestation occurs both in relation to different discourses and through
particular spaces.

CULTURAL SPACES AND MEDIATED IDENTITIES:
YOUNG MUSLIM WOMEN AND THE MEDIA

If, as I suggested above, the media are an important cultural space for the production of 'new ethnicities', there has been little work considering how these ideas are received and transformed by young people. An important recent exception is the work of Marie Gillespie (1995) which examines the role of TV talk among young people in Southall in the construction and negotiation of cultural identities. Gillespie's detailed analysis of conversations about the viewing of soap operas, adverts and local and national news provides considerable insight into the ways in which the media is used productively by consumers both to produce new spaces of identity and to maintain and strengthen boundaries (Gillespie, 1995: 207). In this section I consider how particular cultural spaces – like those provided by television soap operas – are used by the research participants in the construction and the contestation of their identities.

Although use of the media was not a central focus of my research, television emerged as a recurrent theme within many of the group discussions. My analysis of these instances of TV talk parallel many of the more detailed findings produced by Gillespie (1995). Like the respondents in Southall, group discussants enjoyed watching soap operas, particularly *EastEnders*[12] and *Brookside*.[13] What they enjoyed about these programmes was the extent to which they dealt with realistic themes:

> 'Brookside and EastEnders, I think they're the best, they're more realistic, they do cover issues, like Mark on EastEnders with AIDS.'
>
> *(Zakkya)*

Through the representation of such topics on soap operas the young women were able to debate relevant moral questions such as the portrayal of lesbian relationships on *Brookside*:

> *Zakkya*: I mean it is believable it could happen.
> *Sughara*: It probably does happen, because what happened to Beth, that's why she goes after other girls, because she's scared of what her father did to her.
> *Zakkya*: And yeah, because you can understand that.

If soap operas gave participants the opportunity to debate moral questions they were also seen as cultural spaces within which aspects of their own cultural identities were represented:

'The old [i.e. previous] characters on EastEnders, they brought our culture into it, they brought in arranged and assisted marriages . . . they showed the good qualities, where the girls do talk to boys, and the parents trust them more.'

(Eram)

However, while participants could cite some good examples where issues of racism or cultural diversity had been included in drama series they were also critical of the negative ways in which Asian people or Islam were often represented:

Ruhi: It wouldn't be so bad if they showed it once, but everytime they put something on TV about black and Asian people it's always negative, you know, I mean they never ever show anything of the positive side of it.
Eram: If I wasn't an Asian I would probably be prejudiced against Asians. Because the way they show it it's as if one Asian is like that that means that every Asian is going to be like that.

What these comments reveal is what has been termed the 'burden of representation' (Tagg, 1988; Williamson, 1993: 116). Since there are so few images of Asians in the mainstream those that do appear have to carry a greater burden of signification.

This question of representation was most acute when the participants considered discussions of how these images were interpreted by others in different places. Although many respondents enjoyed the drama *Bhangra Girls*,[14] about three Asian girls who formed a band, their enjoyment was mediated by how they expected other girls at school to react to it:

Ghazala: Her mother locked her up in the house and she wanted to go out and perform, so she opened the window, and then she ran, she left her plait on the bed. And then after she had performed, she ran away, up to another town with her boyfriend . . . . it probably does happen, but this country, the rest of them, were probably watching it and having a right laugh.
Husbana: They were thinking, that is typical, that is what Muslims are like.

This extract suggests that the ways in which the media are read and understood by young Muslim women is always contextual. Within the social spaces of school, where identities are negotiated in relation to others, their own reading of the media is always mediated by the ways in which they think other non-Muslims will read and understand the representations given. Thus it is not enough to simply ask whether cultural spaces exist in which positive images of Muslim women can be produced but to consider how all images are interpreted within specific contexts.

This question of the contextual reception of the media was also evident when respondents discussed the work of the performer Apache Indian. For many participants their opinions about Apache Indian revolved around whether or not his music could be called 'authentic'. (Gilroy, 1993a: 82; see also Back, 1996).

> *Ravinder*: He used to sound really traditional Indian. Now he's got all the beats and all the reggae and he's changed his whole image, how he looks and dresses, because he's entered the mainstream
> *Ghazala*: Other bhangra groups they failed because they were singing in Punjabi, the only way you can get in is if you mix the two, like he has done. So he wasn't losing it totally, he just changed it because you can't get pure Asian music into the charts.

At the same time the enjoyment that some participants gained from the music, and their controversial lyrics, was mediated by how they expected others to react to them:

> *Ghazala*: A lot of people hate him because they don't think he should talk about controversial topics like arranged marriages. But he should . . . . Like the song[15] that says I want a girl to cook for me and that, that is like the Asian stereotype of a woman and that's true that is what Asian men do want.
> *Nazreen*: I don't think they should sing about it.
> *Ghazala*: I think they should sing about it, it's true, we laugh about it, we can take it.
> *Husbana*: But if we want to fight racism we have to put on positive images not negative ones.

The disagreements expressed in this discussion illustrate some of the dilemmas for young people of responding to the representations of 'new ethnicities' and hybrid identities produced by artists like Apache Indian. While for some the question may be about the politics of authenticity and the representation of Asian music in the mainstream others are concerned about how the political themes developed in his songs will be understood by others. While Ghazala enjoys the honesty of the parodies suggested in the song 'Arranged Marriage',[16] others argue that such representations should not be encouraged because of the ways in which they will be misinterpreted by others.

What I have illustrated in this brief account of how young Asian

Muslim women talk about the media is the complex and contextual processes by which the media are used in the negotiation of identities. An analysis of how young people talk about both TV, and the media more broadly suggests some of the ways in which young people are involved in processes of 'cultural translation' (Gillespie, 1995: 207). Such processes involve the negotiation of different cultures and different positionings within each culture. Yet such negotiation, as the examples above have suggested, also takes place within specific discursive and spatial contexts.

CONCLUSION

This chapter has demonstrated how young Muslim women seek to define their own identities and resist dominant representations of 'Muslim Women'. Through an analysis of how respondents talked about two different elements of youth culture – dress and consumption of the media – I have illustrated how for young Muslim women the articulation of their own identities requires the negotiation of dominant representations and stereotypes and a challenge to existing discourses. Perhaps the most important theme which was repeated in many different ways by different respondents was that they had to constantly challenge other people's expectations of them. As this chapter has highlighted such challenges draw upon a variety of different discourses and are articulated differently in different places. Robina's quote, which opens this chapter, makes clear that the young women who are represented here want to be seen as individuals who are proud to be British, Muslim and Asian but are seeking ways to articulate these different, and interlocking, dimensions of their own identities on their own terms.

ACKNOWLEDGEMENTS

I am grateful to the editors and to Peter Jackson for comments on an earlier draft.

NOTES

1 In order to protect the anonymity of respondents all names used are pseudonyms chosen by the individuals themselves.
2 The research, carried out between September 1993 and August 1994 involved

interviews and group discussions with 49 young women, aged between 16 and 19, in two schools in a suburban town north-west of London. The majority of the participants (35) were Muslims and most were born in the town with parents of South Asian origin. The support of the Economic and Social Research Council in funding this research is gratefully acknowledged (R00429234082).

3 Apache Indian (Steven Kapur) was born of Punjabi Hindu parents and raised in the multi-ethnic area of Handsworth, Birmingham. He performs in a blend of Jamaican patois, Punjabi and English. His first album *Movie over India* (City to City, SUNREC 001A, 1990) topped both the reggae and bhangra charts in the UK and prompted the new term 'bhangramuffin' (Back, 1995).

4 Back (1996) and Gillespie (1995) provide important recent exceptions, however.

5 It is also important to stress the extent to which the respondents were also negotiating the constructions of their identities in relation to me, the white, non-Muslim researcher. As I suggest later, this is part of the process of the contextual negotiation of identity.

6 There is a considerable literature on this particularly in relation to the veil; see especially Abu Odeh, 1993; Alloula, 1986; Fanon, 1989; Macleod, 1991.

7 For example, the juxtaposition of photographs of Jemima Goldsmith head-lined 'Eastern style' and 'Western style' accompanying an article by her about her conversion ('Why I Chose Islam', *Sunday Telegraph*, 28 May 1995, p. 3). See also 'It'll Soon Be Over For This Maiden' (Vanessa Feltz, *Daily Mirror*, 17 May 1995); 'Fundamentally Fashionable' (Lowri Turner, *Evening Standard*, 18 May 1995, p. 12).

8 *Shalwar kameez* is the name given to the loose trousers and tunic worn throughout the Punjab area of Pakistan and the East Punjab area in India.

9 A parallel here is Valentine's (1993) discussion of how lesbian women adopt 'straight' dress styles to negotiate the heterosexed spaces of the workplace and the street.

10 In Islam the term *hijab* refers to the veil or partition which prevents men from gazing at women. Amongst my respondents the term was used to describe the more complete 'Middle Eastern style' head covering favoured by those who had adopted a more self-conscious Islamic identity.

11 The name given to the long scarf draped over the head and shoulders which is worn with the *Shalwar kameez*

12 *EastEnders* is a soap opera shown on BBC Television which is set in the imaginary East London neighbourhood of Albert Square.

13 *Brookside* is a soap opera shown on Channel Four Television in the UK which centres on a fictional community living in a suburban street called Brookside Close in Liverpool.

14 *Bhangra Girls* (BBC2, DEF II, 4 October 1993. Written by Nandita Ghose)

15 'Arranged Marriage' (Island ID 544)
16 See Back, 1995: 144 for an analysis of 'Arranged Marriage' which suggests
   Apache Indian's own sense of irony about Asian heterosexual masculinities
   expressed in the song.

REFERENCES

Abu Odeh, L. (1993) 'Post-colonial feminism and the veil: thinking the differ-
   ence' *Feminist Review* 43: 26–37.
Ali, Y. (1992) 'Muslim women and the politics of ethnicity and culture in
   northern England', in G. Sahgal and N. Yuval-Davies (eds) *Refusing Holy
   Orders*, London: Virago, pp. 101–23.
Alloula, M. (1986) *The Colonial Harem*, trans. Myrna and Wlad Godzich,
   Manchester: Manchester University Press.
Afshar, H. (1994) 'Muslim women in Yorkshire: growing up with real and
   imaginary values amidst conflicting views of self and society', in M. May-
   nard and H. Afshar (eds) *The Dynamics of 'Race' and Gender: Some
   Feminist Interventions*, London: Taylor and Francis.
Back, L. (1996) *New Ethnicities and Urban Culture: Racisms and Multiculture
   in Young Lives*, London: UCL Press.
—— (1995) 'X Amount of Sat Siri Akal! Apache Indian, reggae music and the
   cultural intermezzo', *New Formations* 27: 128–47.
Bhachu, P. (1993) 'Identities constructed and reconstructed: representations of
   Asian women in Britain', in G. Buijs (ed.) *Migrant Women and Changing
   Identities*, Oxford: Berg.
Brah, A. (1993) '"Race" and "culture" in the gendering of labour markets:
   South Asian young Muslim women and the labour market', *New Commu-
   nity* 19, 3: 441–58.
—— and Minhas, R. (1985) 'Structural racism or cultural difference: schooling
   for Asian girls', in G. Weiner (ed.) *Just a Bunch of Girls*, Milton Keynes:
   Open University Press.
Fanon, F. (1989) 'Algeria unveiled', in *Studies in a Dying Colonialism*, trans.
   Haakon Chevalier [1965], London: Earthscan.
Gillespie, M. (1995) *Television, Ethnicity and Cultural Change*, London:
   Routledge.
Gilroy, P. (1993a) *The Black Atlantic: Modernity and Double Consciousness*,
   London: Verso.
—— (1993b) *Small Acts: Thoughts on the Politics of Black Cultures*, London:
   Serpent's Tail.
Hall, S. (1988) 'New ethnicities', talk given at the ICA, reproduced in J. Donald
   and A. Rattansi (eds) *'Race', Culture and Difference*, London: Sage, 1992.
Hebdige, D. (1991) *Subculture: The Meaning of Style*, London: Routledge.

Hewitt, R. (1986) *White Talk, Black Talk: Inter-Racial Friendship and Communication among Adolescents*, Cambridge: Cambridge University Press.

Kabbani, R. (1986) *Europe's Myths of Orient*, Bloomington: Indiana University Press.

Knott, K. and Khokher, S. (1993) 'Religious and ethnic identity among young Muslim women in Bradford', *New Community* 19, 4: 593–610.

Macleod, A. (1991) *Accommodating Protest: Working Women, the New Veiling and Change in Cairo*, New York: Columbia University Press.

McRobbie, A. (ed.) (1989) *Zoot Suits and Second Hand Dresses*. Basingstoke: Macmillan.

Parmar, P. (1984) 'Hateful contraries: media images of Asian women', *Ten* 8, 16: 71–8

Said, E. (1978) *Orientalism*, New York: Random House.

Tagg, J. (1988) *The Burden of Representation: Essays on Photographies and Histories*, Basingstoke: Macmillan.

Valentine, G. (1993) 'Negotiating and managing multiple sexual identities: lesbian time–space strategies', *Transactions of British Geographers*, n.s. 18: 237–48.

Williamson, J. (1993) 'A world of difference: the passion of remembrance', in *Deadline at Dawn: Film Criticism 1980–1990*, London: Marion Boyars.

....

4

RETHINKING BRITISH CHINESE IDENTITIES

•

David Parker

INTRODUCTION

Recent writing about new ethnicities in Britain (Hall, 1988 for example) has highlighted the redefinitional and prefigurative nature of new British-based youth cultures derived in part from the Caribbean and South Asia. Yet despite the presence of Chinese settlement in Britain for more than two centuries, it is only recently that the question of British Chinese identity has been broached by an emerging generation of young Chinese people. My previous research set out to explore the identities of this generation, concluding that there were signs of an emerging British Chinese sensibility (Parker, 1995). At the time I was keen to place the experiences of young Chinese people alongside those of Black and Asian British youth, together redefining what it might mean to be British. However, I have since been struck by how little significance Britain has in the lives of young Chinese people and the salience of Hong Kong for them.

Due to their spatial dispersal, young Chinese people do not readily fit into what is a mainly urban research framework for studying new ethnicities in Britain (Back, 1996). In addition, the relative prosperity of Hong Kong gives them both an opportunity and an incentive to return to the place of their parents' origin. Several hundred young Chinese people have left Britain for Hong Kong in the last few years on completion of their education.

Drawing on interviews with twenty-five Chinese young people in both

Britain and Hong Kong I show that to understand British Chinese identities the frame of reference needs to be shifted both historically and geographically to encompass the continuing hold of regressive cultural formations in Britain and the possibility for new forms of identification in Hong Kong. In addition the work of young Chinese artists is referenced throughout the chapter to highlight new self-representations of British Chinese identity.

MEDIA REPRESENTATIONS

Whilst preparing this chapter in late 1995/early 1996 my thoughts were regularly punctuated by the stereotypes through which Chinese people in Britain have come to be known. These must be seen against the background of the recurrent fascination and repulsion with which the West has conceived of the East. In the late nineteenth and early twentieth centuries the threat of a so-called 'Yellow Peril' was widely disseminated throughout popular culture by such Orientalist figures as the evil Fu Manchu and Flash Gordon's adversary Ming the Merciless (Clegg, 1994; Marchetti, 1993; Ng, 1992). Alongside these Sinophobic figures, another tradition of Sinophilia expressing admiration for exotic treasures, Chinese wisdom and gentle mysticism has contributed to the sense of distance between East and West (Dawson, 1967). However quaint and archaic such images may seem, they undoubtedly continue to shape contemporary perceptions. These cluster around two themes: firstly a racialised and spatialised criminality associated with male Triad gangs; secondly an effeminising exoticisation of Chinese food and festivals.

Stereotypes of young Chinese men express both an admiration for martial arts prowess and a fear of the potential terror and cruelty coiled within lean bodies. Young Chinese men I talked to had been dubbed 'Bruce' at school in mock reference to the iconic figure of Bruce Lee and teased:

'They think we can do Kung Fu, but most of us can't, I wish I could so I could get them back.'

(Ming, student, aged 21)

Such an image of potent, dangerous young men stems in no small part from the recurrent representation of so called Triad gangs in British media.

It has become a cliché that the increasingly ubiquitous television police series will feature a Chinatown storyline involving drugs, gambling, extortion and protection rackets. Examples include the drama series *Thief Takers* (Carlton Television, February 1996) which had a two part storyline based in Chinatown. *Yellowthread Street* was a thirteen part series for Yorkshire Television in 1990 set entirely in Hong Kong. The TV listings guide for the series 'the cops who tamed Hong Kong' (*TV Times*, January 1990, 13–19) matched the plotlines of British colonial officers pacifying the unruly natives.

Such programmes provide popular newspapers with a repertoire of images that are readily drawn upon when they mention Chinese people in Britain. The murder of London headmaster Phillip Lawrence in December 1995 allegedly by a group of young 'Chinese-looking' men called forth a brief flurry of Triad gang stories in the tabloid newspapers of 11 December 1995. The *Sun* carried the headline 'Teeny Triad Revenge' (Wood and Lauchlan, 1995: 4–5); and page 4 of the *Daily Mirror*: 'Terror Triads Target Schools' (Antonowicz, 1995: 6). These stories have an impact on how young Chinese people perceive their place in Britain. The Triad image can actually deter Chinese young people from making Chinese friends.

'I thought that because this incident was major news and it involved Chinese that the English/whites would become hostile and wary towards us, and the image of peaceful, hardworking Chinese people would be destroyed.'

(*Alison, unemployed, aged 21*)

'You can cook and you know martial arts; if the person's young and he's a lad then he might be in the Triads.'

(*Keith, student, aged 21*)

'When you're isolated from the Chinese community you think up these scenarios and misconceptions about how the Chinese community is. And you don't feel you belong.'

(*Danny, student, aged 21*)

What the Triad conception also fails to appreciate is how 'Triad style' gangs (the vast majority of which have no connection with Triad fraternities) can actually be a form of self-protection from the kinds of constructions Les Back noted in the late 1980s fieldwork he conducted in South London, where Vietnamese Chinese young men were at times marginalised by both black and white peers (Back, 1993).

'When I was young I hung round in a group – there were thirty or forty of us. None of us were in the Triads, but everyone thought we were in the Triads 'cos we were in a massive group . . . . But we saw going round in a massive group as protection for ourselves.'

(*John, store owner, aged 20*)

Stereotypical representations are particularly damaging for a small and dispersed population, with few counter-narratives of their own in popular culture. One series on British television has had a lead Chinese actor – *The Chinese Detective* shown in 1981 on BBC. The star, David Yip, points out that its impact was somewhat marred by the BBC showing old Charlie Chan films shortly after the series ended (personal communication). Such imagery is difficult to challenge as Chinese actors are hard-pressed to find parts other than as gang members. Fifteen years after *The Chinese Detective* it was David Yip himself who played one of the Triad leaders in *Thief Takers*.

REALITIES OF TAKEAWAY LIVES

The second set of representations influencing the conceptualisation of Chinese people in Britain is focused on Chinese food, especially take-away food. Routinely invoked and then dismissed, 'Going out for a Chinese' has become the stuff of urban myths and legends about cats and dogs in the freezer and ammunition for pejorative quips (see Healey and Glanvill, 1996: 82–3). In one bizarre example, the Trade and Industry minister Ian Lang, at a press conference in the wake of the publication of the Scott Report in February 1996, derided a joint press conference held by the opposition ministers Robin Cook and Menzies Campbell, dubbing it the Ming Cook show: 'sounds like a Chinese takeaway, with lots of sweet and sour porkies'. The fact that Chinese culture only appears in the public arena in these ways influences how non-Chinese people interact with the British Chinese population. Such representations are particularly important given that it is mainly young Chinese people who have served British customers over takeaway counters and in restaurants in their formative years.

Every evening in several thousand Chinese takeaway food shops around the country there is a very particular encounter between British and Chinese young people (Song, 1995). The gendered exoticisation of Eastern food attaches a very particular set of meanings to Chinese takeaway food and those serving it across the counter. An overwhelmingly male clientele can all too readily draw on the stock of cultural imagery, in a manner that subordinates the young people, often young Chinese women, taking the orders:

'I don't like working Friday nights because there's this man who comes in and he keeps singing "Suzy Wong, Suzy Wong".'

(*Mei Han, student, aged 19*)

What this means for young Chinese people growing up and serving behind the counter is captured in Pui Fan Lee's autobiographical play *Short, Fat, Ugly and Chinese*, broadcast on Radio 5 in 1992. She remembers her childhood:

> 'But most of all I hated the shop. The Takeaway. Or, as the locals called it, "The Chinky" . . . . Then there was the trouble, the fights, the broken windows, the police . . . [Lee balefully reflects] English kids didn't have to watch behind the counters.'
>
> *(extracts from Lim and Yan, 1994)*

Linda Chui's poem also articulates what it is like on the other side of the counter; in her case from the perspective of having left the family business behind for Hong Kong (Box 4.1).

Such experiences of working most evenings in takeaways hardly constitute fertile ground for the generation of locally grounded Chinese youth cultures in Britain. However, as with young South Asian people, hitherto overlooked cultural products from several thousand miles away are continuing to connect British born generations to Asia (see Gillespie, 1995 and Sharma, Hutnyk and Sharma, 1996 for comparison).

BRITISH CHINESE IDENTITIES

Hong Kong popular music is a multi-million dollar industry, its stars fêted throughout Asia and the overseas Chinese world; and yet the World Music racks of Western record shops are empty when you get to 'HON . . .', and contemporary Hong Kong music has no place in debates about the cultural identities of subaltern populations. Despite being derided as derivative, 'Canto-pop' continues to thrive; HMV in Hong Kong has whole floors of its stores in the territory devoted to both Canto and Japanese pop. Names such as Leon Lai, Emil Chou, Andy Lau and Faye Wong mean nothing to Western youth but dominate East Asian popular culture.

This cultural formation has a real presence in the homes, car stereos and Walkmans of Chinese people in Britain. It is sustained by a network of Hong Kong cultural outlets selling glossy Hong Kong magazines and Hong Kong television serials for video rental, and by the rapidly expanding Chinese cable and satellite channels.

These are all part of a broader Chinatown imaginary and set of cultural practices. Most Sundays see Chinatowns in Britain bustling; a family meal, the children learning Chinese for a couple of hours in a

Box 4.1

ALL QUIET ON THE WESTERN FRONT

Only the contrast of weather
Makes me homesick for England,
Walking in the shadow
Of a Victorian edifice;
The Union building steps
And swing doors take me back
With the sun behind.

Grass lawns and gardens
Life isn't; living in a summer furnace
Warmed by consuming
Midnight oil at 300 degrees;
Recommending the menu
Until you're sick,
Until the lights turn out,
Until the streets are yellow
And dead, and my fingers preserved
In vinegar and salt beneath my nails.

Grass lawns and Eden
Life isn't when you're no more
A fluorescent powered fiction:
Half Frankenstein and half 'chink'
In a white overall
Trying to understand
The foreign poetry
Of insults recited against my kind.
I became myself and myself alone
When after I tried and almost failed
To turn hard like stone,
Turn stone to build
A soundproof chamber.
Out of the deafness and silence,
From the hands and mouths
Of dead men I found
A bastard poetry in a grain
Of salt beneath my nails.

Linda Chui (1995)

supplementary language school nearby; the parents stocking up with provisions and Chinese videos and then returning to the family takeaway.

> 'Every Sunday they all meet up in Chinatown and the younger generations all hang out in Chinatown . . . . They want to keep that identity. And by keeping that identity, you need Chinese otherwise you don't know what your friends are talking about if they talk about pop stars and stuff. And they're always fascinated about what's going on in H.K. . . . The younger ones they enjoy it, or they want to learn Chinese now because they want to read the Chinese songs, the lyrics. And they also want to know in those gossip magazines who's going out with who.'
>
> *(Irene, conference manager, aged 29)*

Chinatowns nurture a rescaling of identity as a combination of both Hong Kong and new British regional identities; especially in the centres of Manchester, London, Birmingham, and to a lesser extent Newcastle, Glasgow and Liverpool.

> 'I'm Manchester born and raised here so I do think it's more like my home, I go out on the street and I know most of the people . . . . If people say "Are you British or Chinese?" I'll say "I'm Mancunian all right?" . . . settles the arguments.'
>
> *(John)*

Manchester and the north-west have become a focal point for Chinese youth in Britain. The only British concert by Hong Kong pop star Jacky Cheung was held at Manchester's NyNex centre in November 1995. On a more regular basis, night clubs in the region cater for young Chinese people on Tuesdays when takeaways tend to have a night off. Chinese night at 'Mr Smiths' in Warrington has attracted young people not just from Greater Manchester and Merseyside, but from as far afield as York, Birmingham and London.

Throughout my discussions with young Chinese people, notable was the lack of investment in British-based national identifications. However, as with John, familiarity with a local Chinatown setting could nurture a regional or city based identity which was strongly felt, but also highly personal. Unfortunately, and in stark contrast to young black and South Asian people, the dispersed Chinese settlement both within and beyond British cities makes it harder for young Chinese people to secure the cohesion that might result from daily contact with their peers. As a result Hong Kong popular culture has not been reformulated or used as the basis to reflect on distinctively British Chinese experiences and thus generate a strong social identity. The notion of a British Chinese youth culture does not come readily to mind; there is no British based Chinese musical form to match Bhangra for example.

A survey of young Chinese people in Birmingham (Parker, Li and Fan, 1996) revealed considerable disaffection with Britain; racism and inequality were seen as ingrained.

'I didn't really mix in with English culture that well, you may think that you're English but because of your colour and family background they don't treat you the same.'

(Lawrence, computer programmer, aged 23)

Hong Kong now has a standard of living and employment opportunities in excess of those in Britain. Together with the continuing cultural connection it is little wonder Hong Kong continues to have an active presence in young Chinese people's lives, exciting them in a way political activity in Britain cannot match; even more so since the handover to China on 1 July 1997. Any understanding of British Chinese identity must recognise the opening up of a new dimension, the return to Hong Kong, which complicates simple notions of either assimilation or British-based cultural hybridity. In fact it is often only when British born Chinese people return to Hong Kong that a self-definition in terms of a national category like British Chinese becomes prominent.

BBCS IN HK

Since 1989 several hundred British born Chinese people (known to each other as 'BBCs') in their early and mid-twenties have left Britain for Hong Kong (Figure 4.1). Dissatisfied with their lack of progress in the British labour market, well qualified young Chinese people returning to Hong Kong are particularly advantaged: they usually have the enviable combination of Chinese language oral skills, local family ties, British education and a British passport which until July 1997 entitles them to a twelve-month stay without a visa.

'People like myself who are fluent in English and can speak Cantonese, they have a role to play with the concept of 1997. So I guess that was one of the reasons for wanting to come out. I felt Hong Kong was also a very fast moving place so for somebody who wanted to progress and wanted to really capitalise on opportunities, I felt it was . . . the right place to be.'

(Mary, retail manager, aged 31)

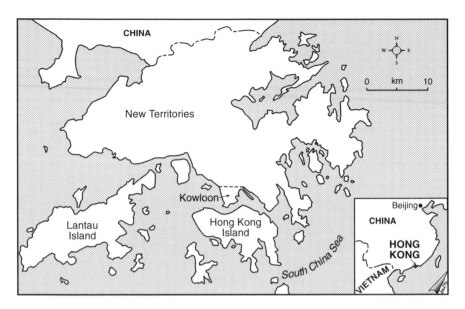

Figure 4.1 The territory of Hong Kong.
Source: Linda Dawes.

For those who migrate to Hong Kong, Asia is the future and they want to be a part of it; Britain is portrayed as the past.

> 'Britain is going down the tubes!'
>
> *(Lily, accountant, aged 25)*

> [of Hong Kong] 'I think it's the buzz and the opportunity if you're prepared to take the risk . . . . And it's a thriving economy. It's almost like a place living on borrowed time and therefore you must get in as much as you can in the time that is left.'
>
> *(Mary)*

Young Chinese people I interviewed in Hong Kong connected this shift in the world's centre of gravity to a change in their own sense of identity. Britain was viewed with rather quaint affection as possessing nice countryside for driving in, but little else.

> 'And my parents are very worried. They're saying "Oh, why are you staying in Hong Kong? '97 is coming. You know, come back to England." I thought "What for? What have I got in England?"'
>
> *(Irene)*

Besides being a haven of job opportunities away from a relatively stagnant and potentially racist British labour market, Hong Kong enables

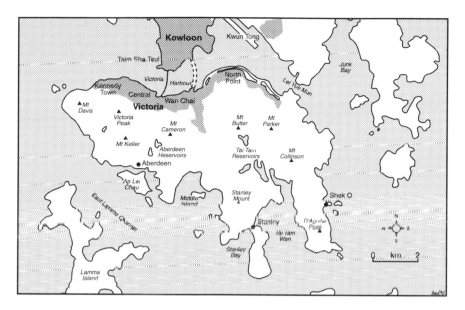

Figure 4.2 Urbanised areas of Hong Kong Island (highlighted) are the focus of night life for young people.
Source: Linda Dawes.

young Chinese people who in Britain would never meet to live within a few square miles of each other.

> 'I've known a lot of BBCs now that I do not know in England because they're from Manchester, Hull, Birmingham, and I didn't go up there. I only knew the ones in London. So that's quite good.'
>
> *(Irene)*

A return to Hong Kong has fostered a new awareness of the commonalties with other British born Chinese. This is encouraged by a spatial concentration in terms of residence in Hong Kong; many returnees live in or near their ancestral villages in the rural New Territories. In addition Hong Kong's night life also has key reference points for overseas Chinese young people (Figure 4.2). Lan Kwai Fong in Central Hong Kong (Figure 4.3 and Plate 4.1) has a special place in the spatialised iconography of Hong Kong as a site of cultural interchange (Choi, 1993). Wan Chai also figures large in the night life of British Chinese in Hong Kong.

> 'It's in Wan Chai . . . where a lot of people host venues like, birthday parties, club events, you know, rave nights and things. In the last year, raves organised by big London based DJ's – they come over here to

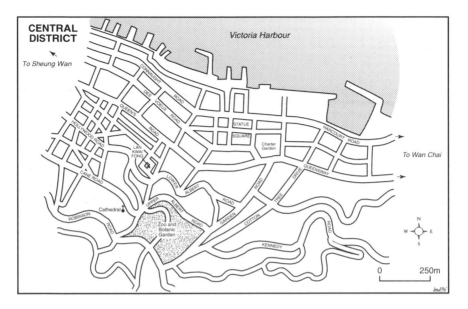

Figure 4.3 Lan Kwai Fong.
Source: Linda Dawes.

organise raves . . . . The BBCs are the ones who know the DJs, who are actually in the circuit. It's sort of like a club circuit.'

(Lori, finance professional, aged 28)

Social gatherings are facilitated by Hong Kong's telephone system where local calls are free of charge.

'If it's someone's birthday, or someone's having a party, the word just gets passed around so quickly that you get loads of people turning up.'

(Michael, graphic designer, aged 25)

However, this gathering together as British Chinese has been supplemented by a further rescaling of identification to encompass the recognition of a wider range of ways of being Chinese. The presence in Hong Kong of other young Chinese who have grown up in North America, Australia and New Zealand has fostered an awareness of the broader identity 'overseas Chinese'. The successful formation of the Overseas Chinese Connection as a social club in Hong Kong by a group of British, American and New Zealand Chinese shows the potential for new links to be forged between young Chinese people who've grown up away from East Asia, with a recognition of the wider Chinese diaspora.

Plate 4.1 Lan Kwai Fong has a special place in the spatialised iconography of Hong Kong.
Photograph: David Parker.

'I'm not local and I'm not English. I am an overseas Chinese it is as easy as that. But to define overseas Chinese is not easy. . . . It's very hard to put yourself in a group . . . . A different identity that's it – I can't describe what it is.'
(Raymond, sales negotiator, aged 24)

Yet because many of the young people who return are fluent in spoken Cantonese, but can't read and write Chinese, they do not always feel completely at ease as overseas Chinese in Hong Kong.

'Oh, we never belong. We'll never belong to Hong Kong because even though as much as we try ourselves consciously, subconsciously you're not. We don't because your face says you're Chinese but your heart is still very much . . . in between because . . . . But when you are neither pure Chinese and neither pure English it's very hard to belong to one country.'

(Vicky, administrative assistant, aged 23)

As the handover of Hong Kong to China approaches, this is not a settled population. Leaving Hong Kong is not synonymous with return to Britain, however, and several interviewees had ambitions to move to Australia, Canada or the United States. Some of the contradictions and uncertainties of being British born Chinese in Hong Kong are captured by another poem of Linda Chui's reflecting on her first year in Hong Kong (Box 4.2).

NEW IDENTITIES

In studying the geographies of contemporary youth cultures, categories such as 'British Chinese' need to be imaginatively reconfigured. British

Box 4.2

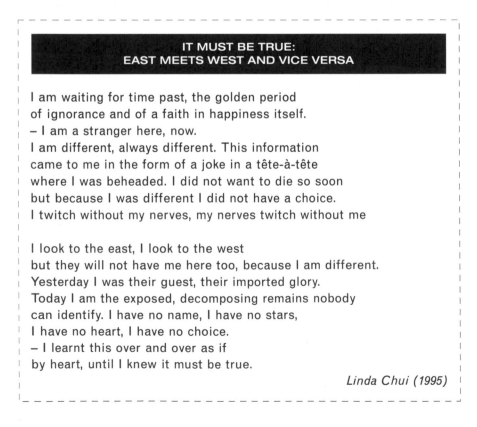

**IT MUST BE TRUE:
EAST MEETS WEST AND VICE VERSA**

I am waiting for time past, the golden period
of ignorance and of a faith in happiness itself.
– I am a stranger here, now.
I am different, always different. This information
came to me in the form of a joke in a tête-à-tête
where I was beheaded. I did not want to die so soon
but because I was different I did not have a choice.
I twitch without my nerves, my nerves twitch without me

I look to the east, I look to the west
but they will not have me here too, because I am different.
Yesterday I was their guest, their imported glory.
Today I am the exposed, decomposing remains nobody
can identify. I have no name, I have no stars,
I have no heart, I have no choice.
– I learnt this over and over as if
by heart, until I knew it must be true.

Linda Chui (1995)

Chinese identity cannot be understood simply by confining research to Britain. It is not that the idea of British Chinese identity should be discarded, merely that it should be seen in the wider perspective of returning to Hong Kong and subsequent migration elsewhere. Whether in Britain or Hong Kong, British Chinese young people have uncertain and mobile futures. The identifications resulting can be doubly non-national: combining local British accents and a mobile transnational sense of 'being Chinese' as expressed by Vicky (a BBC in Hong Kong) in describing how wherever she settles in the future must have a sizeable Chinese population.

'Wherever I go there still has to be a Chinese entity somewhere. I couldn't go where there was no Chinese entity. That would completely lose myself. Because at the very bottom of things I'm Chinese.'

(Vicky)

However, not all young Chinese people can leave Britain for Hong Kong, or have the desire to do so. The political uncertainty of the territory after July 1997 and the fact that British citizens will no longer have advantageous residential rights there may refocus efforts to increase the visibility of Chinese people in Britain. Some young Chinese people have already begun to recognise this.

'I'd like to envisage one day that the Chinese youth in Britain can have and express an identity.'

(Chen, student, aged 22)

It is in addressing these sentiments that a new generation of young Chinese artists has a major part to play in countering the stereotypes of the past and defining the terms on which Chinese young people can contest racist attributions.

The work of two artists in particular confronts the two modes of caricature outlined in the first section of this chapter. Lesley Sanderson's self-portrait in *Fuck the British Movement* (Plate 4.2) is deliberately unsettling and confrontational. The objectification of Chinese women embodied in the figure of Suzy Wong and deployed daily by male customers in Chinese takeaways is directly returned. She aims to 'confront the dominant representation of "ethnic" women as being exotic and readily available for the gaze'.

At first sight Yuen Fong Ling's artwork *Fighting Spirit '96* (Plate 4.3) – an image of *kung-fu* icon Bruce Lee on an appropriately coloured golden T-shirt of the type one might wear when performing martial arts – seems to merely endorse the portrayal noted earlier of potentially violent

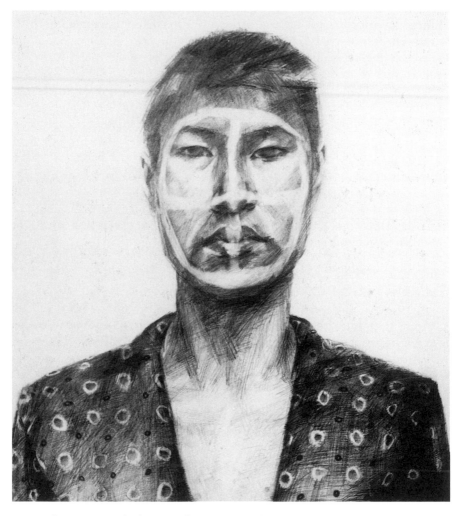

Plate 4.2 *Fuck the British Movement* by Lesley Sanderson (1984).
Photograph: the artist.

Chinese masculinity. Yet only on closer inspection does the work's true
point emerge. The label inside the T-shirt proclaims 'Yuen Fong Ling
Born in England'. As Yuen Fong states of the piece:

> Fighting Spirit '96 addresses the issue of cultural identity head-on through
> its confrontational or KICK ASS approach. The image of Bruce Lee
> embodies my attitude to the experiences of being a British born Chinese.
> His fearless film-screen image also reflects the nature of my cultural
> education, one which has been learned through Western perceptions of
> the Chinese. The 'fighting spirit' that is the title to the work is my quiet
> anger that fuels the passion behind my creativity.

Plate 4.3 *Fighting Spirit '96* by Yuen Fong Ling.
Reproduced by kind permission of the artist.

Only one thing is certain, young Chinese people will need to draw on that fighting spirit in both Britain and Hong Kong over the next few years.

ACKNOWLEDGEMENTS

Thanks to the following for various forms of assistance: Sarah Chan, Justine Fan, James Li, Irene Li and Dr Choi Po King as well as the University of Birmingham School of Social Science and Birmingham City Council.

REFERENCES

Antonowicz, A. (1995) 'Terror Triads target schools', *Daily Mirror*, 11 December: 6.

Back, L. (1993) 'Race, identity and nation within an adolescent community in South London', *New Community* 19, 2: 217–33.

—— (1996) *New Ethnicities and Urban Culture*, London: UCL Press.

Choi, P.K. (1993) 'Introduction', in P.K. Choi and L.S. Ho (eds) *The Other Hong Kong Report 1993*, Hong Kong: Chinese University Press.

Clegg, J. (1994) *Fu Manchu and the Yellow Peril: The Making of a Racist Myth*, Stoke on Trent: Trentham Books.

Dawson, R. (1967) *The Chinese Chameleon: An Analysis of European Conceptions of Chinese Civilization*, London: Oxford University Press.

Gillespie, M. (1995) *Television, Ethnicity and Cultural Change*, London: Routledge.

Hall, S. (1988) 'New ethnicities', in *Black Film, British Cinema*, ICA Document No. 7, London: Institute of Contemporary Art.

Healey, P. and Glanvill, M. (1996) *Now! That's What I Call Urban Myths*, London: Virgin Books.

Lim, J. and Yan, L. (eds.) (1994) *Another Province: New Chinese Writing from London*, London: Lambeth Chinese Community Association and Siyu Chinese Times.

Marchetti, G. (1993) *Romance and the Yellow Peril: Race, Sex and Discursive Strategies in Hollywood Fiction*, Berkeley: University of California Press.

Ng, Chun-ming (1992) 'The image of overseas Chinese in America', in *The 16th Hong Kong International Film Festival: Overseas Chinese Figures in Cinema*, Hong Kong: Urban Council.

Parker, D. (1995) *Through Different Eyes: The Cultural Identities of Young Chinese People in Britain*, Aldershot: Avebury.

——, Li, J. and Fan, J. (1996) *The Needs of Young People of Chinese Origin in Birmingham*, Birmingham: Birmingham Chinese Youth Project.

Sharma, S., Hutnyk, J. and Sharma, A. (eds) (1996) *Dis-Orienting Rhythms: The Politics of the New Asian Dance Music*, London: Zed Books.

Song, M. (1995) 'Between "the front" and "the back": Chinese women's work in family businesses', *Women's Studies International Forum* 18, 3: 285–98.

Thompson, T. (1996) *Gangland Britain*, London: Coronet Books.

Watson, J. (ed.) (1977) *Between Two Cultures*, Oxford: Basil Blackwell.

Wood, M. and Lauchlan, I. (1995) 'Teeny Triad revenge', *Sun*, 11 December: 4–5.

5

REHABILITATING THE IMAGES OF DISABLED YOUTHS

•

Ruth Butler

There are estimated to be 340,000 16–29-year-olds with disabilities in Britain (Martin *et al.*, 1988). Despite their numbers and heterogeneity they have remained largely invisible in representations of youth culture. Images of 'super cripples' overcoming all odds (Figure 5.1), evil 'monsters', bitter and twisted in body and mind, or pitiful charity cases in need of care and attention, account for most billboard, film and other media images of disabled youths (Longmore, 1985; Barnes, 1991).

This chapter is concerned with the strategies disabled youths use in order to cope with the pressures that are placed on them to conform to socially accepted 'norms' of behaviour in public space. It will firstly consider the nature of the stereotyped roles that disabled youths are expected to follow in society. I will draw on discourses about the body to consider how such social images have been constructed and continue to function. Secondly it will discuss the reactions disabled youths have to such expectations of them and their methods of coping with the demands society places upon them. It will finally question what implications these reactions have for the wider disabled population.

Adolescence can be a trying time for any individual. The social and practical problems related to striving for independence from parents, searching for a coherent set of personal values, forming new relationships, finding a job, setting up a home and acknowledging your sexuality create certain developmental needs which are common to all young people, including those with disabilities (Hirst *et al.*, 1991). To

Figure 5.1 Struggling with representations.
Source: reproduced from the cover of J. Morris (ed.) (1989) Able Lives: Women's Experiences of Paralysis *(London: Women's Press) by kind permission of Angela Martin.*

compound matters peer pressures are extreme at this time of life (Smart, 1978).

As they lose the sweet, innocent victim status they often possess as children, disabled youths can face adulthood with a lack of confidence, self-esteem and limited social skills (Hirst *et al.*, 1991). They also miss the respect that is offered to elderly disabled individuals for what they have given to society and/or achieved despite their impairments.[1] Disabled youths face the discriminatory practices of employers when searching for paid work and the misguided beliefs of their peers about their capacities in building more informal relationships. Social beliefs mean that adolescence can be a long, lonely and particularly rough road for this group of individuals. The next section considers why this is so.

REPRESENTATIONS AND SOCIAL EXPECTATIONS OF DISABLED YOUTH

A person's image of him/herself is strongly influenced by the opinions of others and by the representations in language, media images and music of desirable identities. In a self-defensive manner, we tend to see ourselves as 'normal', and hence socially acceptable, and those we view as 'other', as deviant (Shakespeare, 1994). This is one method we employ to distance ourselves from people we think to be on the undesirable margins of society. It produces a social structure where we are all measured against an unspecified yet apparently desirable 'norm'. It is a strong source of motivation for disabled youths to behave and appear as 'normal' as possible. People's fears of marginalisation encourage them to take on board ideas of social 'norms' and build them into their own evaluation of their identity (Goffman, 1963). Young (1990) notes how women's bodies are often treated as objects to be viewed and evaluated. She says:

> The source of this objectified bodily existence is the attitude of others regarding her, but the woman herself often actively takes up her body as a mere thing. She gazes at it in the mirror, worries about how it looks to others, prunes it, shapes it, molds and decorates it.
>
> *(Young, 1990: 66)*

Disabled youths also find themselves objectified by the curious gaze of strangers. Their bodies, however, may be less easy to shape into a socially acceptable image and also into an image that they themselves like and accept. People's lack of comfort with their own bodily image and its openness to interpretation from others have made fashion and beauty products an extremely profitable industry. It is a marketing strategy which disability aids manufacturers are also becoming increasingly aware of and one to which disabled youths are as susceptible as anyone else. To what extent an individual chooses to decorate their body to their own tastes and to what extent they are forced to do so by the consumer culture that we live in is an issue that has divided feminists for years (Baker, 1984). However, young disabled men and women are certainly encouraged by media representations of 'normal' bodies to obscure by dress and bodily decoration what are seen by others as bodily inadequacies.

Youths with disabilities are more reliant on their friends for social activities than people disabled later in life who find their support networks of friends and family already set in place at the onset of the impairment. Whilst it invariably takes time for both the disabled person and the rest of the family to come to terms with a disability arising later in life, the recognition of the individual as a person with value is already intact unlike the situation outside the family for disabled youths. Mobility problems often make it difficult for disabled youths to get to social events, and reliance on family and friends for lifts can be frustrating and embarrassing, and may give the individual a sense of themselves as a burden to others. There may be problems of physically not being able to get involved in some activities, again leading to frustration as well as isolation and fears of being seen as boring (Barnes, 1991). They may feel their 'street cred' is under threat.

All this at a time of life when image is of particular importance as sexuality is recognised and the first tentative steps to more intimate relationships are forged. To 'improve your image', 'maximise your performance' and be 'serious about performance' are issues which the advertisers of wheelchairs and fitness equipment for disabled people take seriously (Figure 5.2). The need for disabled people to make the most of their abilities in order to 'pass' for 'normal' as far as possible seems to be the advertiser's primary concern. The negative connotations of incapacity and dependence are apparent.

Even with a positive social image disabled youths can face problems. Images to be effective must be mediated to those around us. Oliver (1983) recognises how a lack of opportunity for interaction with others coupled with insufficient social and emotional skills results in many disabled youths having problems building friendships and more personal relationships. He notes:

> Disabled people may... find it difficult to initiate contacts in pubs or at parties. To take the initiative and take a seat close to someone who is attractive may be very difficult for someone in a wheelchair, and for visually impaired people it may be impossible. Parents of disabled youngsters are sometimes overprotective and reluctant to allow their children to take the usual teenage risks. Furthermore, disabled teenagers may find it difficult to do things that perhaps they should not (when they go out they

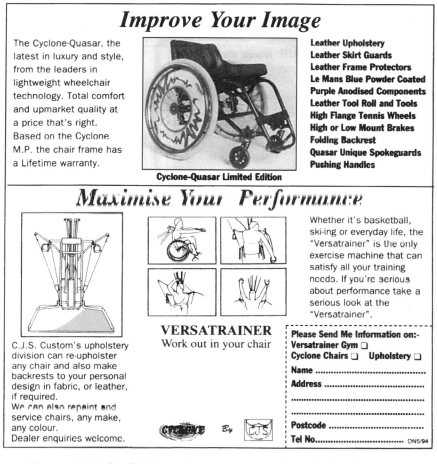

Figure 5.2 Style, fashion and image are used to sell disability aids.
Source: reproduced from Disability Now *(May 1994: 25).*

probably have to be transported by their parents). They therefore can't lie
to their parents about where they have been or who they have been with.

(Oliver, 1983: 72)

Young people's awareness of such problems and the misinterpretation of
their actions by others were apparent in a recent survey of visually
impaired people in Reading.[2] One youth said:

'If it's dark at parties or in the pub I feel really self-conscious that I might
trip over things or knock over the nibbles. I get worried people won't realise
I can't see and will think I'm odd so I decide not to join in in case I make a
mess of things and then people just think you're boring.'

(Female, visually impaired since birth)

> *Simon Weston suffered major facial and body burns in the Falklands/Malvinas war between Britain and Argentina (1982). Following a TV documentary about his rehabilitation he became a national celebrity in the UK.*
>
> People have such misconceived ideas of what's attractive. Finding a girlfriend is going to be twice as hard for me as for someone who looks 'normal', but that's fine, I can live with that; it just means that when I do find someone to settle down with, I'll know I've found a bloody good one.
>
> *(Weston, 1989: 247)*

Another, acknowledging that he sometimes walked into bollards and other obstacles he failed to see, said:

'You feel like something off . . . Oh God what's it called . . . "You've been framed!" You feel a right div.'

(Male, visually impaired since birth)

SOCIAL EXPECTATION

A wealth of literature on society's marginalised minorities has clearly illustrated the role of social expectations and economic marginality in controlling people's behaviour, presentation of self (as discussed above) and simple presence in public space (see for example Goffman, 1963; Liachowitz, 1988; Winchester and White, 1988; Barnes, 1991; Valentine, 1993).

Research by Hirst (1987) showed that only one-third of 21-year-olds with severe physical impairments were in open or sheltered employment. Whilst research has shown that preparation and support services are lacking it has also shown that even when they are in place they have little effect on reducing problems of unemployment (Parker, 1984; Hirst, 1984). This is not because these services are unnecessary, but rather because they are not designed to break down the prejudiced views of employers which are the crux of the problem.

Public reactions to difference can be both negative and positive. They can range from hostility, backed by legal expulsion from space, to pity and acts of unwanted and unnecessary charity. Comments from visually impaired people in Reading illustrate this point. One recalled:

'You do get sympathy and you're treated as if you're mentally deficient and patronised.'

(Female, visually impaired since birth)

Another noted how on his morning walk to work he was regularly shouted at by cyclists:

'Get the fuck out of the way! Are you blind?!'
                                  (Male, visually impaired late in life)

Fear, a desire for superiority and control, distancing and othering processes may all lie behind these reactions. Shakespeare (1994: 283) suggests that disabled people are often used as 'dustbins for disavowal' to relieve able bodied individuals' fears of their biological frailty and ultimate death.

Disabled youths may give displays of physical behaviour which society believes should be kept private (Elias, 1982). A lack of control over their body may result in spasmic movements of limbs or involuntary noises. This coupled with the myth that what is seen as an inadequate body incorporates an inadequate mind (Butler and Bowlby, 1995; Barnes, 1991) has led to disability being seen to represent what Shakespeare (1994: 296) refers to as 'the physicality and animality of human existence'.

The outward appearance of the body has been closely linked to an individual's moral standing. Beauty has for centuries been seen as good, whilst ugliness or abnormality has been related to evil (Goffman, 1963; Synnott, 1993). Failure of an individual to look after their body has been related to a lack of morality and self-discipline (Synnott, 1993).

Bodily presentation is also important in society's expectations of others. While young people who are visibly disabled, 'ugly', find themselves the butt of hostile or patronising behaviour, misunderstandings, often excused by ignorance, repeatedly occur through the invisible nature

> I was desperately trying to live like a normal human being, but I had to look at life in a different way because people looked at me differently. Inside I was normal, but outside I was scarred. I wanted to go places and do things, but everywhere people stopped and stared. The fact was most people felt revulsion when they saw me. They seemed to think I was scarred on the brain, too. I tried not to let it bother me, but inevitably some things could not be shrugged aside. One night outside a disco in Cardiff, a pair of teenage girls stared, pointed and laughed at me. If only they could have known how much more that hurt me than any other kind of pain I had had to endure.
>
> *(Weston, 1989: 198)*

of many disabilities. This point was made clear through discussion with some of Reading's visually impaired inhabitants. One youth noted:

'If you've just been into town you don't want to be hassled by carrying a [white] stick around with you and that makes asking for help difficult. People just think you're stupid, they don't believe you when you say you can't see very well 'cos you look so, so . . . I don't know, normal, capable, I guess.'

(Female, visually impaired since birth)

A further important element of social bodily interpretation is that of its link to nature. Just as women have been seen as closer to nature and our animal ancestors than men (Thomas, 1983), so too have disabled people (Shakespeare, 1994). Men have considered themselves conquerors of and in control of the natural environment and hence of their more 'animalistic' female contemporaries. In a similar way disabled people have been seen as different, dangerous, exciting and generally 'other' to their able bodied contemporaries (Butler and Bowlby, 1995).

PHYSICAL AND SOCIAL STRUCTURING OF PLACES

The social control of disabled people in public space is not necessarily direct, through people's reactions and behaviour towards a young disabled person. It can occur through the structuring of the environments people occupy. Just as employment can mean more than income to a person in terms of social status and self-respect, so the spaces a person frequents and the amenities they make use of can reflect their social standing. The freedom of movement someone can enjoy in any given area and their acceptance in it by others can be highly influential in society's view of them and their own view of themselves.

The physical structuring of the environment can act as an obvious and direct control on disabled youths' actions. As Oliver (1983: 72) notes:

Many social gatherings, such as youth clubs and discos, may simply be physically inaccessible, and in a wheelchair it may not always be possible to participate in that favourite teenage pastime of hanging around on street corners.

The social structuring of the environment we live in and its role in controlling people's movements may not always be as apparent as the physical structuring. Reactions to the presence of undesirable people in public space can vary from hostile stares and verbal comments to physical violence and state legislation to remove them (see for example Brown, 1984; Scarman, 1981 on racial minorities; Davis, 1990; Smith, 1993 on beggars and homeless people). Members of minority groups are often seen as people who have to be distanced from the mainstream in order to reduce 'reminders of the "failures" of society' and in order to reassure the majority of its members that they are not like these unfortunates (Butler and Bowlby, 1995: 2).

A person's assumed biological characteristics are often believed to determine their social status and personal character. Feminists have noted, however, that feminine characteristics come not necessarily from biology, but rather from social conditioning and the power relationships they must enter into with their male contemporaries (Oakley, 1981). In a similar way the hegemonic power relationship between able bodied and disabled people may have a greater role to play in constructing the traits associated with disability than the medical impairments their bodies exhibit.

BODILY CHARACTERISTICS

Recent developments in disability theory and politics have hinged on the awakening of researchers and activists to the power of economic and social relationships in controlling and constructing disabling physical and social environments and hence the problems people with impairments face (see for example Oliver, 1984, 1990, 1996; Swain et al., 1994; Barton, 1996). The direct effects of a bodily impairment on a person's life must not, however, be forgotten (Crow, 1992, 1996; French, 1994).

Biological make-up may mean that a person is unable to use their body in a manner which is socially acceptable or 'normal'. Bodily characteristics can also mean that actions which the person would desire to take can be impossible and that others they would rather not take are impossible to prevent. The pain of aching joints or absence of a limb has little to do with social control, even if the frustration and anxiety impaired people suffer as a result of the behaviour that is expected of them, but not always possible for them, is largely socially constructed. It is difficult to believe that an idyllic society where impairment will be of no importance to a person or to those around them is attainable. Whilst social expectations, negative judgements and economic discrimination presently play the larger role in controlling disabled people's lives, to suggest that the biological functioning of an individual's body is, or could be, of no consequence to their circumstances is to paint a picture of the world that few disabled people can relate to, or believe possible. This is an issue which affects us all, to varying degrees, as we all have physical limitations as well as days when we feel more energetic than others.

The combined effect of impairment and social control, through expectations of ability and social activity, on behaviour was clearly illustrated by one Reading youth who said:

> 'I get pissed off at times when my mates are off to play badminton or something and I can't, but I get more fucked off when there's something I can do and other people try and stop me 'cos they think I'm not capable of it. I mean how the hell do they know what I'm capable of?!'
>
> *(Male, visually impaired since birth)*

It may be argued that frustrations over not being able to join in a game of badminton are due more to social structure than a person's impairment. The game may represent a social event. The activity may be seen as healthy, enjoyable and morally good. Abstention from it may be taken as lazy, unhealthy and suggest moral failings about the person who is unwilling to take care of their body. In an alternative social structure the inability to join in sporting activities may not seem as important. The fact remains, however, that physical pain, illness, exhaustion and the feelings of loss that impairment can bring with it place limitations on an individual regardless of the society they find themselves in. We are all constrained by our physical condition.

IMAGES OF DISABLED PEOPLE

Figure 5.3 outlines some of the misguided expectations society has of disabled youths. How different young people may choose to cope with such images and expectations varies with the nature of the impairment, its severity, time of onset, long term prognosis, visibility to others and so on, as well as a multitude of other economic, political and social circumstances.

They must learn to reconcile the social constructions and lived realities of their bodies. Goffman (1963) has suggested that individuals stigmatised in this way have two main options open to them. They must choose either to accept the images others have of them or to fight against them. It must be noted that these are two extremes of a fluid socialisation process, but they are a useful starting point from which to consider the reactions of young disabled people to the images and expectations others have of them.

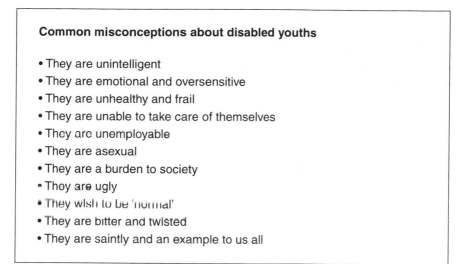

Common misconceptions about disabled youths

- They are unintelligent
- They are emotional and oversensitive
- They are unhealthy and frail
- They are unable to take care of themselves
- They are unemployable
- They are asexual
- They are a burden to society
- They are ugly
- They wish to be 'normal'
- They are bitter and twisted
- They are saintly and an example to us all

Figure 5.3 Common misconceptions about disabled youths.
Source: collated from interviews with visually impaired people in Reading.

RE-PRESENTING THE SELF

Self-image, as already discussed, is strongly influenced by social representations of desirable and undesirable social characteristics. Internalising ideas of how we believe others see us or of how we want to be seen helps us to feel more comfortable with the performances we are expected to give in social interactions (Goffman, 1963; Scott, 1981). It also helps others to predetermine our expected performance in any given situation due to the social, economic and political categories in which they put us (Goffman, 1963; Scott, 1981). This is not a tactic used only by disabled people to survive in the social jungle, but by all people in their social interactions. By performing in socially acceptable ways we also simultaneously create and re-create our social identities. Such performances are sometimes largely unreflective or part of a deliberate strategy of self-presentation.

This tactic of projecting a necessary, socially accepted and expected, if not always desirable, image of self when negotiating the social environment was apparent in interviews with visually impaired people in Reading. When discussing the use of guide canes one youth said:

'I mean if I'm going to a strange place I take my white stick with me, so that people can see I need help. I act kind of more helpless than I am to be honest.'

(Female, visually impaired since birth)

> Like the majority of teenagers, one of my greatest needs at this time was to conform. No one had hinted to me that, for a blind girl, conformity in a world of sighted people would be a near impossibility without some very radical changes of attitude. And so I began a great pretence. I actually pretended sometimes that I could see. I picked up my Braille books, and peered at the invisible dots while feeling them slyly with my fingers. Waiting for a friend, I would look this way and that up and down the road and when we visited art galleries I would try to make intelligent remarks about paintings.
>
> *(Taylor, 1989: 36)*

The importance of a white stick as a symbol of visual impairment and hence as an explanation and justification of the need for help was made clear by another woman who noted:

'I only use it [white stick] when I do need help. And I don't bother with it if I don't.'

(Female, visually impaired since birth)

The need to reverse the act of helplessness to an equally unrealistic one of complete independence often resulted from a not uncommon fear of vulnerability. Such a fear led one visually impaired man to explain how he avoided the use of assistive aids as they underlined his 'weakness' to others. He said:

'I have a telescope thing to read things like the departure board at the railway station, but I don't use it, because if I have my hands full with that and all my attention looking through it I leave myself open to pickpockets and things. I don't risk it.'

(Male, visually impaired late in life)

Carrying a white stick for purely informative purposes and the restrictions that fear of vulnerability place on the use of assistive aids are far from the only ways that adaptation to fit the 'norms' expected of an individual can cause problems. Hegarty *et al.* (1981) have noted that placement in a mainstream school with emphasis on integrating youths into 'ordinary' educational and social settings can result in particular pupils' needs being overlooked as their 'normality' comes to be taken for granted.

FIGHTING THE EXPECTED IMAGE

If an individual feels uncomfortable and/or dissatisfied with the behaviours expected of them they may choose to fight others' expectations by
making clear what they believe to be their true personality and desires.
To this end many disabled youths are finding the courage to contest and
resist the 'normality' with which they are supposed to comply.

In Britain a number of court cases have arisen in recent years as local
authorities have been accused of failing to offer access to and adequate
support in their mainstream educational establishments for disabled
youths (see for example *Disability Now* 1995a, 1995b).

Court cases are an extreme form of confrontation with society's
expectations of disabled youths, however, and one which few choose to
take. To suggest that all or even most disabled youths are quite so
assertive in their political activity would be misleading. Many visually
impaired people in Reading acknowledged the need for self-confidence to
fight others' ignorance and misconceptions. One older visually impaired
woman, when asked if other people's attitudes had ever stopped her from
doing anything she wanted to, noted:

> 'Well not nowadays. There again it's a question of self-confidence. If you
> have self-confidence you can find a way round anything.'
> *(Female, visually impaired since birth)*

As this quote suggests, self-confidence is something which, for most
people, comes with experience, and which may be lacking among
younger disabled people. Many disabled youths find themselves playing
the role of politician not through any conscious decision to do so, but
simply through their attempts to live their own life in the mainstream of
society.

For disabled youths, self-consciously or not, fighting society's images
of them involves fighting for their civil and human rights. Through
projects of mainstreaming in education youths have often found themselves in the position of educator to both their classmates and teachers
alike. The lack of support and forward planning that has often gone with
such schemes have further placed them in the eye of the media and at the
heart of a wider fight for the rights of all disabled people to access
educational, employment and leisure establishments. It is to the broader
lessons which the disability movement as a whole can learn from its
younger members that I now turn.

IMPLICATIONS FOR THE WIDER
DISABLED COMMUNITY

Comparisons with feminism suggest that an issue which will grow in importance for the disability movement is to provide an analysis which successfully integrates the varied experiences and interests of all disabled people, including not only women but also the differences between the old and the young, . . . while retaining the political advantages of recognising some common interests amongst these groups.

(Butler and Bowlby, 1995: 12)

Disability theorists have paid relatively little attention to the needs of disabled youths facing the transition to adulthood. In doing so they have not only failed to recognise the demands of these individuals when influencing policy makers and planners, but they have also left their broader models of understanding of the disabled population as a whole open to criticism.

Lessons can be learned from the experiences of disabled youths in a practical manner. The repeated failure of disabled school leavers to find employment has been blamed on the lack of provision of support services to guide them through the recruitment process (Chapman, 1988). These calls for improved support services have been echoed in community care initiatives, integrated education projects and numerous other areas of disabled people's lives. The same combination of a lack of understanding from others, outright discrimination and inadequate resources and services that disabled people of all ages must face in their social, political and economic lives are responsible for making adolescence a far more trying time for disabled youths than their able bodied peers (Clark and Hirst, 1989).

Disability theory is currently in a state of flux. In the wake of the British Government's 1995 Disability Discrimination Act[3] the disability movement in Britain is reassessing its policies and objectives for future political, social and economic campaigns. Proponents of the social construction model of disability have fought to break down images of disability as a list of medical complaints which isolated individuals must cope with as best they can. This social construction model is central to the thinking of disability activists today. The model has drawn attention to the powerful role society plays in controlling disabled people's lives and recognises the potential of disabled people given the necessary opportunities. In this way it has given disabled people a new feeling of self-worth and a stronger political base than they have ever had before (Crow, 1992). Morris (1993) and Crow (1992, 1996) amongst others have, however, found reason to question the comprehensive nature of a model which fails to recognise the importance of impairment in an individual's life. Whilst some fear that acknowledging the impact an impairment has on an individual's life may mean a return to an individual, medical needs based model of disability, to ignore impairment and other elements which contribute to people's multiple identities, including age, is to leave the movement open to ridicule.

Feminists' experience of politics and debates on the effect of social thought on our experience of our bodies may go some way to confronting the issues disability theorists are now addressing (Butler and Bowlby, 1995). To recognise the differences amongst disabled people and between disabled people and able bodied people including their impairments, ages and a host of other social, economic and political factors does not necessarily mean a return to an individual based model of disability. As Corbett (1991: 260) wrote: 'I like difference. If we rejoice in the fact that we are a marvellous combination of cultures, our lives can be enriched by each other.' The effects of individual circumstances on a person's life cannot be ignored if theory is to reflect reality and avoid criticism from its observers.

Recent media attention given to public demonstrations by the movement have undoubtably raised its profile in the public eye. Disabled activists have led the way in promoting new positive images of disability. They have offered desperately needed, positive role models to disabled people of all ages. They have questioned what is meant by 'able bodied' and pointed to the frailty of the human condition which we all face. The visibility of disability has increased and fears of 'coming out' have been reduced.

The disability movement can learn from disabled youths. For many such media hype is still something distant from their daily experience. It is through integration with disabled people, face to face, that many of the misconceptions people hold about disability are broken down. Disabled youths may draw strength and a feeling of self-worth from the political fight. Nonetheless, without mass media attention, large promotional rallies or networking between each other, they have been and

> People like people, and if you've got a personality, then people will accept you. Keep being you and it'll work. The only way we can get disfigured people – myself included – to be socially acceptable, and bear off this notion that everyone has to look like the front cover of *Vogue*, is to get out there and live our lives in the best, fullest, most active and enjoyable way we can.
>
> (Weston, 1989: 251)

continue to make political progress by simply having the courage to live out their lives in a manner that they see as fit and desirable. For them the personal is most definitely political.

ACKNOWLEDGEMENTS

I would like to thank Sophie Bowlby and Lisa Doyle for their helpful comments on earlier drafts of this chapter.

NOTES

1 The words impairment and disability are used throughout this chapter as they are defined by the Union of the Physically Impaired Against Segregation (UPIAS, 1976): Impairment – 'lacking all or part of a limb, or having a defective limb, organism or mechanism of the body.' Disability – 'the disadvantage or restriction of activity caused by a contemporary social organisation which takes no or little account of people who have physical impairments and thus exclude them from the mainstream of social activities.'
2 Intensive interviews with visually impaired people in Reading are continuing as part of a Ph.D. thesis. For more information see Butler and Bowlby (1995).
3 Between 1982 and 1995 there were '13 unsuccessful attempts to get anti-discrimination legislation for disabled people through the British Parliament' (Barnes and Oliver, 1995: 111). The Disability Discrimination Act (DDA) was passed on 8 November 1995. Its key provisions came into force in November 1996. For the first time disabled people have some power to legally claim their right to equal participation. The DDA does, however, have its limitations. Many within the Disability Rights Movement have dismissed it as a 'sham, a "bigot's charter", without any teeth or real understanding of the operation of disability discrimination' (Gooding, 1996: 1). For more information, see Gooding (1996).

REFERENCES

Baker, N.C. (1984) *The Beauty Trap*, London: Piatkus.
Barnes, C. (1991) *Disabled People in Britain and Discrimination*, London: Hurst.
—— and Oliver, M. (1995) 'Disability rights; rhetoric and reality in the UK', *Disability and Society* 10, 1: 111–16.

Barton, L. (ed.) (1996) *Disability and Society: Emerging Issues and Insights*, London: Longman.

Brown, C. (1984) *Black and White Britain: The Third PSI Survey*, London: Heinemann.

Butler, R.E. and Bowlby, S.R. (1995) 'Disabled bodies in public space', University of Reading, Department of Geography, Discussion Paper No. 43.

Chapman, L. (1988) 'Disabling services', *Educare* 31: 5–20.

Clark, A. and Hirst, M.A. (1989) 'Disability in adulthood: ten-year follow-up of young people with disabilities', *Disability, Handicap and Society* 4: 271–83.

Corbett, J. (1991) 'So, who wants to be normal?', *Disability, Handicap and Society* 6, 3. 259–60.

Crow, L. (1992) 'Renewing the social model of disability', *Coalition*, July: 5–9.

—— (1996) 'Including all of our lives: renewing the social model of disability', in J. Morris (ed.) *Encounters with Strangers: Feminism and Disability*, London: The Women's Press.

Davis, M. (1990) *City of Quartz*, London: Verso.

Disability Now (1995a) 'Schoolgirl fights for her rights', June: 1–2.

—— (1995b) 'School battles', October: 3.

Elias, N. (1982) *The Civilizing Process*, Vol. 1: *The History of Manners*, Vol. 2: *Power and Civility*, New York: Pantheon.

French, S. (1994) 'Disability, impairment or something in between?', in J. Swain, V. Finkelstein, S. French and M. Oliver (eds) *Disabling Barriers – Enabling Environments*, London: Sage.

Goffman, E. (1963) *Stigma*, Englewood Cliffs, NJ: Prentice Hall.

Gooding, C. (1996) *Blackstone's Guide to the Disability Discrimination Act 1995*, London: Blackstone.

Hegarty, S. and Pocklington, K., with Lucas, D. (1981) *Educating People with Special Needs in the Ordinary School*, Windsor: NFER-Nelson.

Hirst, M.A. (1984) 'Education after 16 for young people with disabilities', *Youth and Policy* 2: 37–40.

—— (1987) 'Careers of young people with disabilities between ages 15 and 21 years', *Disability, Handicap and Society* 2: 61–74.

——, Parker, G. and Cozens, A. (1991) 'Disabled young people', in M. Oliver (ed.) *Social Work: Disabled People and Disabling Environments*, London: Jessica Kingsley.

Liachowitz, C.H. (1988) *Disability as a Social Construct: Legislative Roots*, Philadelphia: University of Pennsylvania Press.

Longmore, P.K. (1985) 'Screening stereotypes: images of disabled people', *Social Policy* 16, 1: 31–7.

Martin, J., Meltzer, H. and Elliot, D. (1988) *The Prevalence of Disability among Adults: OPCS Survey of Disability in Great Britain, Report 1*, London: HMSO.

Morris, J. (ed.) (1989) *Able Lives: Women's Experience of Paralysis*, London: The Women's Press.

—— (1993) 'Feminism and disability', *Feminist Review* 43 (Spring): 57–70.

Oakley, A. (1981) *Subject Women*, London: Martin Robertson.

Oliver, M. (1983) *Social Work with Disabled People*, London: Macmillan.

—— (1984) 'The politics of disability', *Critical Social Policy* 4, 2: 21–32.

—— (1990) *The Politics of Disablement*, London: Macmillan.

—— (1996) *Understanding Disability: From Theory to Practice*, London: Macmillan.

Parker, G. (1984) *Into Work: A Review of the Literature about Disabled Young Adults' Preparation for and Movement into Employment*, York: Social Policy Research Unit, University of York.

Scarman, Lord (1981) *The Scarman Report: The Brixton Disorders 10–12 April 1981*, London: HMSO.

Scott, R.A. (1981) *The Making of Blind Men: A Study of Adult Socialization*, London: Transaction books.

Shakespeare, T. (1994) 'Cultural representations of disabled people: dustbins for disavowal', *Disability and Society* 9, 3: 249–66.

Smart, L.S. (1978) 'Parents, peers and the quest for identity', in M.S. Smart and R.C. Smart, *Adolescents: Developments and Relationships*, 2nd edn, New York: Macmillan.

Smith, N. (1993) 'Homeless/global: scaling places', in J. Bird, B. Curtis, T. Putnam, G. Robertson and L. Tickner (eds) *Mapping the Futures: Local Cultures, Global Change*, London: Routledge.

Swain, J., Finkelstein, V., French, S. and Oliver, M. (eds) (1994) *Disabling Barriers – Enabling Environments*, London: Sage.

Synnott, A. (1993) *The Body Social: Symbolism, Self and Society*, London: Routledge.

Taylor, J. (1989) *As I See It*, London: Grafton.

Thomas, K. (1983) *A Man and the Natural World*, London: Allen Lane.

Union of the Physically Impaired Against Segregation (1976) *Fundamental Principles of Disability*, London: UPIAS.

Valentine, G. (1993) '(Hetero)sexing space: lesbian perceptions and experiences of everyday spaces', *Environment and Planning D: Society and Space* 11, 4: 395–413.

Weston, S. (1989) *Walking Tall*, London: Bloomsbury.

Winchester, H. and White, P. (1988) 'The location of marginalised groups in the inner city', *Environment and Planning D: Society and Space* 6: 37–54.

Young, I.M. (1990) 'Throwing like a girl: a phenomenology of feminine body comportment, motility, and spatiality', in I.M. Young (ed.) *Throwing Like a Girl and Other Essays in Feminist Philosophy and Social Theory*, Bloomington: University of Indiana Press.

6

PAPER PLANES

travelling the new grrrl geographies

•

Marion Leonard

INTRODUCTION

Traditionally (male) youth sub-cultures have been associated with particular symbolic sites where sub-cultural activities and meanings are
played out. Conventionally youth has been located in 'the street' (see
McRobbie, 1990: 77; Corrigan, 1993: 103–5). This chapter proposes an
alternative perspective considering how a sub-culture can operate from
within the sphere of the home and in the absence of exterior fetishised
locations. Using riot grrrl as a case study it will consider how a sub-
culture can maintain a sense of 'community' when its participants do not
meet in the collective space of a club or music venue, but are broadcast
over a wide geographical area. This chapter will argue that sub-cultures
should not be considered unified groups tied to a locality, creed or style
but as dynamic, diverse, geographically mobile networks.

This chapter will focus on the development of the riot grrrl network in
print and electronic media in the form of zines. Although riot grrrl has
spread to other countries, this study will concern itself with publications
issued in Britain and the USA. Consideration will be given to how these
texts described and promoted ideas about riot grrrl and to what extent
they shaped the 'movement' itself. It will review how zines facilitated the
geographical spread of riot grrrl, exploring the effects of movement and
dispersal. Moreover, it will examine how riot grrrl's gender politics can
be considered in spatial terms. The chapter will investigate the interplay
between conceptions of public and private space and the apparent contra-

diction of an 'underground community' being disparately geographically located. It will engage with both material and metaphorical notions of space.

TERMINOLOGY

The feminist nature of riot grrrl, and the decision by a number of its members to be straight edge,[1] gives good grounds for rethinking the associations of masculinity and deviancy brought to the term 'youth sub-culture' (see Willis, 1978; Cohen, 1972). Indeed riot grrrl may be understood as the realisation of the 'all-girl sub-culture' that McRobbie looked towards where members might have 'a collective confidence which could transcend the need for boys' and which could 'signal an important progression in the politics of youth culture' (McRobbie, 1990: 80). As Sophie comments in her riot grrrl zine from 1992, 'this summer has taught me alot of things. That it's OK and GOOD to have grrrls as friends, that a grrrl doesn't need a boy around to be cool' (*Notta Babe!*: 46).

Whilst riot grrrl's resistance to the status quo and identification with the 'punk rock underground' gives it grounds to be identified as a youth sub-culture, this chapter will avoid such terminology. The association with youth is misleading for whilst many girls aged 14 and 15 became involved, older women also identified with the initiative. As Tye, publisher of *The Meat Hook* zine, comments: 'my mom and I are doing a zine together (she's 51) and I know she loves the idea of Riot Grrrl!' (Vale, 1996: 72). The term sub-culture tends to suggest a group displaying integrated behaviour, beliefs and attitudes. This chapter hopes to avoid this notion of unity by conceptualising riot grrrl as a 'network'. This is a term used by riot grrrls to describe the informal ways in which they make contacts through letters, friendship books[2] and zines.

GO-GIRL-GO: THE DEVELOPMENT OF RIOT GRRRL[3]

Riot grrrl is a feminist network which developed in the underground music communities of Olympia, Washington, and Washington, D.C. The initiative was promoted by members of the bands Bratmobile and Bikini Kill who sought to challenge sexism in the underground music scene and encourage girls and women to assert themselves. Kathleen Hanna, singer for Bikini Kill, states that the name 'riot grrrl' was inspired in 1991 by the Mount Pleasant riots in Washington, D.C.:

'During that time, Jean Smith, a woman we know who's a writer and musician, said something like, "We need a girl riot, too" . . . . At the same time, Allison and Molly from Bratmobile were also in DC, and they heard this and said, "We're going to start a fanzine called Riot Grrrl".'

(*Hanna, quoted in Juno, 1996: 97–8*)

As women and girls began to identify with this idea, riot grrrl networks spread across the USA and Britain. The realisation of this initiative took several forms. Female audience members began by challenging the traditional division of the gig environment into gendered spaces, where women were largely absent from front of stage. Other grrrls formed bands, wrote zines, arranged meetings and organised events to introduce girls to music making. Activities were not only music related but concerned themselves with a broad range of issues tackling discrimination. This chapter will concern itself with the production and distribution of grrrl zines.

PEN + PAPER + PHOTOCOPIER = ZINE

Zines are self-published, independent texts devoted to various topics including hobbies, music, film and politics. They are usually non-profit-making and produced on a small scale by an individual or small group of people (Plate 6.1). Zines are inexpensively produced, often being photocopied rather than printed. As one writer explains: 'The simplest zine is one sheet of legal-sized paper copied on both sides and folded three times, trimmed and stapled to make a 16-page "mini" zine' (*Homegrown*, quoted in Vale, 1996: 4). These texts can also be found on the Internet in the form of e-zines. However this option is open only to those privileged enough to have access to a computer, telephone line and scanner. Whilst the e-zine has the potential to reach a very wide audience, it does so at the loss of individuality, lacking the personal qualities of paper zines.

The word 'zine' is an abbreviation of 'fanzine' which, in turn, is an alteration of 'magazine'. Nico Ordway dates the origin of the 'fanzine' to the publications produced by science-fiction enthusiasts in the 1950s, although these form part of a lineage of self-published leaflets and newsletters dating from the advent of printing presses (Ordway, 1996: 155). Self-publication has been a method closely associated with several art movements, from the Dada journals to the texts produced by the Surrealists and Situationists. The explosion of fanzines in response to punk rock established these publications as youth culture media. Fanzines could be produced cheaply and quickly on a small scale, articulat-

Plate 6.1 Riot grrrl zines.
Photograph: Rob Strachan.

ing the views of music enthusiasts and providing an alternative discourse to national magazines and newspapers. Punk fanzines had a fairly regular format of editorials, interviews with musicians and record reviews – which has been reproduced by enthusiasts of several subsequent music movements.

Low production costs have resulted in a proliferation of zines, allowing those with minimal wealth, such as children, to have a public voice. V. Vale claims that 'since photocopying became widely available in the 70s, over 50,000 zines have emerged and spread in America alone' (Vale, 1996: 5). The independent character of zines gives them significant value, allowing subjects outside of mainstream fashion to receive coverage. Sara, writer of *Out Of The Vortex*, comments: 'Only by controlling the medium do we control the message. . . . For this reason zines are extraordinarily unique and powerful political tools' (quoted in Vale, 1996: 168).

RIOT GRRRL ZINES

The contracted word 'zine' has greater relevance to riot grrrl publications as their content shows a shift from the role of the fan in documenting and constructing musical taste cultures. This is not to distance zines from music altogether – many make reference to independent groups and record releases – but merely to state that they have a broader project. Their content extends beyond music – addressing a range of topics from racism and queer politics to the narration of personal experiences. Riot grrrls create zines which reflect their thoughts and experiences: '[we] know that we're tired of being written out – out of history, out of the "scene", out of our bodies . . . for this reason we have created our zine and scene' (*Riot Grrrl! #3*: 1).

As with other fanzines, the production style of riot grrrl publications confers a tone of intimacy. The text is often hand-written and frequently includes hand-coloured pages or decoration with stickers. The zines use many stylistics of punk fanzines which, as Hebdige comments, convey a sense of urgency as 'Typing errors and grammatical mistakes, misspellings and jumbled pagination were left uncorrected in the final proof' (Hebdige, 1979: 111). Other similarities with earlier self-publications can be noted in the use of collage techniques and the detournment of magazine images. However, riot grrrl zines are noteworthy as, in both America and Britain, the selected images concern themselves with the construction of femininity. Writers offer a visual critique of beauty standards and expected behaviour by juxtaposing text and illustrations taken from comics, advertising, teen magazines and fashion photography.

Riot grrrl zines employ small scale production, with issues sometimes numbering only twenty copies. Most are not sold through retail outlets but distributed at gigs or by mail order. Whilst attendance at a gig at least suggests shared musical tastes, ordering through the post may seem a more distant and impersonal exercise. In practice, however, the reverse seems to hold true. Acquiring the addresses of zines involves tapping into the informal friendship networks active within riot grrrl. Whilst some grrrls produce contact lists, other addresses are printed in review sections of zines or enclosed with the publication in the form of small 'flyers'. Thus lines of connection are made through ordering and collecting zines. Zines despatched by post are always accompanied by a letter or note encouraging the reader to write back with comments. This request for personal interaction is in keeping with the zine content. A large number of grrrl zines, particularly those produced in America, position the reader as confidante by disclosing experiences of rape or child abuse.

ARTICULATING SPACE AND SPACES OF ARTICULATION

The language of riot grrrl zines indicates how the network is conceived of spatially by its participants. Contributors involve themselves in a discourse negotiating the ideological location of riot grrrl, exploring notions of public and private space. Several writers picture riot grrrl as opening up frontiers and allowing girls and women access to places from which they were previously excluded. One zine writer characterised the birth of riot grrrl as 'revolution summer, breaking down every wall' (*Quit Whining! #1*: 31), whilst another declared 'anything that ever was/ is exclusive to boystown has had all previous right removed. The world is yours, so do what you want' (*Riot Grrrl #7*: 17).

The rhetoric of occupation and conquest is offset by the importance placed on being a subterranean network. Whilst writers celebrate the multi-dimensional trajectories of riot grrrl initiatives, many express resentment when these are documented in national publications: 'seeing ourselves described by these mainstream writers puts boundaries in our minds. . . . We can counteract it by keeping alive the "underground" aspect of riot grrrl' (*Riot Grrrl #8*: 14). Zines thus present a complex, if not contradictory, concept of riot grrrl as an underground network which rejects public intrusion yet claims to be open to all: 'every girl is a riot grrrl' (ibid.: front cover). The band Huggy Bear exploits a sense of insiderness in their zine whilst simultaneously refusing to mark out any defining characteristics: 'for a change you can't just blag your way into this "scene" . . . all peasy easy by nonchalantly leaning over at the bar and enquiring after its signifiers' (*Reggae Chicken*: 7).

The text of these publications thus offers an articulation of ideas about space and place. However, these publications can be understood as creating 'rhetorical spaces', to employ Lorraine Code's spatial terminology, where individual voices and particular feminist ideas can be articulated (Code, 1995). The contributors first position themselves as gendered subjects within the text and then proceed to explore the issues and interests relevant to them as women. As the writers of *Hotskirt* argue, 'we want to create a voice, for ourselves and for anyone who is concerned about woman's situation in society today. . . . To show that girls can write zines, play in bands, set up shows, live up to their own expectations instead of the media's' (*Hotskirt #1*: 1). By writing themselves into the text, through relating personal experiences and concerns, riot grrrls have expanded the discursive parameters of the fanzine.

Although this is a somewhat poetic use of geographical terminology, the subject is far from abstract. In considering the fanzine as a 'textured location where it matters who is speaking and where and why' (Code, 1995: x), one begins to approach the significance of the existence and proliferation of grrrl zines. Fanzines have traditionally been produced within a male culture and have concerned themselves with aesthetic judgements. Whilst the discussion of feminist concerns was not prohibited by fanzine editors, it was nonetheless absent as there existed no rhetorical space for its articulation. These fanzines effectively excluded female voices and concerns by their failure to acknowledge their relevance to the texts under discussion.

To theorise the zine writer/producer requires a shift away from the association between sub-culture and the street. Textual production necessarily locates youth indoors and, most often, in the private space of the bedroom. Again this is not a neutral location for it is often regarded as the feminised space of teen consumerism (see McRobbie and Garber, 1993: 209–22; Brake, 1980: 143). However, through the production of a zine, this private space becomes a site of activity. Moreover, the private sphere is integral rather than incidental to the text itself. Whereas previous music fanzines have been responses to an external (street or club) culture, here the culture is produced in the very act of writing. The multiplicitous nature of riot grrrl and the accompanying rhetoric of inclusion – 'We are *not* a club and there are no rules' (*Riot Grrrl #8*: 14) – encourages anyone to identify with the network and contribute to its expansion. As one zine writer comments. 'riot grrrl is . . . BECAUSE every time we pick up a pen, or an instrument, or get anything done, we are creating the revolution. We ARE the revolution' (*Fantastic Fanzine*, reproduced in *Persephone's Network*: 26).

Zines allow young women to voice their participation in the network from a safe space, encouraging participation by those who would be precluded from involvement in a pub or club environment. A number of teenaged women I have interviewed about their involvement in riot grrrl said they often could not attend live performances due to concerns over safety, travel arrangements and access to venues. One zine writer from Hertfordshire explained: 'I was about 15 at the time . . . and couldn't go travelling on the underground on my own' (personal communication, 29 December 1995).

BORDER CONTROL: DISTRIBUTION, SCALE AND PUBLICITY

The style of the zines, their scales of production and systems of distribution are critical to notions of riot grrrl as an underground network. The presentation style and small scale of production supports a sense of alliance with the reader. A remark made by one zine writer from Minneapolis illustrates how she credits her readership with a level of trustworthiness: 'I print around 200 an issue . . . looking at that number scares me because I am really vulnerable in my zine. . . . Feedback has been amazing though, it's what keeps me going' (personal communication, 26 August 1996). Thus, whilst these zines may reach people in other countries, their content and scale of production give the impression of conversing with a close group of friends.

The system of distribution informs the way in which the interiority of the network is maintained as these texts are not 'publicly' sold or advertised. By encouraging feedback from readers, zine writers create a sense of engagement rather than acting as impersonal distributors:

> I want to encourage people not to just order zines like they were any commodity but to write to anyone whose zine you feel inspired by or have a critique of. It would truly bumm me out if this turned into a commodification of 'girl zines' where if you have the cash you can have access to whatever you want.
>
> (Riot Grrrl Press Catalogue, *July 1993*)

In a very real sense riot grrrl has grown, not in spite of its participants being disparately located but because of the freedom and opportunity created by not being bounded within a particular locale. Zines act as a means of accessing like minded people who do not live locally. As one 17-year-old zine writer comments: 'none of my friends around me here in Norfolk became involved in riot grrrl at all, they'd never even heard of it' (personal communication, 30 November 1995). Zines became a tool for empowerment allowing geographically isolated people to correspond with each other and share a common sense of identity. One 16-year-old from Liverpool remarked: 'I've made lots of new friends through riot grrrl and it's helped me a great deal to feel more accepted and more in control of my life and happier with myself' (personal communication, 6 September 1996).

National press interest in riot grrrl in 1992/3 disrupted this sense of a subterranean discursive community.[4] The relationship between riot grrrl participants and the various strands of the international media is a complex one, but it is not the focus of this chapter. Certainly the publicity does not fit a simplistic exploitation or moral panic model. Flows of information about riot grrrl in the national media were both regulated by those involved and ungoverned; accounts were variously informed, negative, inaccurate and supportive. Whilst arguments can be made as to the benefits and drawbacks of this exposure, the interest here is in how it reflected the discursive space of the zine.

Several writers seemed to view the public exposure as an invasion and responded by printing textual defences against the intrusion. A number of American zines criticised the publicity afforded by teen magazine *Sassy*:

> Let it be known that I, Margaret Rooks DO NOT like *Sassy* magazine, and am annoyed at its attempts to infiltrate the underground music scene. . . . I do not appreciate *Sassy* latching on to something they think is hip, then spoonfeeding it to the mainstream.
>
> *(Quit Whining! #1: 22)*

Zine writers in Britain expressed similar indignation at the usurpation of their cultural terrain and argued against such public trespass: 'We're growing, we're underground, and we're denying their power by not talking to them . . . we don't need the corporate press. We're truly independent, we'll pay our way, we'll use our own channels' (*Ablaze! #10: 15*).

A SLIPPERY, SHIFTING NETWORK . . .

One repeated concern of participants was that journalistic involvement might rigidify riot grrrl. Reporters frequently attempted to present a user's guide to the 'phenomenon', plotting its lineage and describing its aims as if participants adhered to a central manifesto. Gwen, a zine writer from Northamptonshire, argues press attention resulted in 'the music press defining the idea for us, claiming it to be something it necessarily isn't and putting their own expectations on it rather than giving girls/people the information and letting them interpret/use it as they wish' (personal communication, 24 December 1995). Attempts to contain riot grrrl within an explanatory grid ran counter to its very ethos as an expansive and changing network.

Those involved in riot grrrl stress that no one viewpoint or cultural product (be it a zine or record) can be taken as representative or indicative of the whole. Zine writers repeatedly emphasise they can only offer personal comment: 'I won't offer a definition because it wouldn't be fair to other grrrls to whom riot grrrl may mean something totally different. I will however offer my insight on what I have seen happen' (Spirit, 'What is a Riot Grrrl Anyway?'). In this sense riot grrrl is truly multiplicitous. In order to clarify what I mean by this I defer to Elizabeth Grosz's explanation:

> A multiplicity is not a pluralised notion of identity (identity multiplied by *n* locations), but is rather an ever-changing, nontotalisable collectivity, an assemblage defined, not by its abiding identity or principle of sameness over time, but through its capacity to undergo permutations and trans-formations, that is, its dimensionality.
>
> *(Grosz, 1994: 192)*

Those involved in the network delighted in the possibilities opened up by allowing different people to produce their own interpretations of riot grrrl: 'take the ball and run with it' (*Riot Grrrl*: 2).

Deleuze and Guattari's notion of a rhizome is useful when considering the nature of riot grrrl and its zine network. They reject the dominant metaphor of the tree in Western thought with its suggestion of unity, a centralised core and binary thinking ('arborescent pseudomultiplicities') (Deleuze and Guattari, 1988: 8). Instead they propose a celebration of the nature of rhizomes, the underground stem of plants branching out in different directions without a singular or central trunk. This is a very different concept to arborescent thinking 'which plots a point, fixes an order' (ibid.: 7). Instead the rhizome:

> is composed not of units but of dimensions, or rather directions in motion. It has neither beginning nor end, but always a middle (milieu) from which it grows and which it overspills. . . . The rhizome operates by variation, expansion, conquest, capture, offshoots.
>
> *(Deleuze and Guattari, 1988: 21)*

This is an effective way of conceptualising multiplicity and is a useful explanatory model for riot grrrl. The image of a root-like structure matches the idea of an underground culture multiplying via lines of connection which are not controlled from a primary location: 'Riot grrrl isn't centralised, it's not organised, we've no leaders, no spokeswomen' (*Ablaze! #10*: 15).

Riot grrrl calls then for a complication of the notions of a youth sub-culture. Whilst critics have been keen to argue culture as a process, riot grrrl splits apart any concept of this as a unified progression. Whilst those who identify with the initiative may use a collective name, riot grrrl is polymorphous. In 1992 Karren, author of *Ablaze!*, speculated gleefully on the way this diversity and contradiction would be reflected in zines:

> kids will kick over the news stands and build their revolutionary methods of communication in the form of 'fanzines'. . . . We will confuse them by disseminating different pieces of literature under the same name, and there will be no sense in which any is more 'authentic' than any other.
>
> *(Karren, in 'Girlspeak', GirlFrenzy #3: 31)*

SLIPPING THROUGH THE NET

The appearance of riot grrrl pages on the Internet further complicates notions of a unified community as the potential audience is equal to the number of on-line subscribers.[5] Of course the actual number of 'visitors' will be relatively small yet the ease of accessing any site disrupts the concept of an underground community. The decision to use the word 'grrrl' in the titling of web sites to some extent filters the audience by separating it from the mass of material aimed at men. This point is made by Chrystal Kile, who publishes the *Pop Tart* site:

> a very practical reason grrrls/geeks/nerds use these codewords in titles of our site is to make clear that we're not naked and waiting for a hot chat with you! I mean, just do an Infoseek search using the keyword 'girl' or 'woman' and see what you find. Cybergirl. com [not to be confused with Cybergrrrl] is a nekkid – chick. gif site or something.
>
> *(Kile, quoted in DeLoach, 'Making Two Negatives A Positive')*

The title grrrl thus helps to differentiate a site and identify it with certain feminist attitudes; 'Girls are not girls, but grrrlo, super kewl (cool) young women who have the tenacity and drive to surf the net, network with other young women on-line and expand the presence of young women in new and emerging technologies' ('Friendly Grrrls Guide To The Internet – Introduction').

Whilst riot grrrl on the Internet disables any notion of a clearly identifiable audience, it also highlights the folly of constructing riot grrrl as a single strand of youth culture. Use of the word grrrl suggests a certain feminist affinity; however, it cannot be assumed that authors of sites will identify themselves as a riot grrrl. RosieX, writer of e-zine *geekgirl*, comments that:

> except for friends like St. Jude, I actually wasn't familiar with [others in the movement such as] the Guerrilla Girls or Riot Grrrls, in fact we still ain't made acquaintances. But I guess I've always liked the grrrowl in grrrl.
> *(RosieX, quoted in 'Making Two Negatives A Positive')*

This demonstrates how riot grrrl terminology has spread and is being used in new and different contexts. The Internet allows international access to the riot grrrl initiative and a migration of its ideas across new political and ideological terrains. As Amelia DeLoach comments, 'Like the Riot Grrrls, the grrrls on the Web don't have a neatly defined central purpose. In many ways, both the online and offline movement are like the web itself – diffuse' (DeLoach, 'Grrrls Exude Attitude').

There is a sense in which ideas of geographical location are removed by the very communication medium of the Internet. The immediacy of e-mail and accessibility of virtual 'sites' around the world remove notions of distance; the desire for effective, direct communication results in short summary Web pages that flatten differences, and the regular typed print and box format gives the reassurance of the familiar rather than the foreign or exotic. However, despite this loss of identifiable locality, the rhetoric of geography is pervasive. If the individual is, to some extent, divorced from their real world context, authors have been quick to offer a virtual world as substitute. Those posting e-zines on the Internet stake out their own virtual territory through the language of geography. Clara Sinclair describes her Internet connection as allowing her to 'surf the Web from my own turf' and talks of her excitement in exploring 'the booming new region of Cyberland' (Sinclair, 'Net Chick Clubhouse').

Admittedly the terms 'site' and 'home' page are common on the Internet; however, e-zines often play with this idea of place. Megan Larson, writer of 'ratgrrrls' hideout', creates a sense of insiderness in her opening text:

> welcome to the new, improved ratgrrrls' hideout. I've rearranged the furniture a bit, but it's still the same old place, honest . . . stay as long as you like – you don't even need to know the secret handshake to get in (but if you're really nice, I might teach it to you).
> *(Larson, 'ratgrrrl's hideout')*

Clara Sinclair evokes a similar atmosphere in her e-zine 'Net Chick Clubhouse'. The introductory page features a picture of the (virtual) clubhouse, the message 'you're at the front door' and the invitation to explore 'the room of your choice'. The rooms 'inside' contain biography details, book reviews, selected links and Clara's 'secret' diary. The author trades with ideas of selectivity and secrecy by blocking access to the diary until you pass a sartorial style challenge, 'I don't let just any voyeur read my personal memoirs' (see Figures 6.1 and 6.2). This simple exercise, demanding a very basic shared cultural knowledge of what clothes to select, supports the illusion of being inside a 'gang'.

This chapter has considered how riot grrrl reworked the fanzine format to create paper texts that reflect female concerns and experiences. Similarly e-zines were a response to the absence of women on the Internet, encouraging women to involve themselves in technology. RosieX explains that the pervasive masculinism of the Internet motivated her to create a cyberzine: 'I was appalled at the Internet "boys club". . . . I created geekgirl as a vehicle to express my ideas, about feminism/cyberfeminism' ('RosieX explains . . . '). She stresses the ideological importance of her Internet site in the statement 'I consider the WWW my political home' (ibid.). Whilst the Internet cannot offer the privacy of a limited publication and distribution, it does offer a welcoming sense of anonymity. A Web page is a public space where women can talk without having to disclose their home address or even use their real name. As Amelia Wilson, author of *NerdGrrrl!*, comments 'The web provided the perfect forum for a shy girl with something decidedly unshy to say' ('Amelia explains . . . ').

CONCLUSION

This chapter has explored how riot grrrl operates using different ideological conceptions of public and private space and internal and external location. Grrrls have invited involvement by anyone, whilst holding fast to the idea of 'the underground'. Zines have been considered as spaces of articulation, promoting a rhetoric of inclusion: 'This will result in a worldwide society, in which all exist freely and in wonderment, unbounded by old laws of how to live' (*Ablaze #10*: 17). Using riot grrrl as a case study, this chapter has illustrated the complexity of spatial notions operating within a sub-culture. However, there is a need to temper riot grrrl's celebration of possibility with an acknowledgement of its limitations. Perversely, the call by riot grrrls to unite resulted in some women feeling their differences were being denied or ignored. As zine writer Sisi, who is of Mexican origin, comments: 'everyone tried to

Wanna read my secret diary? Well I won't give you the key unless you give me some clothes! (Hints: 1. Don't make me look like a dork. 2. I need the right combo of 3 items.)

Figure 6.1 Before access is granted to the secret diary one has to dress a 'paper doll' in suitable clubhouse clothes by clicking on various options with the cursor.
Source: *http://www.cyborganic.com/People/carla/Rumpus/Toychest/Doll*

create this *utopia*: "We should all love each other" cuz we're girls. It doesn't matter what class, race or religious background we have" . . . it's just not true. There *are* disparities between us; there are differences' (Vale, 1996: 54). Whilst some young women of colour were active in the network, the grrrl revolution tended to be confined to white, middle-class

Not bad!

I guess I gotta give you the key to my diary now...

Figure 6.2 If suitable selections are made you'll be able to unlock the mysterious book and read on (Net Chick Clubhouse).
Source: http://www.cyborganic.com/People/carla/Rumpus/Toychest/Doll/Minibootstank.

women. Although the network connected many women, the difficulty in accessing zines should also be considered. Zines are a transient medium where P.O. box and e-mail addresses rapidly obsolesce and small scales of distribution mean access to back issues is often impossible.

The appearance of e-zines raises the question of placing boundaries around sub-cultures. Whilst some e-zines declare themselves as riot grrrl, others, used in this chapter, do not. By employing the term 'grrrl', they create a relationship with the network but present themselves as an off-shoot. This is in keeping with the model of a rhizome mentioned earlier, where movement and growth results in rupture and metamorphosis. In spatial terms, Deleuze and Guattari ally the rhizomatic network with a map: 'the rhizome pertains to a map that must be produced, constructed, a map that is always detachable, connectable, reversible, modifiable, and has multiple entryways and exits and its own lines of flight' (Deleuze and Guattari, 1988: 21). This conveys the dynamism of riot grrrl as 'a world-wide network' (*GirlFrenzy* #3: 30), encouraging women to enter and contribute new lines of connection. Perhaps this chapter can be understood as a view of the industries of grrrl cartographers. As Karren, author of *Ablaze!*, enthusiastically declares: 'New Girl Geography: We're construct-ing new road maps of our territory, the globe, which is unbounded by walls and unmarked by flags' (Karren, *Ablaze! #10*: 17).

NOTES

1 Straight edge is a lifestyle choice of no drugs, meat or alcohol.
2 Friendship books and 'Slam' books are small hand-made booklets comprised of several squares of paper stapled together. Each 'book' contains a list of names and addresses and a brief outline of the respective entrant's likes/ interests. When full, the book is posted to the addressee on the front page.
3 'Go-Girl-Go' is a reference to 'Clock Song (Go-Girl-Go)' by the female band Scrawl. The song appears on the compilation album *International Pop Underground Convention* – a live recording of the convention which took place in Olympia, Washington on 20–25 August 1991. The album is released by K records of Olympia, Washington. The convention was important in promoting riot grrrl. On the first night of the event, named 'Girls' Night', female musicians performed and riot grrrl zines were handed out. The words 'Go-Girl-Go' were printed in some zines, motivating readers to become involved in riot grrrl.
4 Riot grrrl articles appeared in a variety of publications including *Rolling Stone, New York Times, Ms* and *LA Weekly* in the USA and *Daily Star, The Wire, Guardian* and *Melody Maker* in Britain.
5 The Friendly Grrrls Guide To The Internet calculates this as 40 million.

REFERENCES

Brake, M. (1980) *The Sociology of Youth Culture and Youth Subcultures: Sex and Drugs and Rock 'n' Roll?*, London: Routledge and Kegan Paul.

Code, L. (1995) *Rhetorical Spaces: Essays on Gendered Locations*, London: Routledge.

Cohen, S. (1972) *Folk Devils and Moral Panics*, London: MacGibbon and Kee.

Corrigan, P. (1993) 'Doing nothing', in S. Hall and T. Jefferson (eds) *Resistance Through Rituals: Youth Subcultures in Post-War Britain*, London: Routledge.

Deleuze, G. and Guattari, F. (1988) *A Thousand Plateaus; Capitalism and Schizophrenia*, trans. Brian Massumi, London: Athlone Press.

Grosz, E. (1994) 'A thousand tiny sexes: feminism and rhizomatics', in C.V. Boundas and D. Olkowski (eds) *Gilles Deleuze and the Theater of Philosophy*, New York: Routledge.

Hebdige, D. (1979) *Subculture: The Meaning of Style*, London: Methuen.

Juno, A. (ed.) (1996) *Angry Women in Rock, Volume One*, New York: Juno Books.

McRobbie, A. (1990) 'Settling accounts with subcultures: a feminist critique', in S. Frith and A. Goodwin (eds) *On Record: Rock, Pop, and the Written Word*, New York: Pantheon Books.

—— and Garber, J. (1993) 'Girls and subcultures', in S. Hall and T. Jefferson (eds) *Resistance Through Rituals: Youth Subcultures in Post-War Britain*, London: Routledge.

Ordway, N. (1996) 'History of zines', in *Zines! Volume One*, San Francisco: V/Search.

Vale, V. (ed.) (1996) *Zines! Volume One*, San Francisco: V/Search.

Willis, P. (1978) *Profane Culture*, London: Routledge and Kegan Paul.

WEB PAGES

Amelia explains . . .
http://www.december.com/cmc/mag/1996/mar/delame.html
(28 August 1996)

DeLoach, Amelia, Grrrls Exude Attitude
http://www.december.com/cmc/mag/1996/mar/deloach.html
(28 August 1996)

DeLoach, Amelia, Making Two Negatives A Positive
http://www.december.com/cmc/may/1996/mar/delrec.html
(28 August 1996)

Friendly Grrrls Guide To The Internet – Introduction

http://www.youth.nsw.gov.au/rob.upload/friendly/fintro.html
  (28 August 1996)
Larson, Megan, ratgrrrl's hideout
http://gladstone.uoregon.edu/~meganl/
  (20 August 1996)
RosieX explains . . .
http://www.december.com/cmc/mag/1996/mar/delros.html
  (28 August 1996)
Sinclair, Clara, Net Chick Clubhouse
http://www.cyborganic.com/people/carla/index.html
  (7 November 1996)
Spirit, What is a Riot Grrrl Anyway?
http://www.columbia.edu/~rli3/music_html/bikini_kill/girl.html
  (20 August 1996)

ZINES

Ablaze! #10 (Leeds).
GirlFrenzy #3 (London).
Hotskirt #1 (Little Rock, Arkansas).
Notta Babe! (Richmond, Virginia).
Persephone's Network (Cleveland, Ohio).
Quit Whining! #1 (S. Hadley, Massachusetts).
Reggae Chicken (London).
Riot Grrrl (Washington, D.C.).
Riot Grrrl! #3 (Massachusetts).
Riot Grrrl #7 and *#8* (Washington, D.C.).

....

two

matters of scale

Geography as a subject has been relatively late in 'discovering youth' and to date young people are largely missing from the Geographical imagination. In this section five authors employ different understandings of geographical scale in their examination of aspects of young people's lives and youth cultures.

In her chapter on the 'Spatial construction of youth cultures' Doreen Massey argues that youth cultures are not closed local cultures, nor are they undifferentiatedly global cultures, but rather that they are in fact complex 'products of interaction'. Using examples from Mexico to the UK, Massey demonstrates some of the ways that young people carve out local spaces for themselves, while also highlighting the undeniable interconnectedness of these spaces and youth cultures with others across the globe.

The fact that global processes impact upon young people in very different parts of the world in complex ways is highlighted by Cindi Katz. In her chapter she describes the ways in which global economic restructuring and other 'globalising' forces have profoundly damaged the local living environments of young people in places which we may consider to be worlds apart – namely, Harlem, New York, USA and 'Howa', Sudan.

In the following chapter Tim Lucas focuses on Santa Cruz, California, USA. Here he explores the ways in which local concerns about gangs and gang violence are fuelled by, formulated in response to and engage with, national-scale moral panics. Rather than local–national intersections, Birgit Richard and Heinz Hermann Kruger are concerned with global-national interconnections. In this chapter they use the example of 'rave' to discuss the ways in which global musical and dance forms have been interpreted in specific ways in German youth cultures.

In the final chapter of this section Luke Desforges examines the way in which at the very time young travellers from the 'West' are forging their identities through contact with others around the world they are also participating in the process of 'othering' and constructing 'First World' representations of the 'Third World'. From '"Checking out the planet": global representations/local identities and youth travel' we learn that the cultural capital which young people gather from their independent travel experiences is often then effectively converted into economic capital back home in the workplace. Thus through their engagement with the 'Third World' young people can earn a privileged position in the 'West'.

7

THE SPATIAL CONSTRUCTION OF YOUTH CULTURES

•

Doreen Massey

A few years ago I was filming in the Yucatán in Mexico. I was inter-viewing a group of women in their home. We sat on stones on an earthen floor, the light from an open fire flickering across the baked walls and thatched roof. The women were making tortillas (corn bread) and talk-ing about their lives. This was the way bread had 'always' been made they said, as they slapped it from hand to hand before throwing it on to the red hot grill. Outside, the late afternoon faded into evening, and the noises of the night grew in intensity among the trees in which the house was set. It seemed a picture of a 'truly indigenous' culture preserving its customs, its clothing, its corn bread and its beliefs.

The interview over, we talked on for a while and then walked back to our jeep. As we approached the road our ears were assaulted by a racket of electronic noise. In a pool of light from another building – this one wired for electricity – a dozen or so youngsters were urgently playing computer games. Machines were lined up around the walls of the flimsy shack and every one was surrounded by players. Electronic noises, American slang and bits of Western music floated off into the night-time jungle.

Of course, we filmed the youngsters too. But the question is what to make of this contrast: was the youth culture of the Yucatán countryside the entry-point for external influences into (maybe the eventual break-up of) this inherited Mayan culture which had lasted so long and endured so much? Are the cultures of the young more outgoing and international-ised than those of older generations? Indeed, how would one define the culture of these lives which we caught for such a brief moment of evening play? Was this a local culture, or were these youngsters part of an emerging global culture of youth?

The answer to these questions depends on how one conceptualises 'culture' and this is something which has been a matter of fierce theoretical debate. That roomful of computer games was certainly a link between this small cluster of houses in eastern Mexico and something that might be characterised as 'global culture' (though already some questions should be registered: is it 'global' or, say, American? Is it 'global' in the full sense that everyone, everywhere, has access to it?). And there were plenty of other links too – the T-shirts with slogans in English, the baseball caps, the trainers, the endless litter of cola cans. And yet of course the 'youth culture' here was also quite different from that of, say, San Francisco, or a small American mid-west town, or Redditch, England, or Tokyo. In each of these places the T-shirts and the computer games are mixed in with locally distinct cultures which have their own histories. The very meaning of the 'global' elements themselves will change: what is and what is not a status symbol, how particular slogans or music are interpreted. And they will be embedded in a host of particularities – Mayan family relations, an understanding of an ancient cosmology, a particular attitude to the USA which comes from being its southern neighbour, a vague consciousness of 'Latin America'. In each place the mix of 'local' and 'global' will be different. So, in partial answer to some of the questions which were posed above: local specificity – such as local variations in youth cultures – can be constantly reinvented even while international influences are accepted and incorporated.

The local youth culture of the Yucatec Maya, then, is *a product of interaction*. It is certainly not a closed, local, culture, but neither is it an undifferentiatedly global one. And such interactions could be exemplified in a million ways. The spatial openness of youth cultures in many if not all parts of the world is clear. Across the world even the poorest of young people strive to buy into an international cultural reference system: the right trainers, a T-shirt with a Western logo, a baseball cap with the right slogan. Music draws on a host of references which are fused, rearticulated, played back (see, for instance, Meegan, 1995). Anti-roads protesters and young ecologists link up via the Internet with environmental battles worldwide (internationalism can mean interconnection without implying sameness). The youngest generations of diaspora societies wrestle constantly to find an enabling interlocking of the different 'cultures' in which they find themselves: it is a struggle indeed to build another, different – 'hybrid' – culture.

And yet the evidence seems to be that all youth cultures – and not just those more obvious cases such as the children of diasporas – are hybrid cultures. All of them involve active importation, adoption and adaptation. And indeed this touches the heart of some recent debates about how to conceptualise culture. Indeed what is at issue is the *geographical constitution of cultures*. In a long accepted formulation, 'cultures', and certainly 'local cultures' were understood as locally produced systems of social interaction and symbolic meanings. Cultures which felt themselves to be under threat would conduct a kind of archaeology in search of origins, a search for what was 'authentic' and essential to that cultural formation. Imagined geographically, such a culture was understood as preserving its authenticity through closure. It was not invaded by cultural intrusions (foreign elements) from outside. In such a geographical imagination of a culture there would be a clear distinction between what was 'local' and what came from outside. So the authenticity of the Mayan women in that house would be seen as being under threat from the computer games their children played on endlessly. And quintessential Englishness is interpreted as in danger of defilement by the 'pouring in' of 'immigrants'.

Such a view of cultural authenticity, and especially of an authenticity which depends on purity and closure is now increasingly argued to be inappropriate (see, for instance, Hall, 1995). For 'Englishness' did not somehow grow out of the soil but rather is a complex product of all the peoples who over the centuries have settled that part of the British Isles, of all their contacts and influences. The quintessential cup of tea could not be sipped without plantations in India, Opium Wars in China and – if you take sugar – a history of slavery in the Caribbean (Massey, 1991). And those women in the Yucatán? – on the earthen walls was pinned a picture of the Virgin of Guadaloupe, and the language we were chatting in around that fire was not Mayan (though Yucatec Mayan is still used) but that of the Spanish who conquered Mexico 500 years ago. Appreciating this means reimagining the geographical constitution of cultures: here they are not closed but open, not ingrown ('pure') products of relative isolation but the outcome of incessant processes of social interaction. Here it is much more difficult to distinguish the local from the global. This is a reworking of the geographical imagination of culture which has been well captured in the formulation, from 'roots' to 'routes'.

•

There are three points which can be drawn out of the argument so far.

Firstly, while youth cultures certainly leap geographical scales in the search for influences and references to tap into, this openness of cultural formations is not specific to the young. 'Hybridity' is probably a condition of all cultures. So the young Maya may be adding another element to an already mixed brew.

Secondly, if it has been necessary to rethink the geography of cultures then this process in itself has encouraged a challenge to the manner in which we think about space. One of the ways, both in popular discourse and academic analysis, in which it is customary to represent the organisation of space is by making use of the concept of *scale*. It has indeed on occasions been claimed that 'scale' is one of the central concepts in geographical thinking (see, for instance, Smith, 1993). Thus many analyses work – implicitly or explicitly – with a spatial framework which is organised into a nested hierarchy of different levels, or scales. Smith, for instance, has suggested the sequence 'body, home, community, urban, region, nation, global' (1993: 101). The list itself, as Smith agrees, is immediately open to debate and challenge: why are the 'scales' of the continent or the workplace excluded? Some of the 'scales' (such as community) are not necessarily specifically spatial; a 'region' of Luxembourg is of a very different size (scale?) than a region of China; and so on. But the analysis of (youth) culture above raises even more fundamental doubts about what is meant by the concept of scale and how it can be used in geographical analyses.

What emerged from the discussion of changing conceptualisations of culture (from roots to routes) was a notion of space as organised, not into distinct scales, but rather through a vast complexity of interconnections. A so-called 'local' youth culture was argued to be not a closed system of social relations but a particular articulation of contacts and influences drawn from a variety of places scattered, according to power-relations, fashion and habit, across many different parts of the globe. Social spaces are best thought of in terms of complicated nets of inter-relations in which each particular culture is differently located, but as networks which are certainly not tidily organisable into distinct 'scales'. Moreover, as was also argued, the different 'scales' influence each other – the 'global' is inside the 'local', for instance (indeed often itself has 'local' – e.g. American, Caribbean – origins), and the 'local' affects the character of the 'global'. Finally, the geographies of cultures themselves cut across many of the most commonly accepted hierarchies of scale. The interconnections which bind together and internally differentiate a diaspora culture, for instance, cut across regions, nation states and continents, linking local areas in, say, a British city to a Turkish region of Cyprus, to a particular island in the Caribbean or a village in India. Such links take no notice at all of the neat packaging of space into a hierarchy of scales. What they suggest is that the social relations which constitute space are not organised into scales so much as into *constella-*

tions of temporary coherence (and among such constellations we can identify local cultures) set within a social space which is the product of relations and interconnections from the very local to the intercontinental. This is a view of social space which recognises its enormous complexity and which refuses prematurely to tame it into any hierarchy of neatly ordered boxes labelled urban, regional, national. We shall return to this issue later.

But before that there is a *third* point which it is important to draw out from the opening argument. This is that all these relations which construct space, since they are social relations, are always in one way or another imbued with power. That is to say, such relations are not just neutral 'connections' between one cultural constellation and another, or others; they reflect in their form and their direction geographical differences (uneven development) in cultural influence, fashion, economic power, the spatial structure of the media industry, the traces of migrations perhaps centuries ago, trade routes, the access to ownership of computers, the dominance of Hollywood – and a host of other phenomena. What's more, these relationships deeply affect the meanings of cultural influences and cultural contact. When, say, young people in Guatemala sport clothing marked clearly as 'from the USA' (or – ironically – with an 'American' logo and trademark emblazoned upon it but in fact quite likely made in Guatemala, a T-shirt quite likely sewn up by the mother of the Guatemalan kids themselves) they are tapping into, displaying their knowledge of, their claimed connection with, that dominant culture to the north. The social relations (both cultural and economic) embedded in this flow of cultural influence (and thus in the particular moment of the wearing of this T-shirt) are complex but they are clearly to do with the subordination of the Guatemalan culture and economy to the greater power of the United States of America. For, say, a middle class white youngster in those United States to wear the brightly coloured textiles of Guatemala (a wrist-band or a jacket perhaps) has a very different meaning and embodies and expresses very different social relations. It may be that Guatemalan textiles are seen as 'exotic', as tapping into the otherness of a (perhaps rather romanticised) vision of 'unspoilt' indigenous culture. At its worst this can be read as the children of the relatively wealthy 'West' brightening up their lives by tapping into less powerful and less 'modern' cultures – the 'Third World' as exotic decoration. And yet again it could be far more than that. Many lines of cultural connection around the world are expressions in one way or another of solidarity or of a desire to belong to something believed in. An awareness of Central America among US youth might extend to solidarity with its people's resistance to persistent US intervention. (Accusations of US involvement with right wing Guatemalan death squads has long been a political issue.) The colours of Rasta, sported by millions on both sides of the Atlantic, from Boston to Rio, from London to Cape Town, were a deliberately visible sign of belonging, maybe even of commitment. Or again, on the banks of the Niger in Mali musicians are borrowing music from Cuba to weld into their own, well aware that some of that music is itself a long historical development of instruments and melodies first taken across the Atlantic in the opposite direction, in the slave ships which left West Africa some centuries ago.

It would be a fascinating exercise to try to map even just a few elements of this complex of cultural influences and the different kinds of social relations and social power which they involve and express. A map of the world would certainly reveal some parts of the world as foci of more powerful influences than others. More modestly, the exercise might be tried for a particular youth culture, in order to capture the geography of influences (both inward and outward), their evolution over time, and the power relations which they embody.

●

And yet, if it is important to begin by conceptualising space in terms of a complexity of interacting social relations, it is also important to recognise that within that open complexity both individuals and social groups are constantly engaged in efforts to territorialise, to claim spaces, to include some and exclude others from particular areas. At the political level the carving up of the world into nation states (a relatively recent, and already embattled, phenomenon) is probably the most obvious example. What it is important to recognise is that the boundaries of nation states, and the integrity of the territories which they enclose/define/claim, are social constructions – they cut across a far more open intermeshing of the social relations which construct space. An excellent critique of how political scientists assume nation states as natural (that is: take them as given to their analyses) has been written by Walker (1993). It is a critique which can be applied much more widely to the ways in which we think about social space.

Our attempts to territorialise space can have a range of different motivations. At one level, representing space as essentially organised into compartments – at the extreme as organised into nested scalar hierarchies – seems simply to be an attempt to tame the unutterable complexity of the spatial: it is a way of gaining some control – even if only in our heads – by constructing an ordered geographical imagination through which to frame our worlds. In more material practices, fencing off particular areas may be part of wider strategies to protect and defend particular groups and interests. Fencing off space may also, on the other hand, be an expression of attempts to dominate, and to control and define others. What *is* clear is that such strategies of spatial organisation are deeply bound up with the social production of identities. (A general discussion of these issues can be found in Sibley, 1995.)

The construction of youth cultures, certainly in 'Western' societies, exhibits many of these phenomena. The design, definition and control of spatiality is an active ingredient in the often contested social processes of such construction.

On the one hand, a range of 'authorities' in wider society invent and implement rules for the spatial ordering of the population in terms of age. So teenagers are not allowed into children's playgrounds (which are reserved for people younger than them) or into certain clubs, drinking places and cinemas showing certain films (these places being reserved for people *older* than them). Such rules have a number of evident rationales – the protection of toddlers (assumptions here about the potential behaviour of teenagers in children's playgrounds), or the protection of teenagers themselves from contact with influences they are deemed not yet sufficiently mature to cope with. Even such 'ordinary' rules are bound up with assumptions about identity and attempts to construct socially acceptable identities. And indeed the very drawing of age lines and the definition of the spaces where particular age groups are allowed, is part of the process of defining an age group in the first place. The control of spatiality is part of the process of defining the social category of 'youth' itself.

It is also part of the process of defining what is to be deemed as acceptable behaviour on the part of that group. Recent decades in Britain have seen a number of hotly contested struggles over this. Perhaps the most widely analysed have been attempts by the Conservative government to control what was evidently – to them – a disturbingly high degree of mobility and lack of desire to 'settle down' on the part of significant numbers of young people. The battle over the Criminal Justice Act, with notions such as 'aggravated trespass', is a classic case in point. A host of issues coincide here. There was clearly in play a conservative resentment of people not growing up to respect, and to conform to, what conservative supporters wanted to see as 'normal' lifestyles and ambitions. These young people apparently did not *want* to own their own 'nice home' in some salubrious avenue (or, at least, not yet); they appeared to reject the strivings of the already established. Attempts to forbid behaviour which is different have always been part of the armoury of those who insist on establishing their own behaviour as 'normal' and 'natural', and therefore by some leap of logic to be conformed to by everyone. Moreover in this case the attempt to control and define hinged centrally on spatiality. The anxiety over mobility was evident. By politician after politician, New Age Travellers were contrasted with the rest of society ('the public is fed up with New Age Travellers'; 'they disturb ordinary people going about their normal business'). Such statements serve to define not only New Age Travellers (not part of 'the public', not 'ordinary people') but also 'ordinary people' themselves: the implication is that the ordinary public does not (should not) wander about the countryside. And indeed, as Sibley shows, the nature of the countryside and what kinds of behaviour were to be deemed (in)appropriate to the countryside were central stakes in this battle. Here, then, control of spatial behaviour was part of a social struggle over what are/are not acceptable forms of youth culture. (A host of other instances could be cited: benefits legislation aiming to keep young people in the parental home, arguments over whether or not homeless people – and particularly those living up trees on roads protests – should have the vote, come immediately to mind.)

But the attempts to territorialise or regulate space are not all one way. The very mobility of so many young people in the 1980s and 1990s was an attempt to undermine the dominant assumption of settledness as the better option. And youth cultures claim their own spaces too, and may be as excluding and defensive about them as any nation state. The gang cultures of graffiti, sworn loyalties, and heavily marked and bounded territories, are probably the most celebrated examples, but similar processes go on in much more ordinary, day-to-day, ways. From being able to have a room of one's own (at least in richer families), to hanging out on particular corners, to clubs where only your own age group goes, the construction of spatiality can be an important element in building a social identity.

In her book *Goliath*, Campbell (1993) discusses the TWOCers whose behaviour caused such media fury for a few years in the 1980s/1990s. TWOCing was 'taking cars without the owner's consent'. Campbell writes of one group of young men on a particular estate who used such cars to demonstrate their driving skills to their admiring peers. From 10 at night they transformed what was otherwise thought of as the square near the shopping centre into an arena of dramatic performance. Outrageous speeds, handbrake turns, sudden stops, dramatic exits. After 10 at night this space belonged to them; everyone else kept well clear. And the claiming of this space was integral to the identity the young men were striving to establish: the fact that other sections of the community would stay away was part of the point. After 10 this was young men's territory.

And yet it was not simply a closed space. For that space near the shopping centre after 10 at night was also the meeting place of cultural references drawn from a wide range of other places. Some of the cars were (at least in part) foreign made; the clothes that were worn made all the correct up-to-the minute references to a wider youth culture; the fact that this kind of activity was getting coverage in the national media cannot have been unimportant; and so on.

In this apparently local arena, then (just as in that small Mexican room crowded with electronic games), the different sides of the spatiality of the construction of youth cultures (indeed, perhaps, *any* cultures) come together. On the one hand, the apparently endless process of the carving up of space and the claiming of it for one's own, and on the other hand the undeniable interconnectedness of any space, or any culture, with others even on the other side of the world.

REFERENCES

Campbell, B. (1993) *Goliath: Britain's Dangerous Places*, London: Methuen.

Hall, S. (1995) 'New cultures for old', in D. Massey and P. Jess (eds) *A Place in the World?*, Milton Keynes and Oxford: The Open University and Oxford University Press, pp. 175–213.

Massey, D. (1991) 'The world of food production', Unit 1 of Society and Social Science: A Foundation Course, Milton Keynes: The Open University.

Meegan, R. (1995) 'Local worlds', in J. Allen and D. Massey (eds) *Geographical Worlds*, Milton Keynes and Oxford: The Open University and Oxford University Press, pp. 53–104.

Sibley, D. (1995) *Geographies of Exclusion*, London: Routledge.

Smith, D. (1993) 'Homeless/global: scaling places', in J. Bird, B. Curtis, T. Putnam, G. Robertson and L. Tickner (eds) *Mapping the Futures*, London: Routledge, pp. 87–119.

Walker, R.J.B. (1993) *Inside/Outside: International Relations as Political Theory*, Cambridge: Cambridge University Press.

8

DISINTEGRATING DEVELOPMENTS

global economic restructuring and the eroding of ecologies of youth[1]

•

Cindi Katz

INTRODUCTION

Global economic restructuring and other forces of 'globalisation' have cut a swath through the everyday environments of young people in settings as different as New York City and rural Sudan, creating rents and tears in both places that are strikingly similar. My work focuses on the nature of these cultural-ecological and political-economic changes, their implications for children and young people, and the practical responses to them by children, youth, and adults in the course of their everyday lives. I will discuss the systematic disruption of social reproduction – the ways young people do not receive the knowledge and skills necessary for the world in which they will come of age – in two sites deeply affected by economic restructuring, rural Sudan and New York City. By making connections between changes 'on the ground' in both sites I hope to encourage a broader, and potentially more empowered, understanding of global change and its local workings.

I want to accomplish five things: to provide texture to the notion of global change by demonstrating some of its local consequences and causes; to reveal the ties that bind Sudan and New York as (not so) different 'locals' of the 'Third' and 'First' Worlds; to address some of the consequences and meanings of these changes – local and global – for young people, such as disruptions of social reproduction and ruptures in the ecology of childhood and youth. In doing so, I hope to point to the rarely acknowledged centrality of young people in debates about global change, and thereby to the creation of a transformative transnational politics of experience (Mohanty, 1987). Finally, I want to broaden the notion of 'environment' as it is used in discussions of everyday life, and suggest its importance to a reinvigorated politics of change.

GLOBAL CHANGE

The notion of global economic restructuring suggests the movement of capital from one form of investment to another, usually with a geogra phical aspect. Depending on historical geography, it includes such well known processes as de-industrialisation, the dismantling of fordist forms of manufacturing, the growth of the service sector, and the devastation of subsistence economies. It also includes transnational shifts in investment that have led to particular forms of industrialisation in previously unindustrialised countries, the development of high technology service and 'knowledge based' economies in many urbanised areas, and massive shifts in the location, organisation and social construction of employment and the meaning of work. These changes, which are neither uniform nor unidirectional, rework the scale of uneven development. Taken for granted political-economic realities and relationships have been shattered and reworked – tens of thousands have been left un- or under-employed where once were robust manufacturing economies, and elsewhere, members of groups not traditionally part of the paid labour force have become factory workers in droves. These operations are well documented in the literatures of political economy, regional economics, planning, sociology, and geography (e.g. Bluestone and Harrison, 1982; Harvey, 1987; Piore and Sabel, 1984; Pred and Watts, 1992; Sassen, 1988; Sawers and Tabb, 1984; Wolf, 1992), and are not of explicit concern here. My concern is the disintegration of possibilities for meaningful work and the broader social contracts that have been associated with it. These entailments of global economic restructuring and its partner, structural adjustment, will be discussed with greater specificity below.

Alongside these broad political-economic changes is an increasing globalisation of cultural production (and especially consumption) through television, film, videos, music, electronic media, and products that reflect 'style'. The reach of these products has led to a transnational burgeoning of desire and a breathtaking heterogeneity of means to satisfy it. 'Globalisation' also has created unprecedented possibilities for communication and brought about means of information sharing that take global simultaneity for granted. At the same time, relatively inexpensive production technologies hold out the possibility of democratising cultural production, which in some areas has been realised.

These global changes – the political-economic and the socio-cultural – have myriad consequences for young people's everyday lives as well as their futures. Some of these are discussed below.

ERODING ECOLOGIES OF CHILDHOOD AND YOUTH

In many parts of the so-called 'Third World', capitalist relations of production and reproduction, and with it the logics of modernity, are visited upon people through agricultural development projects (e.g. Katz, 1991; Pred and Watts, 1992).[2] In Howa,[3] the village where I have conducted field research since the early 1980s, the state sponsored and internationally financed Suki Agricultural Project transformed social relations of production and reproduction as well as the local political ecology. With the Project, peasant cultivators became tenants as agricultural production shifted from the dryland cultivation of sorghum and sesame to the irrigated cash cropping of cotton and groundnuts. Prior to the Project, Howa was characterised by a largely subsistence economy. Villagers' limited needs for cash were met in the main through the sale of sesame, other agricultural products, and on occasion, livestock. Some men earned money as agricultural labourers and even tenants in the private cotton schemes along the Blue Nile. With the Suki Project, the village was entwined increasingly in the national cash economy and money gradually came to predominate all exchange.

The Suki Project also transformed the local political ecology. To establish the Project, approximately 1,050 hectares in the vicinity of the village that had been characterised by a mix of woodland, fields, and pastures were levelled and cleared for the irrigated cultivation of groundnuts and cotton in 100 per cent rotation. Chemical fertilisers, pesticides, and herbicides were required to sustain the intensification of land use. Moreover, the removal of so much land from mixed use to cultivation alone put tremendous pressure on remaining pasture and wooded areas. By the time of my study, ten years after the inception of the Project, the area surrounding the village was severely degraded. By the time of my last visit in 1995, the nearby pasture areas could not support the numbers of livestock they had just a decade before, and most commercial woodcutting had moved elsewhere because of local environmental degradation.

The alteration of the landscape together with the transformation of the social relations of production and the tightening of ties to the cash economy were associated with significant shifts in the allocation of household labour. For example, with de-vegetation, the time it took to procure an adequate domestic fuelwood supply stressed and some cases exceeded the labour capacity of many households. While elsewhere in Africa women are largely responsible for fuel provision, in Howa children and teens bore greater responsibility because of the preference associated with Muslim practice for women to remain secluded if at all possible. Because of the changed labour demands associated with the Project and its effects, only young women who were just wed or in the midst of bearing and caring for young children did not engage in work outside of the household compound. In a similar vein, the labour requirements of tenant farming were greater than those that had prevailed when sorghum and sesame were the major crops. This situation also led to increased family labour time from all quarters, including women in many households. Only the minority of households that could afford to purchase the labour time of others were not implicated in these shifts. As a result, many formerly free or commonly held goods including wood and in some cases water were becoming commodities. With commoditisation, of course, the need for cash increases. In Howa this shift exacerbated the demand on teens' and children's time, because parents needed their labour contribution either to earn cash or to provide goods and services that would otherwise have to be purchased. Somewhat paradoxically then, it appeared that Howa's incorporation in a so-called 'development' project had reduced young people's 'free' time and with it, for many, the possibility for school enrolment and attendance.

One of the central arguments that arose from my research was that because of the changes under way in Howa, children and adolescents were not learning what they were likely to need to know in their adulthoods. This issue was of enormous consequence for the young people themselves, and, writ large, for the social formation of Sudan as a whole. In brief, my research suggested that because of the problems outlined above, children and youth were working long hours in a variety of subsistence and cash-oriented activities. They were thus learning quite well how to participate in the socio-economic and cultural-ecologic frame that obtained in the village, but this was itself under erasure. Moreover, because of the land tenure relations associated with the agricultural project, in which a fixed number of tenancies (250) was allocated to the village, most young people were unlikely to have access to productive land when they came of age. This situation was aggravated by the prevailing demographic circumstances in Howa. Specifically, couples were generally young when they began childrearing, and thus would be likely to remain economically active as their older children reached maturity and required land of their own. In addition, the average household had six children, which was likely to create competition between siblings for land rights, *if* parents were to cede their tenancies. Under these circumstances, young people's considerable knowledge of farming, animal husbandry, and the use of local environmental resources would be rendered dormant if not entirely useless as they reached adulthood. Rural de-population is often the outcome of agricultural intensification, but the stagnation of the larger economy of Sudan will leave these displaced young people with disturbingly few options. The burgeoning population of 'street children' in the towns of Sudan in recent years is but one outcome of this developing crisis. Under these conditions, the prospect of meaningful work for the population coming of age grows dimmer each year.

The situation is not much different in New York City and elsewhere in the United States. In New York, for instance, disinvestments in manufacturing, shipping, and warehousing along with declines in construction and infrastructural maintenance, and the shrinkage of government posts (including the military) have dimmed the prospects for reasonably well paying, stable employment, and most certainly for meaningful work, of many working class young people. At the same time, the transition to a high-tech service economy associated in large measure with the finance industry has bifurcated the labour market in New York, creating extremes of pay, skill, and stability. Unemployment among teens 16–19 years old in New York City, for instance, increased from 18 to 36 per cent in the five years from 1988 to 1993 (Citizen's Committee, 1995: 47). Biases around race, gender, and ethnicity, among other things, further divide the labour market, leaving some segments – most notably young African American men – with disastrously few prospects for employment. As in Sudan, political-economic shifts (many associated with global processes) render the knowledge acquisition of many young people moot.

The question of knowledge leads to the issue of schooling, which – especially in the inner city – has deteriorated in the gutting of state and local budgets tied integrally to urban disinvestment and the renewed segregations of race and class with which it is associated. Concomitant with disinvestment in production there has been massive public disinvestment in social reproduction as well. The two are obviously related. Most crudely, it recalls the old farmers' adage, 'if you don't need the milk, why pay for the cow'. There have been steep declines in public outlays for housing, health care, social welfare, education, job training, job creation, child care, recreation, and open space over the last two decades in many parts of the United States and elsewhere in the industrialised West. In the US these declines and their consequences have been most severe in the major cities which have been punished by calculated federal neglect and (white) suburban taxpayer revolt. With an unresponsive and hostile state, individual and private resolutions are sought to the problems engendered by these cutbacks. Privatisation, the order of the day, is already taking its toll on those most in need of public services and socio-economic support. Privatisation is at best double-edged; it can lead to some improvements in service provision but there is less public accountability and greater unevenness in their pattern of distribution.

Apart from the well documented consequences of declines in education, health care, and social services for children's and teens' well-being, and the obdurate problems posed by the dearth of jobs for those with limited education and training, there are less well known but equally pernicious effects created by the decay and outright elimination of public environments for outdoor play and recreation. With public space deteriorated and perceived as unsafe from a variety of perspectives both social and physical, young people have fewer opportunities for autonomous outdoor play or 'hanging out' (Plate 8.1). This lack has implications for many aspects of their healthy development related to such diverse processes as gross motor development, the building of culture, and the construction of identity. In the absence of safe outdoor spaces young people become prisoners of their homes, often isolated with only the television or worse for companionship. The restricted access to the public environment, and with it many opportunities for forging and negotiating peer culture and acquiring the various social skills associated

Plate 8.1 The deterioration of public space in New York has left few spaces for young people to hang out. This schoolyard in upper Manhatten had few features for play. Its renovation was stymied for lack of funds.
Photograph: Cindi Katz

with these negotiations, is generally worse for girls than boys (Katz, 1993). The construction of subjectivity and identity formation are inflected and seriously compromised by these uneven socio-spatial relations in the contemporary urban environment.

In urban New York and rural Sudan – vastly different settings in many ways – the ecology of youth has eroded definitively in recent years. Without setting up some mythical halcyon past ecology of childhood and youth, it is fair to say that the local manifestations of global economic restructuring have had serious and deleterious consequences for young people in both settings only pointed to here. These changes are associated with serious disruptions in the material social practices of social reproduction that can lead to crisis, a renovation of the social relations of production, or continued erosion of social life.

The cases discussed are but two examples of degradation in young people's everyday lives. However, as important as it is to be critically aware of this situation, and to formulate a political and material response to it in all of the settings where it is manifested, it is theoretically, politically, and practically important to resist not only the degradations themselves but the collapse into 'dismalism' that these conditions engender. The fortitude to resist is becoming increasingly difficult, and that much more important.

FIGHTING EROSION

My concerns about young people enable, and even require, me to resist the increasingly common collapse into hopelessness and despair that I am calling 'dismalism'. Focusing on youth compels me at the very least to grapple with the future, with possibilities. From my work with and among young people I have gleaned that one of the key constructions of childhood and youth, in theory and in practice, is that of a time or space 'without walls', when and where all futures seem possible, even as everyone knows that they are not. I want to invoke that notion of childhood/youth here in a non-innocent but strategic way; to suggest that it may be useful politically to think of childhood/youth as a metaphorical site for a new kind of politics in which everyday life is a theoretical site. Building such a project would draw on Lefebvre's (1987) notion of everyday life as a critical concept, where rupture is immanent in the routine. Knowledge is, of course, not simply made obsolete by new structures and forces of production and reproduction, such as those associated with global economic restructuring. The acquisition and deployment of knowledge underwrites resistance as well as reproduction, and can help to change as well as maintain the social relations of production and reproduction.

In the practical sites that are the focus of my work, for instance, we can find changes that at least offer a glimpse of the new politics I am suggesting is possible. In a return visit to Howa two years after my study, I found standpipes were about to be installed throughout the village. No small feat, the acquisition of the piping and its installation was an expensive undertaking that was entirely a self-help initiative. The anticipated labour savings were enormous, and as suggested by my study, likely to accrue in large measure to children and teens, especially girls. The local population was clearly mindful of this, because their next self-help initiative – already planned – was the construction of a girls' school Unlike government and international bureaucrats who construct schools and wonder why enrolment is so low,[4] the village population understood the crucial labour contribution of girls to the procurement of domestic water supply. Thus they constructed the means to procure household water without the constant labour of young people, and so freed their daughters to attend school in significant numbers (Mascarenhas, 1977).

The standpipes were dependent upon the operation of one of two diesel powered artesian wells. These were unreliable for lack of fuel, spare parts, and the well operator, who it turned out had a wife in a neighbouring village. Thus the standpipes were less labour saving than anticipated. When I returned in 1995, eight hand pumps had been constructed throughout the village under the aegis of UNICEF (Plate 8.2). The wells de-centralised the water supply, making the trek to fetch water shorter for almost all households. Reliant only on human labour, the pumps were 'open' all the time. Young people and others no longer needed to fit their work schedules around the short opening hours of the well. Moreover, the proximity of the pumps to most households increased the likelihood that a broader range of household members, including women, would do the task. The new relative ease of fetching water that resulted from the hand pumps, freed more young people, especially girls, for other activities. Not coincidentally, the finishing touches on a high school (for girls and boys), constructed in part with self-help funds, were under way in the village in 1995. In both periods the ties between water supply, young people's time, and schooling were striking.

These efforts were part of a politics of survival. The local population saw that their children were being displaced by the socio-economics and political ecologies associated with the irrigation project, and developed a means to provide them with new skills and knowledge. Their efforts were not isolated instances of action. The first, for instance, was preceded by a period of resistance and concerted opposition on the part of

Plate 8.2 The construction of hand pumps in Howa liberated young people from the chore of trekking long distances to wells for water.
Photograph: Cindi Katz

the farmers' union to win the right to cultivate sorghum, the staple food crop, on Project land. Standpipes, schools, and food security are of a piece. They help ensure local vitality in the face of imposed change. Such strategies, significant in themselves, can be the beginnings of a more transformative politics.

Similar strategies of survival and empowerment are found in the US urban context. While the despair and anger that led to the 1992 uprisings in Los Angeles have no doubt increased given the absence of any meaningful response by either the federal or state government (except for increased policing), there has been a popular response in neighbourhoods and city-wide (e.g. Davis, 1993a, 1993b; Leavitt, 1992). Most notable, in terms of the concerns of young people, has been the gang unity that has largely held in the years since it was instituted just before the uprisings (Davis, 1993b: 34–5). Not only has the relentless black on black, youth on youth gang violence abated, but gang members were radicalised before, in, and as a result of the process, all to the chagrin of the Los Angeles Police Department (Shakur, 1993; Davis, 1993a, 1993b).

As in Howa, the population oppressed and marginalised by the larger political-economic structures and the material social practices of the capitalist state, was well aware of what must be done. The united members of the Crips and Bloods developed their own plan for Los Angeles' recovery, in which a call for youth employment and improved schools figured centrally, and with Maxine Waters, the Congresswoman from South-Central LA, have struggled to make jobs the main political issue in the area (Davis, 1993b: 36).

Similarly in Harlem, New York, Jacqueline Leavitt and Susan Saegert (1990) found that residents of landlord abandoned buildings organised to oppose their displacement, and formed limited equity co-operatives. They were able to regain control over their homes and buildings in part by developing strengths rooted in those very sites. Women leaders emerged from these struggles to reclaim their spaces, and using the logics and communicative strategies associated more with the home than with the public sphere, were able to build a broader politics around issues such as child care, neighbourhood safety, education, and jobs (Leavitt and Saegert, 1990). Likewise, in my work on the participatory redesign of two schoolyards in central Harlem, parents, students, teachers, school staff, local youth, and the broader local community collaborated for more than five years in a frustrating climate of neglect and disinvestment to see to it that two neighbourhood schoolyards were transformed into workable and inviting public spaces. Their efforts were geared to providing children with safe and interesting play spaces, teens with recreational facilities and a secure place to 'hang out,' adults and young people with pleasant sitting areas, and a multi-function schoolyard and garden where teens and children could engage in performances, gardening, sporting events, and other activities during and outside of school hours (Katz and Hart, 1990). Finally, in East Harlem, Rina Benmayor, Rosa Torruellas, and Ana Juarbe studied the participation of Puerto Rican and Latina women in the El Barrio Popular Education Program as a means to confront the conditions created by poverty and the ongoing immiscration of their community. They use the theoretical framework of cultural citizenship, that is, 'practices that affirm the right to equal participation in the society through the right to cultural difference' (Benmayor *et al.*, 1992: 3) to demonstrate how these women asserted their claims to education. In the process they transformed themselves, fought their marginalisation concretely, and were empowered to take on other issues that wove together the relations of production and reproduction in their everyday lives. Their practices negotiated the knife edge between 'the politics of difference and the politics of inclusion' (Benmayor *et al.*, 1992: 74). These findings were reinforced in the course of my own work in East Harlem with the CAMEO (Community, Autobiography, Memory, Ethnography, and Organization) Project, where several of the women participating in our community ethnography chose to document the workings of the Centre for Popular Education.

Like the school, standpipes, and sorghum in Sudan, these strategies, focused on jobs, housing, public space, and education, are hardly incendiary. (Mike Davis [1993b: 36] calls the LA gang members 'tendential social-democrats'.) They are, however, crucial to any broader movement towards transformation. They allow people to reweave the strands of their lives – to survive – but by struggling for and attaining these goals, people 'resist the economic and ritual marginalization they suffer and [can] insist on the minimal cultural decencies of citizenship' (Scott, 1985: xviii).

CONNECTING THE FIGHTS

The above are modest success stories in an increasingly dirty multi-locale war of attrition. The events in rural Sudan and urban America are connected. They are, in part, local effects of the tendencies of global capitalism that have produced a future with little stable or meaningful work in both sites. The common cause of many of the pressing problems in both settings calls for strategies that connect oppositional practices in each site, including a search for avenues in which young people can be connected city to country, across cities, and transnationally. The inspiration to kids and others of the LA gang unity movement, and the growing influence of politicised gang members (often radicalised in prison) underscores the possibilities in connecting youth movements.[5] The issue at hand transnationally is to produce historical geographies that foster self-determination, social participation, and growth and change at all scales, and to connect these 'local' struggles to one another and more broadly. I will sketch out the contours of what this might mean in the remaining paragraphs.

The political ecology of youth is both a constellation of ideas and a set of material circumstances that frame young people and the environment as a pivotal social, political, economic, and cultural relation. What kind of environment is needed to produce a political ecology of youth and childhood that fosters self-determination (the development of capacities and the construction of identity); participation in a social web that offers both 'cultural citizenship' and meaningful work; and growth and change from the individual to the global scale? At a minimum, an environment that provides for young people's health and well-being; safety; physical, emotional, and intellectual development; and future, broadly understood, is called for. This is basic stuff. Health and well-being, whether for young people or old, depend on adequate food, shelter, health care, sanitation, social services, and protection from environmental pollution and hazards.

Clearly, these concerns suggest a notion of environment that is as much social as physical. While much of the attention of environmental activists is focused on the deterioration of putatively 'natural' environments, young people's well-being world-wide might be better served by the provision of adequate and affordable housing. Housing, too, is an 'environmental' problem, and it is time environmentalists recognised that (Jackson, 1991: 51). Along analogous lines, young people's safety is a concern with physical and social dimensions. Addressing physical harm and social predation together helps avoid the traditional separation between public and private domains, and gets at all of the environments in which children and youth are injured or subjected to violence whether that be from parents or strangers, on account of war or other systematic violence, because of carelessness or lack of regulation, or because of crime and neglect. For these reasons and others, young people are at extraordinary physical and psycho-social risk all over the world (see, e.g., Stephens, 1995).

Young people's growth and development depends upon environments that provide stimulation, allow autonomy, offer possibilities for exploration, and promote independent learning and peer group socialising. These criteria are important in all settings, not just those designed specifically for teens such as schools, leisure environments, and teen centres.

Here again, the artifice of the distinction between nature and culture is invoked. Stimulating environments are not limited to 'natural' settings as is often assumed by young people's advocates. The privileging of rural environments as settings for young people is as deeply troubling as it is wrong considering the large and growing numbers of children and teens world-wide who live in cities, and the diverse fascinations of urban environments. In all settings, whether urban or rural, it is poverty, disinvestment, violence, and privatisation that threaten environments nurturing growth and development, not something intrinsic to the environment.

Finally, the environments of youth speak to them of the future – both their own and across generations. It is important to take seriously what the absence of work – what 'no future' – means to them as they construct their identities and make place. Young people see and viscerally experience these limits, absences, walls and foreclosures all around them. At the extreme there are street youth, abandoned children, young people with HIV, and homeless children and teens. But these absences are also manifest in the mundane interactions of many young people's everyday lives. They witness them firsthand in poor schools, broken down playgrounds and recreational facilities, in the lack of child care or teen programmes, in the din of messages that convey the badness and repugnance of children and youth in the public imaginary, and in the criminalisation of young people simply because they are young. A chilling sign of this phenomenon in US (and many European) cities and suburbs can be witnessed in the signs on a growing number of convenience stores limiting the number of young people that can enter the shop at one time. Youth have got to be among the best customers of these establishments, yet they are banned from entering freely because of assumptions of their criminality.

The adult world is often inattentive to youth except to oversee them or to exhort consumption, conformity, and 'good' behaviour. There are, of course, class, national, gender, ethnic, and racial dimensions to these problems. For some young people the best that they can hope for themselves is rather narrow, and this is more punishing now than ever before because of how much more they have learned to want thanks to the reach of globalised cultural production. These issues, which are environmental in the broadest sense, have deep, and largely unexplored, effects upon young people's constructions of identity, in how they see their 'place' in the world, and ultimately, in how they produce the world to come.

These questions, with their broad political, ethical, social, and economic ramifications, return me to the metaphorical site of childhood, and its importance to an effective politics of change. Recognising this site, and its invocation of the future, may not only help children and teens, but may reinvigorate politics more generally; moving away from the pessimistic backward-looking gaze of being 'post' everything, and facing the future and the transformations yet to be. My argument is that the metaphorical, practical and theoretical sites of youth both demand and create positions from which and for which the future is not only faced, but indeed, constructed.

NOTES

1 An earlier version of this chapter was published as 'Textures of global change: eroding ecologies of childhood in New York and Sudan', in *Childhood: A Global Journal of Child Research* 2 (4) (1994): 103–10. Reprinted with kind permission of SAGE Publishers.

2 While the internecine webs of international finance, and before that colonialism – if the two can be separated – have long entangled east, west, north, and south in 'debt traps' of dependency and deprivation, the effects of these relations in countries of the so-called 'Third World' were not generally

discernible *as such* in daily life, especially in the countryside. Of course, the uneven developments associated with colonialism, imperialism, and transnational capitalism have long had pernicious effects in these areas through the grind of immiseration and impoverishment, but these do not in themselves change local relations of production and reproduction as an incursion such as an agricultural project does.

3 The name of the village has been changed to protect the privacy of its inhabitants.

4 The village school at the time of my original stay was open to young people of both sexes, but was overwhelmingly male. Only 4 per cent of the school aged girls attended while 42 per cent of the boys did. Apart from the extensive demands upon children's labour time, the low levels of female school enrolment were also the result of fairly widespread resistance to co-education. The construction of a girls' school in Howa was clearly a means to reach girls, and in that way, at least a tacit recognition that new forms of knowledge would be required of them in the future.

5 Many politicised gang members and other young people are not particularly leftist. Many for instance, find inspiration in the black nationalism of Louis Farrakhan which has both patriarchal and entrepreneurial tendencies (Katz and Smith, 1992).

REFERENCES

Benmayor, R., Torruellas, R.M. and Juarbe, A.L. (1992) *Responses to Poverty among Puerto Rican Women: Identity, Community, and Cultural Citizenship*, New York: Centro de Estudios Puertorriquenos, Hunter College.

Bluestone, B. and Harrison, B. (1982) *The Deindustrialization of America*, New York: Basic Books.

Citizens' Committee for Children of New York (1995) *Keeping Track of New York's Children*, New York: Citizens' Committee for Children of New York.

Davis, M. (1993a) 'Who killed LA? A political autopsy', *New Left Review* 197: 3–28.

—— (1993b) 'Who killed Los Angeles? Part Two: The verdict is given', *New Left Review* 199: 29–54.

Harvey, D. (1987) 'Flexible accumulation through urbanization: reflections on "postmodernism" in the American city', *Antipode* 19, 3: 260–86.

Jackson, W. (1991) 'Nature as the measure for sustainable agriculture', in F.H. Bormann and S.R. Kellert (eds) *Ecology, Economics, Ethics: The Broken Circle*, New Haven, Conn.: Yale University Press.

Katz, C. (1991) 'Sow what you know: the struggle for social reproduction in rural Sudan', *Annals of the Association of American Geographers* 81, 3: 488–514.

—— (1993) 'Growing girls/closing circles: limits on the spaces of knowing in rural Sudan and US cities', in C. Katz and J. Monk (eds) *Full Circles: Geographies of Women over the Life Course*, London and New York: Routledge.

—— and Hart, R. (1990) *The Participatory Design of Two Community Elementary Schoolyards in Harlem P.S. 185 and P.S. 208*, New York: Children's Environments Research Group, the City University of New York.

—— and Smith, N. (1992) 'L.A. intifada: interview with Mike Davis', *Social Text* 33: 19–33.

Leavitt, J. (1992) 'Los Angeles: the community view', *City Limits*, Aug./Sept: 14–15.

—— and Saegert, S. (1990) *From Abandonment to Hope: Community-Households in Harlem*, New York: Columbia University Press.

Lefebvre, H. (1987) 'The everyday and everydayness', *Yale French Studies* 73: 7–11.

Mascarenhas, A. (1977) *The Participation of Children in Socio-Economic Activities: The Case of Rukwa Region*, Dar es Salaam: University of Dar es Salaam BRALUP. Research Report 20-1.

Mohanty, C.T. (1987) 'Feminist encounters: locating the politics of experience', *Copyright* 1: 30–44.

Piore, M. and Sabel, C. (1984) *The Second Industrial Divide*, New York: Basic Books.

Pred, A. and Watts, M.J. (1992) *Reworking Modernity: Capitalisms and Symbolic Discontent*, New Brunswick, NJ: Rutgers University Press.

Sassen, S. (1988) *The Mobility of Labor and Capital*, New York and London: Cambridge University Press.

Sawers, L. and Tabb, W.K. (1984) *Sunbelt Snowbelt: Urban Development and Regional Restructuring*, New York and Oxford: Oxford University Press.

Scott, J.C. (1985) *Weapons of the Weak: Everyday Forms of Peasant Resistance*, New Haven, Conn.: Yale University Press.

Shakur, S. (a.k.a. Monster Kody Scott) (1993) *Monster: The Autobiography of an L.A. Gang Member*, Boston: Atlantic Monthly Press.

Stephens, S. (ed.) (1995) *Children and the Politics of Culture: Risks, Rights and Reconstructions*, Princeton, NJ: Princeton University Press.

Wolf, D. (1992) *Factory Daughters: Gender, Household Dynamics, and Rural Industrialization in Java*, Berkeley: University of California Press.

9

YOUTH GANGS AND MORAL PANICS IN SANTA CRUZ, CALIFORNIA

•

Tim Lucas

> They're everywhere: hanging out in front of coffee shops on Pacific Avenue, violating city law by sitting on curbs, smoking cigarettes around the corner, having unprotected premarital sex, spraying graffiti all over the place, and pretty much heading for a quick trip to nowhere. Nowhere, that is, if you look at them through a lens of mainstream media negativity.
> (The Fish Rap Live, *14 February 1996*)

Young people, as this quote taken from a local student publication neatly captures, are the subject of popular suspicion and anxiety in Santa Cruz, California. While the demonisation of youth has a long history and has taken many forms, in recent years attention in Santa Cruz has focused largely upon the presence of gangs and in particular Latino street gangs. The issue has become increasingly politicised locally in the 1990s as elected officials and law enforcement representatives have sought to demonstrate their willingness to tackle the city's so-called gang 'problem'. Popular concern has resulted in a number of public meetings in the city and the local daily newspaper, the *Santa Cruz Sentinel*, ran a week long series of in-depth articles in October 1994, entitled 'Gang Life'.

The problematisation of young people is by no means unique to Santa Cruz. Across the US, contemporary 'Teenaphobia' has reached an all time high/low (*San Jose Mercury News*, 26 April 1995). Institutional responses have included attempts to re/introduce 'teen curfews' in large cities and smaller communities alike. The 1994 Violent Crime Control and Law Enforcement Act, the most sweeping crime bill in US history,

GANG VIOLENCE!

ON WESTSIDE OF SANTA CRUZ

"OPEN FORUM"

Saturday, June 25,
Natural Bridges
School Gym
10:00 AM

Please join us to find out what is happening in our local neighborhoods. A concerned parent group of neighbors, parents of students attending our elementary, junior high and high schools on the Westside, and local business owners, want to alert you to the increasing gang activity here in our neighborhoods, fast food restaurants, and schools. Please pass the word and bring a friend.

WE CAN MAKE A DIFFERENCE!

Figure 9.1 National problem, local panic?
Source: Tim Lucas.

also placed special emphasis upon the prosecution of gang members, and of juveniles as adults. While generational conflicts are nothing new, one of the primary factors behind current fears has been an increasing concern about the omnipresence of youth gangs (Figure 9.1). As 'gang' has become a code-word for 'race' in the United States (Muwakkil, 1993), so current representations of gangs also draw upon the problematised issues of drugs, gun violence, graffiti and gangsta culture (Cross, 1993). Concern that the measures outlined above are an over-reaction, disproportionate to the threat and the result of political and interest group exploitation, has led some authors to argue that a moral panic is in progress (Barak, 1995; Chambliss, 1994; Giroux, 1995; Krisberg, 1994; Platt, 1995).

This chapter will detail some of the processes by which the city's gang 'problem' has been racialised and represented within the local media. One of the major weaknesses of existing literature about moral panics has been the assumption that localised panics are mere inflections of more widespread social cleavages and scares surrounding marginalised populations. Attention will therefore be drawn to some of the processes by which moral panics are constructed, negotiated and contested at a series of interconnected socially constructed geographic scales.

GANGS IN SANTA CRUZ CITY

The local moral panic about youth gangs is identifiable by a number of factors. Official police figures reproduced in local news media sources have exaggerated the number of gang members, largely as a result of the misconception that a consensual definition of gang, gang member and gang activity exists (Ball and Curry, 1995). It is, however, in the selective framing of gangs and the racialised construction of youth violence that the moral panic is identified. Denying the presence of gangs can be as harmful as the overreaction evident at present; however, it is not contradictory to represent the gang as both the problem and the solution to a range of problems facing large numbers of young people.

The dominant representation of Santa Cruz's gang problem has been achieved through a variety of discourses. Of most interest here are those processes by which the city's situation is explained through reference to specifically local events, actors and locations. Central to the construction of the gang problem in Santa Cruz and its constitution as a problem resulting from one ethnic peer-group are the local geographies evident in representations of street gangs. Most important in this process has been the reporting of the city's Beach Flats area (Figure 9.2), where Eastside gang members live and meet. A number of geographers have drawn attention to the fact that the construction of difference and deviancy is linked to the stereotyping and moral construction of place (Smith, 1986; Cresswell, 1992). The neighbourhood is a nine acre triangle of land adjacent to the city's major tourist attractions. Once the site of seasonal

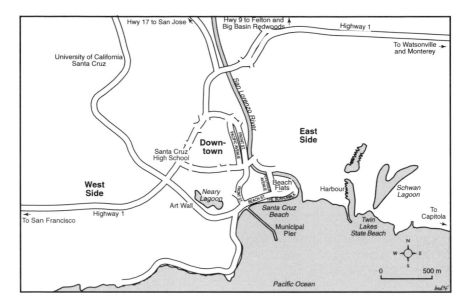

Figure 9.2 Santa Cruz, California.
Source: Linda Dawes.

tourist chalets, it is now a permanent home to approximately 1,200 residents. With 78 per cent Hispanic residents, a median income level significantly less than the city average and 65 per cent of the apartments in the area failing to meet minimum standards (City of Santa Cruz, 1990), the neighbourhood has become synonymous with a range of social problems including drug dealing, prostitution and gangs.

The stereotypes of gangs and of the Beach Flats neighbourhood are mutually constitutive, and evidence of what Sibley (1995: 14) describes as 'the way in which group images and place images combine to create landscapes of exclusion'. The social and spatial distancing of the neighbourhood has largely focused upon descriptions of the community as an eyesore, dirty and unhealthy. The use of metaphors of illness has also been consistent in the coverage of the neighbourhood. Given the technical difficulties in reporting gangs and the attempts of the police to prevent the glamorisation of gangs, pictures of actual gangs or gang members rarely appear in print. Photographs of graffiti have therefore become one of the most common means by which both gangs and the Beach Flats neighbourhood have been represented. Interpretations of and reactions to graffiti are contested (Figure 9.3) and specific to their locational and temporal context (see Cresswell, 1992; Ley and Cybriwsky, 1974). In this example, however, graffiti are a clear indicator of social disorder.

BAY STREET "MURALISTS" IN FULL SWING.

Figure 9.3 Graffiti: representing gangs.
Source: Santa Cruz Sentinel, © John Weiss.

Douglas (1966) argues that when moral boundaries are transgressed and an established order disturbed, by labelling the deviant factor as filth, boundaries between the socially acceptable and intolerable are re-established. Local media representations of the Beach Flats display a distinguishable 'language of defilement' (Sibley, 1995: 55) through which the neighbourhood and its residents are stereotyped and the necessary means of purification are framed. As the gang member has come to personify the problems of the Beach Flats, so the distancing and margin-alisation of the neighbourhood are an important element in the othering of gang members.

The construction of the neighbourhood is significant also in that the parameters of and language used in designating a problem are crucial for framing and influencing its resolution. The implications and actions which have been taken in order to 'clean up' the Beach Flats reflect these images of the area as an alien space (Wacquant, 1994) and its residents as social and moral polluters. Whilst by their own admission, the City Council have done little to improve the area (SC Sentinel, 27 June 1993), the Flats are often subject to 'sweeps' by various agencies. This includes the Immigration and Naturalisation Service (e.g. 'INS Sweeps Beach Flats; Six Arrested', *SC Sentinel*, 31 March 1992) and by local police and county anti-narcotics agents ('50 Arrested in Drugs Raid: Sweep Targets Beach Flats', *SC Sentinel*, 13 September 1991).

> It's time that we crack down on prostitution, rampant burglary, . . . it aint gonna be easy to get that council moving, they've had 15 years of progressives rule to do something about the Beach Flats, and the slum area of Beach Flats has spread under their tenure . . . busts surely only put a dent in it. We've got to start sweeping the streets clean of this kind of garbage.
>
> *(Bryan Maloney, speaking on the Bryan Maloney Show, KSCO talk,*
> *6 May 1995)*

The above quote also emphasises the fact that concern is largely based upon a fear that the pollution associated with the neighbourhood is 'spreading' to other parts of the city. This implies that the problems of the Beach Flats have previously been contained (at least in local imaginings) within this part of the city: 'Police say another misconception is that gangs are contained within the Beach Flats and lower Ocean Street neighborhoods' (*SC Sentinel*, 5 April 1992). This construction denies the fact that the problematisation of youth and concerns over youth, crime and drugs have never existed solely in this one location. The notion that the problem has been confined to the Beach Flats serves to represent the residents as compliant with the activities of gangs and raises important issues of political representation and policing of minority communities. It also goes some way to explaining the fixation with the 'Other' which, as McRobbie (1994) argues, the moral panic has failed to address. Exploring the construction of racial stereotypes, Bhabha (1983) argues that in the search for knowledge of the racial 'Other', an ability to speak as an expert on the characteristics of the 'Other' requires a constant surveillance.

The containment thesis allows the normalised community to exoticise, to know and to speak of the Beach Flats and of gangs, whilst retaining a reassurance of social and geographic distance. Thus whilst the 'community has been the subject of countless surveys, studies, plans (news reports) and proposals' (*El Andar*, August 1994) and its Anglo residents called upon to propose solutions to the local gang problem, little attempt has been made to represent the opinions of the neighbourhood's residents. The employment of such frames at both local and, as will be shown shortly, the non-local scale, are means by which the police demand that actions be taken to maintain moral boundaries and halt a potential spread of gangs beyond the imagined borders of their containment.

A LOCAL OR NATIONAL MORAL PANIC?

Whilst arguing that the city's gang problem is constructed with reference to local actors, practices and events which take place within the city of Santa Cruz, the local panic is also one element of a wider scare concerning youth across California and the US. Interpreting the framing of the gang problem in Santa Cruz must therefore account for the uniqueness of the local, as well as the influence of structures and processes operating on broader levels. Attention will now be drawn to some of the connections between the city, state and nation.

It is impossible to understand the fears surrounding gangs in Santa Cruz and the criminalisation of minority youth, without reference to the current political and cultural debates about crime, justice and race in the US. Whilst these contextualise events and practices in Santa Cruz, their more direct impact upon criminal justice and its representation in the city can also be developed with reference to specific examples. A number of articles published recently in academic and popular journals have described current law and order discourse and policy in the US as a moral panic on a national scale. These arguments are summarised by the National Criminal Justice Commission, a coalition of academic, policy, law enforcement and citizens representatives. In their report *The Real War On Crime* they argue that:

> Crime is a real threat, but at least some of the tremendous fear Americans have is the product of a variety of factors that have little or nothing to do with crime itself. These factors include media reporting on crime issues and the role of government and private industry in stoking citizens' fear.
>
> (Donziger, 1996: 63)

Much of the extra attention paid to crime in recent years has focused on the apparent rise in juvenile crime and the impact of violence upon young men and women. Concern over increasing levels of teenage violence has centred on firearms, popularly represented in the spectre of gang-related drive-by shootings. FBI reports indicate, however, that youth gang shootings only 'account for a small percentage (3.6 per cent) of all homicides' (Pacific Center for Violence Prevention, 1995). Homicide rates, especially amongst minority youth have risen dramatically, to the extent that between 1986 and 1992 firearm deaths of children jumped by 144 per cent compared to a 30 per cent increase for adults (Donziger, 1996). Despite an escalation of attention towards the number of juveniles involved in the most serious crimes of violence, Barry Krisberg (1994: 39), President of the National Council on Crime and Delinquency, states that '[j]uveniles represent a small and declining part of serious crime in America'. A September 1994 Gallup Poll found that American adults already hold 'a greatly inflated view of the amount of crime committed by people under the age of 18', with the most salient reason being 'news coverage of violent crime committed by juveniles' (Males, 1996: 7).

The nation-wide panic identified here, is directly relevant to events in Santa Cruz and the framing of the gang problem locally. Santa Cruz residents are familiar with many of the criminal justice issues discussed above. Deviancy and crime control are staples of national news discourse. Such stories are reported in the local media from both nationally syndicated press sources and in the form of local stories and editorials which place a local angle or hook to criminal justice practices in the county, state and nation. At the same time editorials in local newspapers address criminal justice policy in other US cities and states. These concerns and the publication of federal crime statistics all inform local residents' perceptions of crime and of youth.

It is not just in the mediation of crime and juvenile violence that the national panic over gangs and juvenile crime is of local relevance. The Violent Crime Control and Law Enforcement Act 1994 demonstrated the direct relationship between national and local panics and that policy at the Federal scale has direct consequences upon the construction and reaction to crime in Santa Cruz. The legislation allowed for the creation of 100,000 new police officers nationally and the Santa Cruz Police Department applied for and received funding for two additional officers as part of the COPS FAST program (*The Fish Rap Live*, 26 April 1995). These will be used locally to increase levels of 'community policing' in the Beach Flats area. This is largely in response to the locally constituted, racialised construction of the neighbourhood and of youth gangs discussed previously.

It is within the practices of US criminal justice that minority youth have been increasingly criminalised. Incarceration rates represent just one example of the current, racialised politics of law and order, however. As the police and academics alike have sought to explain the increasing numbers of gang members in the US, 'one of the terrifying visions of white suburbanites is that of the migration of drive-by-shooting gang-bangers in South-Central (LA) to the more prosperous (white) peripheral neighbourhoods' (Dumm, 1994: 185). In order to understand such fears surrounding the presence of gangs in the US, it is also necessary to further interrogate the connections between the cultural and the political and more specifically reactions to gangsta rap in the early 1990s. Giroux (1995: 13) argues that in an increasingly violent culture, 'representations of violence are largely portrayed through forms of racial coding that suggest that violence is a black problem outside white suburban America'. He describes the current moral panic as a 'white panic', rooted in fears and hopelessness among the white middle class created by declining economic, social and political standards.

Mediated scares about gangsta rap cannot be disaggregated from the wider fear of the spread of actual gang members from larger metropolitan areas to smaller communities previously believed to be free of the problems of the inner city. Law enforcement representatives have been keen to advance the gang migration theory. An article from Associated Press which appeared in the *SC Sentinel* (3 April 1995) reported the actions of more than 800 law enforcement officers who 'swept through gang infested neighborhoods' in Los Angeles. An FBI special agent argued that the raids would have a national impact 'Los Angeles based street gangs are spreading across the United States, and it's absolutely essential that we hit the source. That's right here.' Academic research has challenged police perspectives in a number of cities. Despite Skolnick's (1989) argument that gang culture facilitates but does not direct the migration of gang members, other studies found no evidence of satellite gangs in distant locations and questioned the role of gangs in the trafficking of drugs (Hagedorn, 1988; Huff, 1989; Klein and Maxson, 1988). The representation of gang members as social and geographical 'outsiders' does, however, serve to divert attention from local economic conditions and avoids engagement in challenging social issues such as changing race and age demographics (Cummings and Monti, 1993; Enriquez, 1990; Hagedorn, 1988; Moore, 1993; Zatz, 1987; Zevitz and Takata, 1992).

THE GANG MIGRATION THEORY

Local gang activity has been explained as a result of the actions of young Latino males from Santa Cruz. Constructions of the 'indigenous' gang member have, however, had the effect of alienating and excluding these individuals socially and geographically. Many of the processes raised in discussing the distancing of the Beach Flats also apply to representations of the gang problem as a result of the movement of gangs and individuals from outside the city into Santa Cruz. The use of metaphors, selectively employed for this purpose, frequently draw upon geographic imagery. Such constructions also rely upon the representation of Santa Cruz as an island into which social problems have been imported.

Firstly, there are examples where the presence of gangs and gang members within the city are explained by the presence and influence of gangs from other parts of the state. The relationship has on occasions

---

GANGS: THREE VIEWS

Talk radio show host
(*Bryan Maloney Radio Show, KSCO-AM1080*)

I was downtown, disgusted to park in the parking lot at Cathcart and Cedar and if there aren't 50 young people there from the ages of 10 up to 22 just hanging out, doing drugs with their dead-head clothes on, the entire hippie with the hippie wagons what the heck is going on? They're hanging out there all day, they're not from Santa Cruz.

Detective
(*Detective Bailey, Santa Cruz Police Department*)

So when they say 350 that would include people who are associated, who we feel are associated frequently with gang members. Maybe they only meet one or two of the criteria that are established, you know to be able to, quote, 'classify them' as a confirmed gang member, so that 350, about 350 to 400 member figure . . . . Probably out of that . . . probably looking at about 25 per cent of being confirmed gang members. . .there's probably some in there that can be boosted, but probably a good 25 per cent are the actually confirmed.

Community worker
(*Kathy Dominguez, Barrios Unidos*)

People who don't know or who aren't aware of the culture, see someone in baggy jeans and a blue sweatshirt, whatever, and the thought is 'Oh he's a gang member'. And a lot of times kids can't help it . . . . Because even if you're not a gang member, for example if you live in the Villa San Carlos. It's a given you're gonna wear loose jeans

---

> with a blue sweatshirt ... because you're not gonna go walking around the VSC wearing red .... So if somebody sees this kid, 'Oh this kid's a gangbanger, this kid's a Sureno' or whatever claiming VSC [affiliation with the Villa San Carlos gang], and that's not true. It's the environment that the kids live in. This kid may not be gangbanging and that's the problem with the cops. A lot of cops think that the kids are gangbangers. We know they're not, and the cops assume they're gangbangers because of the way they dress. You know, we had some cops going into the VSC and having one of the boys lift up his shirt to see if he had any tattoos and take pictures of him and this guy isn't even a gangbanger you know yet he was wearing a shirt

been described as one of direct causation. In 1992, Detective Sean Upton of the Santa Cruz Police Department argued that:

> nationally affiliated gangs have infiltrated the area and established drug dealing networks. . . . Upton says Southern California's Mexican Mafia and Northern California's Nuestra Familia have infiltrated the city, and that the powerful Crips and Bloods gangs, which operate throughout the nation, have drug trafficking connections in Santa Cruz.
>
> (SC Sentinel, 5 April 1992)

Similarly, on a number of occasions, crime stories have appeared which have identified the presence of Los Angeles based Crips gang members in Santa Cruz. In 1989, when four teenagers were stopped by a sheriff's deputy, he was reported to have asked:

> . . . why a black LA street gang would want white teenagers. The youths told him that the gang was *expanding* and opening up its ranks to all

races. . . . According to a deputy on the Heroin Task Force in the Beach Flats, Crips recruitment posters have been appearing in the area recently.
(*SC Sentinel, 20 March 1989*)

Geographic comparatives have been employed in an attempt to warn Santa Cruz residents of the existence of gangs locally. They were simultaneously used to inform residents that local conditions were favourable when contrasted to those in other Californian cities. There is, however, evidence that officials can, through the use of metaphors, warn of the potential for conditions locally to deteriorate as a result of the continued growth of gangs in the city. Again a comparison can be drawn with representations of the Beach Flats and the propensity of its social problems to spread, polluting the rest of the city. Upon the establishment of a Countywide Task Force to control gangs, Bud Frank, Director of County Criminal Justice Panel, put forward the following argument:

> Area law enforcement agencies have traditionally been effective in controlling youth gangs but current resources and approaches could be overwhelmed if emerging gang problems 'spill over' from the more heavily impacted areas of the state.
>
> (SC Sentinel, *1 October 1989*)

The media watchdog organisation, Fairness and Accuracy in Reporting, have criticised news organisations for employing such images of crime 'invading sanctuaries'. 'This theme suggests that communities once considered safe are now being hit by a spreading epidemic of violent crime' (Donziger, 1996: 73). The *Report of the National Criminal Justice Commission* refutes such claims and argues that crime is actually falling in the suburbs. Locally, the police have been responsible for attempting to apply similar invasion metaphors. The most obvious example of this attempt to forewarn Santa Cruz residents followed the beating of a student in 1991. Detective Upton argued that 'the attack was senseless and is a sign of urban gang problems spreading into Santa Cruz. Similar acts of violence will occur as gangs increase locally' (*SC Sentinel*, 10 May 1991).

BARRIOS UNIDOS: CONTESTING MORAL PANICS

As the moral panic is constructed through a range of geographically interconnected scales, so are attempts to contest the panic and bring about the social change necessary to reduce gang violence. An example of this is Santa Cruz Barrios Unidos, the local chapter of the National Coalition to End Barrio Warfare, part of a wider resistance and new urban peace movement (Childs, 1994 and 1995) largely ignored or criticised by academics and the mainstream media (Garcia, 1993). The organisation works at a variety of levels, from the personal to the international, in its attempts to bring about social change. At the street and neighbourhood level, Barrios Unidos is involved in a variety of community outreach activities at three sites in the city, Beach Flats, Neary Lagoon and Villa San Carlos. These include kids' clubs, youth groups and parent groups as well as street outreach and work in local schools and Juvenile Hall. Barrios Unidos works with a range of individuals and institutions within the local community but also in the county, state and nation-wide. The vision of Barrios Unidos has allowed the organisation to move beyond the local scale to address national issues and to organise nationally and internationally with various organisa tions. The result has been an attempt to unite Sureno and Norteno gangs from across California and the establishment of Barrios Unidos chapters in eight states across the US. Politically, the organisation has sought to promote neighbourhood initiatives at the national scale, by working to formulate an 'Organisational and Community Development Plan'. Through national 'peace summits' held in Santa Cruz and other US cities, the group has formulated the 'Cezar E. Chavez Peace Plan' which was presented to the Californian legislature and US Congress in 1996.

In 1993 Barrios Unidos was also invited by the World Council of Churches to be represented at a global conference about violence prevention, at the UN in New York. Such actions represent concern that violence amongst young Latinos was not being addressed on a national scale in any way other than by incarceration. As the organisation's founder Nane Alejandrez argued in 1994, 'Hispanics must be represented at anti-violence conferences' (*SC Sentinel*, 17 August 1994). They are also symbolic of the acknowledgement that social and economic processes and policies at the national scale were adversely impacting upon Latinos. The organisation is now evidence of the fact that local initiatives can and do affect national discourse and policy.

Conscious of the importance of the media and the need to promote the organisation and its objectives, Barrios Unidos has also become increasingly active in this field. Whilst attempting to influence the framing of gang issues in local news media sources, the organisation is also keen to attract national and international coverage for their work in Santa Cruz and at the national scale. As such Barrios Unidos is an example of what McRobbie (1994) describes as the 'deviant as media expert'. The organisation is interested in reaching as broad an audience as possible geographically, socially and culturally:

We need some wire services to pick us up. . . . When you're on a wire AP and UPI and almost every newspaper subscribes to it, that means that we will be known all over the country and that our cause will be known about. What's happening with the youth killing themselves and about the solutions. We're talking about the solutions and we want the media to know that we're going to Washington DC to present the Cesar Chavez Peace Plan.

(interview with Ella Seneres, Barrios Unidos, 1 April 1995)

CONCLUSION

The themes discussed in this chapter (gangs, moral panics and their geographies) are closely connected. Events at the local scale offer an insight into wider social processes through which these practices are themselves interpreted and influenced. Moral panics about the presence of gangs and the associated ills of violence and drugs are also indicative of a more general social polarisation. The everyday news presentations of the Beach Flats and Latino youth in Santa Cruz result in real and symbolic distancing. In many parts of California and the US, similar fears of gangs, drive-by shootings and inner city crime invading purified suburbs have been important in explaining processes of suburbanisation, white flight, fortified residential and business developments and increasing levels of private surveillance, policing and incarceration (Davis, 1990; Holston and Appadurai, 1996). As Caldeira notes:

> Contemporary urban segregation is complementary to the issue of urban violence. . . . Everyday discussions about crime create rigid symbolic differences between social groups as they tend to align them either with good or with evil. In this sense they contribute to a construction of inflexible separations in a way analogous to city walls.
>
> *(Caldeira, 1996: 324)*

Social exclusion at the neighbourhood and city levels has been replicated in the passage of legislation such as Proposition 187 in California which sought to deny state health and education provision to the children of 'illegal' immigrants (Davis, 1995).

The continual presentation of the Beach Flats as a site of disorder and deviancy is also informed by and speaks of issues raised by debates about the North American suburb and fortified communities as sites of homogeneity and order (Sennet, 1970; Davis, 1990). Given the current period of ethnic change in the city of Santa Cruz and state of California, and the contemporary racialised panics about gangs across the US, such attempts at social and spatial distancing allow moral boundaries to be reinforced. In so doing the normalised community seeks both conformity and deviance, 'for without deviance, there is no self-consciousness of conformity and vice-versa' (M. Davis, quoted in Sibley, 1995: 39).

REFERENCES

Ball, R.A. and Curry, G.D. (1995) 'The logic of definition in criminology: purposes and methods for defining "gangs"', *Criminology* 33, 2: 225–45.

Barak, G. (1995) 'Between the waves: mass-mediated themes of crime and justice', *Social Justice* 21, 3: 133–47.

Bhabha, H. (1983) 'The other question: the stereotype and colonial discourse', *Screen* 24, 6: 19–35.

Caldeira, T.P.R. (1996) 'Fortified enclaves: the new urban segregation', *Public Culture* 8, 2: 303 28.

Chambliss, W.J. (1994) 'Policing the ghetto underclass: the politics of law enforcement', *Social Problems* 41, 2: 177–94.

Childs, J.B. (1994) 'The value of transcommunal identity politics: transcommunality and the peace and justice gang truce in Kansas City', *Z Magazine*, July/August 1994: 48–51.

—— (1995) 'Street war and new urban peace movement', paper available from the author, Department of Sociology, University of California, Santa Cruz.

City of Santa Cruz (1990) *General Plan 1990–2005*.

Cresswell, T. (1992) 'The crucial "where of graffiti": a geographical analysis of reactions to graffiti in New York', *Environment and Planning D: Society and Space* 10: 329–44.

Cross, B. (1993) *It's Not About a Salary . . . Rap, Race and Resistance in Los Angeles* London: Verso.

Cummings, S. and Monti, D.J. (eds) (1993) *Gangs: The Origins and Impact of Contemporary Youth Gangs in the United States*, Albany: State University of New York Press.

Davis, M. (1990) *City of Quartz: Excavating the Future in Los Angeles*, London: Verso.

—— (1995) 'California Über Alles?', *Covert Action Quarterly* 52: 15–19.

Donziger, S. (ed.) (1996) *The Real War on Crime. The Report of the National Criminal Justice Commission*, New York: HarperCollins.

Douglas, M. (1966) *Purity and Danger*, London: Routledge and Kegan Paul.

Dumm, T.L. (1994) 'The new enclosures: racism in the normalized community', in R. Gooding-Williams (ed.) *Reading Rodney King/Reading Urban Uprising*, London: Routledge.

El Andar, August 1994.

Enriquez, V.G. (1990) *Hellside in Paradise: The Honolulu Youth Gang*, Manoa: University of Hawaii, Center for Philippine Studies.

Garcia, L. (1993) 'Mainstream Media Bashes Peace Summit', *People's Tribune*, 24 May 1993: 5

Giroux, H.A. (1995) 'White panic', *Z Magazine*, March 1995: 12–14.

Hagedorn, J.M. (1988) *People and Folks: Gangs, Crime and the Underclass in a Rustbelt City*, Chicago: Lake View Press.

Holston, J. and Appadurai, A. (1996) 'Cities and citizenship', *Public Culture* 8, 2: 187–204.

Huff, C.R. (1989) 'Youth gangs and public policy', *Crime and Delinquency* 35, 4: 524–37.

Klein, M.W. and Maxson, C.L. (1988) *Gang Involvement in Cocaine Rock Trafficking*, Los Angeles: Center for Research on Crime and Social Control, Social Science Research Institute, University of Southern California.

Krisberg, B. (1994) 'Distorted by fear: the make believe war on crime', *Social Justice* 21, 3: 38–49.

Ley, D. and Cybriwsky, R. (1974) 'Urban graffiti as territorial markers', *Annals, Association of American Geographers* 64: 491–505.

McRobbie, A. (1994) *Postmodernism and Popular Culture*, London: Routledge.

Males, M. (1996) 'Wild in deceit: why "teen violence" is poverty violence in disguise', *Extra!*, March/April 1996: 7–9.

Moore, J.W. (1993) 'Gangs, drugs and violence', in S. Cummings and D.J. Monti (eds) *Gangs: The Origins and Impact of Contemporary Youth Gangs in the United States*, Albany: State University of New York Press.

Muwakkil, S. (1993) 'Ganging together', *In These Times*, 5 April 1993: 22–3.

Pacific Center for Violence Prevention (1995) *Preventing Youth Violence: Reinvesting Our Resources. Policy Paper*, San Francisco: Pacific Center for Violence Prevention.

Platt, A.M. (1995) 'The politics of law and order', *Social Justice* 21, 3: 3–13.

San Jose Mercury News, 26 April 1995.

Santa Cruz Sentinel, various editions.

Sennet, R. (1970) *The Uses of Disorder*, Harmondsworth: Penguin.

Sibley, D. (1995) *Geographies of Exclusion: Society and Difference in the West*, London: Routledge.

Skolnick, J. (1989) *Gang Organisation and Migration*, Sacramento: California Department of Justice.

Smith, S. (1986) *Crime, Space and Society*, Cambridge: Cambridge University Press.

Wacquant, L.J.D. (1994) 'The new urban color line: the state and fate of the ghetto in post-Fordist America', in C. Calhoun (ed.) *Social Theory and the Politics of Identity*, Oxford: Blackwell.

Zatz, M.S. (1987) 'Chicano youth gangs and crime: the creation of a moral panic', *Contemporary Crises* 11, 2: 129–58.

Zevitz, R.G. and Takata, S.R. (1992) 'Metropolitan gang influence and the emergence of group delinquency in a regional community', *Journal of Criminal Justice* 20, 2: 93–106.

10

RAVERS' PARADISE?

German youth cultures in the 1990s

•

Birgit Richard and Heinz Hermann Kruger

This article concentrates on the characteristics of the Techno sub-culture in Germany which marks a new era in the history of youth culture. The post-punk era of the mid-1980s was defined by a patchwork of fragmented sub-cultures that were clearly separated from each other – for example the punk, gothic, heavy metal and revival styles of 1960s and 1970s music and fashion. However, in the mid-1990s (1993–5) a pure dance style – Techno – became a mass movement for the first time, uniting nearly two million German youngsters and post-adolescents in the so-called 'rave nation', which has changed the shape, aesthestics and traditional structures of contemporary German youth culture.

Techno is the first sub-culture to use and exploit all the computer-generated forms of the digital age, including desktop and electronic publishing, computer graphics and animation, producing an explosion of new symbols and designs which have infiltrated everyday life (Plate 10.1). For example, Techno magazines have led the way in magazine layout, and Techno clubwear and streetwear styles have trickled slowly into haute couture. More significantly, Techno has revived the importance of dance. Not since disco in the 1970s has music been purely for dancing.

What has made Techno so big in Germany is the fact that it offers different musical styles ranging from the extreme-fast hardcore (not identical with the British understanding of hardcore) or gabber, to the soft and psychedelic (ambient music that produces calm sounds and Goa or Tribal music which contains ethnic sounds). This is possible because the music is produced completely electronically, enabling artists to create a

Plate 10.1 Techno is the first sub-culture to exploit computer-generated
forms of the digital age.
Source: 'Front Page', June 1996.

variety of new elements and sounds quickly and easily. The result is that
Techno music contains a special electronic expression for nearly every
mood, its heterogenous rhythm uniting dancers regardless of their race,
class, age, gender or sexuality. In the words of the Minister of House:

'I am, you see, I am the creator and this is my house. And in my house there
is only house music. . . . You may be black, you may be white, you may be
Jew or gentile – it don't make a difference in our house.'

(Minister of House, 'Mister Fingers', 1989)

The rave experience is ephemeral, however. After the weekend, daily routines are re-established and the ideals of the rave nation – love and peace – are quickly forgotten as social divisions re-emerge.

Raves take place largely in post-industrial landscapes, transforming rundown warehouse sites into timeless, de-localised and de-realised spaces, where obsolete industrial infrastructure is juxtaposed to state of the art technology to create a surreal, almost virtual world – a fun factory. In this environment, dancing becomes an exciting new form of work – the sweating body on the dance floor, symbolically replacing the exertions of the factory floor. For example, 'work your body' is often used as a vocal sample in house tunes. The Techno kids are doing their job on the dance floor, a job which may last from eight hours to three days. It is a provocative form of expression because this kind of work does not produce a living but constitutes pure relentless pleasure and, as such, offers young people an escape from the isolation and problems often experienced by Western youth.

Techno exaggerates the ideal of a well trained body by working it day and night into a state of hyper-fitness, often verging on hyper-exhaustion (akin to endurance sports like aerobics or long-distance running). To endure the three day dance marathon, many ravers turn to the drug Ecstasy (E) to overcome the body's 'natural' desire for sleep. According to McRobbie raves and E go hand in hand: 'The scale is huge and ever increasing, the atmosphere is one of unity, of dissolving difference in the peace and harmony haze of the drug Ecstasy' (McRobbie, 1995: 169). It would be wrong, however, to assume that Techno and House are music styles which are purely driven by Ecstasy, just as nobody would describe beat or hippie styles as deriving only from marijuana and hashish misuse, or assume that punk is a product of excessive beer drinking.

NATIONAL AND GLOBAL STYLE

German post-war youth culture has always been heavily influenced by British and American fashion; indeed, until the emergence of punk German youth culture was often no more than a poor imitation of transatlantic style (Richard and Krüger, 1995: 94). Punk spawned the *Neue Deutsche Welle* (New German Wave), including bands such as as Der Plan, die Krupps, DAF (one of whose former members, Robert Görl, is now a well known Techno musician). These groups adapted New Wave and created a unique national sound, although their lyrics did not often translate well into the German language. Since then, German music has become more adept at developing its own national inflections of international trends (Figure 10.1). For example, the geographical roots of the

Figure 10.1 Germany, post-1989.
Source: Linda Dawes.

current Techno scene in Germany lie in the black community dance scenes of New York, Chicago and Detroit.

House music reached Germany via Britain – where it was known as Acid House (influenced by Spanish Balearic music and dance styles coming especially out of Ibiza). When it first appeared in Germany in 1988, it was not initially popular because the sound was deemed too 'artificial' and because Germans were slow to organise and recognise the significance of raves which were crucial to the success of this musical style in Britain. As McRobbie has argued: 'Where other youth sub-cultures have focused on street appearances, or have chosen live rock performances for providing the emblematic opportunity for the display of style, in rave everything happens within the space of the party' (1995. 168).

However, a small and dynamic underground movement did develop in Berlin and Frankfurt, and from these clubs grew the German form of Techno or Tekkno at the beginning of the 1990s. The fall of the Berlin Wall played an important part in its development. In the following period of political uncertainty, ravers were able to claim politically and commercially unmarked space for illegal raves. From this underground scene an informal network emerged with several, sometimes very quickly changing illegal locations and parties, the most famous being the 'TRE-SOR' or the 'E-WERK' in Berlin.

The German Techno scene has developed a differentiated sub-structure: varieties like Hardtrance, Trance, Acid, Goa and Tribal have emerged as distinct musical styles whereas in the UK these styles have become more hybridised. In the UK this German Techno is known as hardcore (which is not to be confused with Dutch Rotterdam Hardcore, called gabber). Whereas in Britain raves were often held illegally in the countryside, in Germany raves are more often commercially organized events, being held throughout the year, often indoors. For example, the MAYDAY mega event (Berlin, Dortmund and Frankfurt) attracts well-known DJs from around the world, and Techno also takes to the street in the Berlin LOVEPARADE where the number of participants has taken off in just two years, from 100,000 in 1994 to half a million in 1995 (Plates 10.2 and 10.3). The 'Union Move' in Munich and the 'Night Move' in Cologne are other occasions where ravers proudly and visibly take over the streets for their parade.

The neutral matrix of electronic sound which is the basis of Techno means that it can be understood in any culture or language. Different national music scenes are then able to add their own interpretations to this internationally recognisable basic style – usually in the form of voice samples in their own language. House music as a form of dance music has been bouncing back and forth between Europe and America for several decades in the process of which the sound produced has been

Plate 10.2 Youth take to the Berlin streets en masse
in the LOVEPARADE.
Photograph: Sonke Streckel.

interpreted and reinterpreted. There is a special connection between black Afro-American dance styles and German musical technologies. House music has been deeply influenced by the pure synthetic music pioneered by Kraftwerk, Can, Tangerine Dream, Klaus Schulze, for example. (Kraftwerk also influenced early forms of hip-hop like electric boogie) produced out of Germany in the 1970s (Reynolds, 1995). In the 1990s, the popularity of German House music attracted American DJs like Jeff Mills and Frankie Knuckles who revived their waning careers in Europe. Now House and Techno are being reimported into the US; for example, the Berlin Club 'Tresor' plans to open a club in Detroit.

RAVE: CREATING NEW SOCIAL RELATIONS

Dance is where girls were always found in sub-cultures. It was their only entitlement.

(McRobbie, 1995: 169)

The history of dance styles is strongly gendered. For girls and young women, dancing has always been a form of erotic self-expression and

Plates 10.2 and 10.3 Youth take to the Berlin streets en masse
in the LOVEPARADE.
Photograph: Sonke Streckel.

sexuality, whereas men have traditionally preferred to watch, only venturing on the dance floor in order to try and seduce a partner (Mungham, 1976). Acrobatic dance styles originating in the black communities,
such as breakdancing, or Jungle dancing style, offered men new athletic
ways of expressing themselves on the dance floor (McRobbie, 1985;
Willis, 1991). But in doing so these dance cultures excluded women
because to dance in this way would be to lose their femininity in the
eyes of young men. Hip-hop is a typical example of a style totally based
on male competition through its different expressions, breakdance,
graffiti and rap. While the emergence of Techno and House with its
emphasis on the sheer fun of dancing and on body movement, has also
attracted men onto the dancefloor, it is a less misogynist culture (Fritsch,
1988). Techno is a more democratic dance movement. The unequal

positions between the male voyeur and the female dancer dissolve because the emphasis of Techno is on the addictive bodily pleasures of dance rather than dating and sex.

> The trope of masculinity is visually one of largely white unadorned, anti-stylish 'normality'. But laddishness has been replaced by friendliness. Indeed the second irony of this present social moment is that working-class boys lose their 'aggro' and become 'new men' not through the critique of masculinity which accompanies the changing modes of femininity . . . but through the use of Ecstasy.
>
> *(McRobbie, 1995: 169)*

However, Ecstasy alone cannot be held totally responsible for the emergence of these new masculinities on the dance floor. Rather the rave scene has created an environment in which men are becoming more attuned to their bodies. For example, young men are placing more emphasis on the sensual, rather than the traditional physical aspects of their bodies (Plate 10.4).

Rave culture has also transformed femininity. In particular, while the Techno and House scene is associated with hyper-sexual dress codes (Plate 10.5), women wear these forms of clothing while simultaneously subverting their meanings as McRobbie describes:

> One solution might lie in cultivating a hypersexual appearance which is, however symbolically, sealed or 'closed off' through the dummy, the whistle, or the ice lolly. This idea of insulating the body from 'invasion' is even more apparent in the heavy duty industrial protective clothing worn by both male and female fans of German techno music. In both the body signifies sociability and self-sufficiency.
>
> *(McRobbie, 1995: 169)*

Despite the hyper-sexual dress code, the unwritten rule of the rave is that it is not a sexual event. Raves offer a space where women can dance without fear of sexual harassment. According to McRobbie this is only possible because rave is a culture of sexual avoidance. She claims that: 'The orgiastic frenzy of dance culture also hints at the fear of AIDS among young people . . . . The culture is one of childhood, of a pre-sexual, pre-oedipal stage' (McRobbie, 1995: 169).

However, this is to oversimplify a more complex situation as Techno and House are not characterised by total avoidance of sexuality. Rather this culture transforms eroticism into a dance style, sexuality is expressed in a ritual form. The dance itself becomes a form of sexual intercourse where beats and rhythms imitate different stages of orgasm.

Plate 10.4 New masculinities on the dance floor.
Photograph: Sonke Streckel.

The dancers experience virtual sex on the dance floor, releasing their sexual tension through ecstatic shouts.

Despite some of the positive implications for young women, the commercialisation of raves is still reproducing gender inequalities. For example, there are very few female DJs and producers with the notable exceptions of DJ Marusha and Miss DJAX (DJ and founder of the record label DJAX UP BEATS). This under-representation in the music

Plate 10.5 Playing around with femininity.
Photograph: Birgit Richard and Heinz Hermann Kruger.

business also means that young women are less likely to be found in record shops listening to and buying new records.

THE RAVERS' PARADISE: ALL-DAY POLITICS OR JUST COMMERICAL HYPE?

> There are so many dangers (drugs, sex, alcohol . . .) . . . so many social and political issues . . . that rave turns away from this heavy load head-long into a culture of avoidance and almost pure abandonment.
>
> (McRobbie, 1995: 172)

An obvious criticism of rave is that it is pure escapism from the burden of responsibilities, not least because the rave event is a suspension of everyday rules. In addition rave's ideals of democracy are not transferred from the virtual space of the rave into everyday life which leads to the question whether a dance movement like Techno could potentially have a political impact? While Techno culture asks its fans to 'shut up and dance' (McRobbie, 1995), its potential to deliver a political message lies in its power to influence patterns of love and friendship.

In Germany, youth culture has undoubtedly been affected by the experiences of Techno music and rave events – there are messages in the music, messages in the dance. However, its production is defined as hedonistic rather than political, self-indulgent rather than agentic. Social science analysis in Germany tends to assume that movements emerging from within *Kulturindustries* cannot have the power of political statement (Adorno, 1989). This means that the production and effect of Techno and rave can be read in several ways. However, we would argue that Techno and rave can have a political element which may run parallel to the consumerist/hedonistic elements. We also acknowledge that not all rave participants will recognise our political interpretations. The most pertinent example of the politicisation of Techno is the on-going dispute between the administration of Berlin and the organisers of the LOVEPARADE concerning the parade's political status. The LOVEPARADE has to be described as a political demonstration otherwise the organisers have to pay the local authority for policing the event and cleaning up afterwards. Hence the event would then be excessively commercialised as participants would have to be charged to cover these costs. The organisers turn the LOVEPARADE into a political event by claiming that the ravers are not just occupying public space for fun but are demonstrating for peace, that a kind of everyday politics is at play. This is not just a matter of semantics in order to save money. The ravers physically take control of Berlin's most culturally and geographically important thoroughfares, in 1995 the Kurfürstendamm and in 1996 the Strasse des 17 Juni between Ernst-Reuter Platz and the Brandenberg Gate which until 1989 was part of East Berlin and could only be seen by West Berliners from the top of the heavily guarded Berlin Wall (Figure 10.2, Plate 10.6).

Despite the politics of this particular public event there is no doubt that Techno is a very commercialised form of youth culture. It is a consumer-led style, reflecting the excessive consumption of contemporary Western societies. Everything from clothing (clubwear, TechnoKit) (see Plate 10.7) and food (energy drinks) to special sports (snowboarding) or holiday camps are designed specially for ravers (Richard, 1995). However, for youth purists committed to Techno as a genuine German sub-culture Techno remains under ground with a non-commercial club scene. Indeed, Techno is a very democratic and productive style because it offers – as punk did – possibilities for young people to produce and sell their own music (although as discussed above, it is a gendered style). In Germany a new independent national system of producing, distributing and selling electronic dance music has developed: small labels produce limited editions of vinyl records for the direct use of DJs in clubs.

For youth culture of the 1990s the hope for a loving, peaceful and unifying community remains one of the few powerful, utopian ideals left. International rave culture transcends the harsh realities of everyday

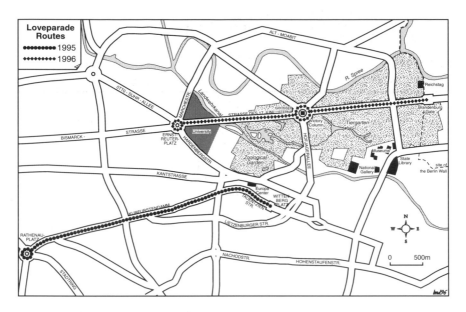

Figure 10.2 Ravers proudly take to the streets for the Berlin LOVEPARADE.
Source: Linda Dawes.

Plate 10.6 The LOVEPARADE: demonstrating for peace.
Photograph: Sonke Streckel.

Plate 10.7 TechnoKit: a kind of suitcase, a collection of items from the raver's
world, including clothing, CDs and energy drinks.
*Photograph: Birgit Richard – Project TechnoKit, Techno and House Archive at
the Art and Design Department, University of Essen, Germany.*

life. In the equal, loving space of the rave, young people are creating a
potential blueprint for the whole of society to follow.

REFERENCES

Adorno, W. (1989) 'Kulturindustrie', in M. Horkheimer *Dialektik der Auf-
klarung*, Leipzig, pp. 139–88.
Chisholm, L., Büchner, P. and Krüger, H.-H. (1995) *Growing up in Europe*,
Berlin/New York: de Gruyter.
Fritsch, U. (1988) *Tanz, Bewegungskulture, Gesellschaft*, Frankfurt: Berluste
und Chanen.
McRobbie, A. (1985) 'Tanz und Phantasie', in R. Lindner and H.H. Wiebe (eds)
Neues zur Jugendfrage, Frankfurt, pp. 126–38.
—— (1995) *Postmodernism and Popular Culture*, London: Routledge.
Mezger, W. (1980) *Diskokultur. Die jugendliche Superszene*, Heidelberg.
Mungham, G. (1976) *Working Class Youth Culture*, London
Reynolds, S. (1995) *The Sex Revolts. Gender, Rebellion and Rock 'n' Roll*,
London: Serpent's Tail.
Richard, B. (1995) 'Love, peace and unity. Techno, Jugendkultur oder Marketing
Konzept?' *Die Deutsche Jugend, Heft 7–8*: 316–24.
—— and Krüger, H.-H. (1995) 'Vom "Zitterkäfer" (Rock 'n' Roll) zum
"Hamster im Laufrädchen" (Techno)', Streifzüge durch die Topographie
jugendkultureller Stile am Beispiel.
von Tanzstilen zwischen 1945 und 1994', in W. Ferchhoff, U. Sander and R.

Vollbrecht (eds) *Jugendkulturen: Faszination und Ambivalenz*, Weinheim, Munich: Juventa, pp. 93–109.

——, Neuke, A. and Klanten, R. (Spring 1996) *Techno-Icons*, Berlin: Die Gestalten (German/English).

Willis, P. (1991) *Jugend-Stile*, Berlin: Argument.

•••••••••••••••••••••••••••••

11

'CHECKING OUT THE PLANET'

global representations/local identities and youth travel

•

Luke Desforges

INTRODUCTION

'Before I went away I thought "Right, I have to go on a long haul trip, otherwise I'm not going to be the sort of person I want to be . . . . I feel like I ought to experience that sort of thing".'

(Ceri)

[T]ourism is not an aggregate of merely commercial activities; it is also an ideological framing of history, nature, and tradition; a framing that has the power to shape culture and nature to its own needs.

(MacCannell, 1992: 1)

Modern travel has been associated with youth from the seventeenth century grand tours of aristocratic sons (Urry, 1990: 4) to the *Wander-vogel* generation that formed the 1920s German youth movement (Janik and Toulman, 1973: 204). The tradition continues today amongst British youth in the form of the 'year out' from higher education or time off from flexible 'McJobs' and short term contracts. Ceri, in her mid-twenties and just returned from a month in Peru, shows how important a part travel can play in youth identity, leading to questions about the sort of person that travel enables her to be, and the sort of travel experiences she needs if she is to feel as though she has the identity she wants. This chapter is about the links between travel and identity, looking at how Ceri and other young travellers 'frame' the places they visit as containing experiences. It is also about how they bring these experiences back home to use in the narration of identity. Travel is one way in which youth identities 'stretch out' beyond the local to draw in places from around the globe. But, to follow the quote from Dean MacCannell, the 'framing' used by travellers forms part of the complex power relations between the West and its global Others in the 'Third World', and the 'framings' used by young travellers are no exception.

The power of Western representations of the 'Third World' has been recognised in geography for some time, particularly in the context of nineteenth century British colonialism. Although Western imperial domination of the colonised world relied on the military power to conquer, it also drew on an imagination of the world which supported and legitimised Western control. Western colonial projects were based in the power to define Others as different. Understandings of such Othering emerged largely from Said's study of the representations of Orientalism, which, he argues, justified European hegemony over the Orient by representing it as 'backward' or 'irrational' compared with Europe's 'maturity' and 'objectivity' (1978: 7). The West's representations are perhaps not as homogeneous as Said suggests (Driver, 1992: 33), and subsequent studies of colonial discourse have uncovered other versions, for example Nochlin (1991) describes the mysterious and erotic East painted by nineteenth century French artists. What these representations share in common is firstly an imagination of the 'Third World' as Other and different from the self, and secondly an assumption that this difference can be understood, described, known and re-presented in an objective fashion by the West. Two general points to draw from colonial discourse theory are, firstly, that representations of the Other and their fixing of difference are somehow linked to the interests of Western power, and secondly, that they have material consequences for the place represented because representations are lived as reality, informing the practices of Westerners in the 'Third World'.

In this chapter I argue that one way in which young travellers 'frame' the world is as a series of differences to be known and experienced. But twentieth century travel is clearly not the same as nineteenth century colonialism, and today's travellers do not have the same projects as Said's colonialists. So what do youth travellers have to gain from travel? Munt (1994) suggests that contemporary travel is part of a process of distinction, in which Western cultures use consumption to mark differences between class fractions. Munt draws upon the work of Bourdieu (Munt, 1994: 106–7), arguing that class power is based on the ability to define 'taste', through 'classificatory struggles' surrounding lifestyle and identity, such as food, education and culture. Munt argues that today's middle class youth (Bourdieu's 'new petit bourgeoisie') define 'alternative travel' as a means to 'stoke up on cultural capital' (Munt, 1994: 109), which they can then convert into economic capital. Travelling, as long as it is done the 'right' way, becomes an entry qualification for some professions, such as working for the charity sector, overseas development or the travel industry itself. As Munt puts it '"travelling" has emerged as an important informal qualification with the passport acting, so to speak, as professional certification; a record of achievement and experience' (Munt, 1994: 112). Through intellectualising travel into the collection of knowledge and experiences rather than, for example, sitting on a beach or going to Disneyland, young travellers define themselves as middle class, gaining entry to the privileges of work, housing and lifestyle that go with that class status.

In this chapter I look firstly at how travellers frame the 'Third World' as a place to be 'collected', and the travel practices they use. Secondly I look at how travellers 'bring it back home', using their accumulated cultural capital to narrate new identities on return to Britain. My account of youth travel and identity is drawn from a larger research project where I interviewed fifteen travellers who had recently visited Peru. This included seven young independent travellers, aged in their twenties and early thirties, as well as six group travellers aged between 35 and 60, and two bespoke travellers in their fifties. I interviewed each traveller twice for ninety minutes, about their latest trip, their previous travel experiences and the way travel fitted into their lives. The interviews were conversational, and although I prompted certain lines of discussion I also attempted (and at times failed) to keep the topics of dialogue open. Given the complex nature of constructing identity, in-depth interviews with a relatively small number of people gave the opportunity of thoroughly analysing the uses of travel in the lives of the interviewees. Excerpts from the interviews are reproduced as 'vignettes' in Boxes 11.1–11.8.

'MOVING ON': COLLECTING TRAVEL PLACES

'I've got a relatively good education but I've never enjoyed studying, I've always done it as a means to an end. And I've always really, really had to slog at it . . . . When I go travelling it's like this knowledge it stays in my mind so easily, because I'm interested in it. And I find to me the whole point of life is, . . . I always want to find out something more, look at something more, know something more.'

(Sarah)

Talking about travelling often brings up ideas about education. For Sarah, an avid traveller since her early twenties, travel offers the opportunity of gaining knowledge which contrasts with the difficulties of formal education. Justin, a trainee lawyer in his late twenties (Box 11.1), sees travel as providing new knowledge in a way which work cannot. Ceri, in her mid-twenties, does not oppose formal education and travel, but sees visiting the shanty towns of Lima as adding to the knowledge she gained studying urban geography at school and university (Box 11.1). Discourses of education point towards a 'framing' of the world in which knowledge and experiences can be gathered or 'collected' through travel. Put in another way by Matt (Box 11.1), who has been on long trips to India, Indonesia and South America, travel is about 'checking out the planet'.[1]

'Collecting places' is built upon a representation of the world as a series of differences from home. Difference is defined by the traveller, as something which can be known, understood and experienced through

Box 11.1

Justin	'I think the thing to be gained from it is really the experience, the way it broadens your mind, education, and you know you get so many great memories from it. And you feel as though you're really living I think, you're actually living life you know. Different things happening all the time, it's not like being at work where, it's not that things aren't happening but it's all within the work sphere. And you're in the same place all the time so your environment never changes. You know, I can think of going off travelling, every day there's something to remember sort of thing.'
Ceri	'What interested me most was the subject of shanty towns and people coming in from the countryside setting up a home, and it was the nearest I think my A level [examination] got to Social Geography . . . so I went and saw Lima and the shanty towns around. I don't know if these are places on the map but they're in the textbook. I just thought "Well go and find out [laughs]. Bring it alive."'
Matt	'So the idea of getting down to South America was to get round that part of the world. See what was going on . . . I mean there are things that we wanted to check out, see what's happening and generally get around a bit, checking the planet out.'

travel. The project of building up systematic knowledge about the world is referred to in a colonial context by Pratt (1992: 29–30) as a form of 'planetary consciousness'. For Pratt, the European colonial encounter with its 'Third World' Other was organised by the planetary project of constructing an institutional written record of the world, for example in the mapping of the world's coastlines. Long-haul travel today is different in that it is organised around the pursuit of individual knowledge and experiences, recorded largely in personal memory or cultural goods such as souvenirs and photographs. Justin (Box 11.2) gives an example of the global nature of this project as he chooses where to go for his next trip. Once Justin has 'collected' a place, he has to move on to new experiences, as he shows in his reluctance to return to Asia because he has already been to the Himalayas. On a more regional scale, travellers represent destinations as containing a number of collectable places, for example to 'do' Peru for some means using a popular division of the country into desert, mountain and rainforest as the basis for a trip (Plate

Box 11.2

Justin	'When you look round the world at places where you could go and do some trekking, where we'd want to go, there was the Himalayas, we could have gone to Nepal, but I'd been there about six or seven years ago. So I didn't want to go there again. So that really brought it down to South America, unless we went to somewhere like Africa, ... I think there are some good mountain ranges but I don't think they're as good as you get in the Andes say. And I suppose there was North America as well, . . . but again I'd been there and I also wanted to go somewhere which had got a completely different feel to it. Like I went to India a few years ago and I thought that was absolutely brilliant because it was just a completely foreign environment. I'd never been anywhere like it.'
Jo	[Arriving in a new place] 'I mean it's a little bit, not even difficult but I find that exciting in going to a new city with a map and like "Yeah" . . . I mean of course there's an adjustment but I find that more like quite exciting and you've got that feeling of being somewhere different that you've never been to before.'

Plate 11.1 Collecting places: gazing at Machu Picchu, Peru.
Photograph: Luke Desforges

11.1). It is the association between movement and new experiences which provides the thrill of arrival described by Jo (Box 11.2), an Anglo-Australian in her mid-twenties, who bases herself in Britain, transforming the difficulties of arriving in a strange city into the exciting new discoveries of beginning to know and experience a place. Travel as 'collecting places' is then a form of relating to the Other which is an individualised version of 'planetary consciousness'.

'Collecting places' forms the basis for travel practices, and in particular the authenticity of places and experiences is important to young independent travellers. Travellers face the anxiety that the places they visit are not 'really' different, that they are not the 'real' India or the 'real' Peru. There is no point experiencing a new place if it is not really new, and any claim to really know a place means claiming to get 'behind the scenes'. Frow (1991) calls this 'authentication', the ascription of authenticity to places and experiences by labelling them as, for example, 'traditional' or 'natural'. 'Collecting places' demands travel practices which young travellers use to authenticate their knowledge and experiences.

Matt and Kim (Box 11.3) construct authenticity through a familiar division between tourists and travellers. They see tourists as people who go on holiday in hotels where they are separated from the 'real' locals. 'Collecting' the new experiences which Matt and Kim are looking for means going to places which are not organised for tourists but for locals

Box 11.3

Matt	'You wouldn't find yourself chatting away in a restaurant or café with some sixty-year-old. . . . Especially if they're like sixty-year-old tourists.'
Kim	'Well they're not in the sort of area that you're in. Cos you're in the ones where you're more mixing with the locals. And they're in the ones where it's been more arranged or they eat in the hotel or they stay in certain areas. Whereas like we'd go into the midst of where the locals go. So you don't really meet many people like that.'
Matt	'It's trying to get more experience out of your trip rather than going on holiday and just having a break.'
Justin	'And Santa Teresa was much more out of the way wasn't it. It clearly didn't have many tourists there. I mean getting into the mountains, it wasn't like starting a trek like the Inca Trail where you get off the train and there's sort of thirty others. So it was very nice to be sort of on our own. . . . I really like being away from other tourists, you know I felt like I was really travelling when I was there . . . . I mean I'd really like to go where no man's [sic] been before. But it's increasingly difficult. That's probably the sort of thing I really strive for. You know I'd like to go somewhere like Central Asia, the Pamirs. Not a lot of people go there I don't think. So you go to places where they don't see Westerners very often at all.'

For Justin, trekking in the Peruvian high Andes is the only way to travel 'properly' because it gets you away from other travellers, who he says stick to the more popular treks such as the Inca Trail. Justin's travel practices are based on a desire for authenticity which goes beyond just avoiding other travellers, in that he wants to visit places which are completely Other (Box 11.3), where nobody goes (by which Justin means no other Westerners). The presence of other travellers, who Justin labels as 'tourists' because they use the travel industry to organise their trips to a greater extent than he does, interferes with the Otherness and difference of places.

Young independent travellers are not only concerned that the places they visit are authentic, but that their experiences of them are as well. Travelling 'properly' for Jo means going individually (Box 11.4), because

Box 11.4

Jo	'. . . especially with nature and landscape I like to just sit and observe it by myself for a while in peace and quiet, instead of having six or seven people around and chatting all the time. And I think when you've got other people with you all the time, they'll say what they're thinking about it, and it kind of influences what you were going to think about it without you even realising . . . . Whereas if you're by yourself, all the opinions that you'll have'll just be totally what your opinions are.'
Jenny	'We'd been round and we were going up to the watchman's hut to take pictures of the rest of it from a hut, a viewpoint. But we didn't quite make it in time. We were sitting looking and watching. All of a sudden there was loads of people there. In their thousands. It was [pause] it just took something away from it somehow. And there were all these people. They hadn't walked there. They were all wearing their nice white trousers and their posh cameras and all looked pristine. I mean [pause] I somehow felt cheated because I'd walked there and I'd found something and discovered something about myself and about the place they were trying to capture arriving by coach or whatever they did. And they were invading.'

travelling as part of a group means that others get in the way of her personal reaction to the landscapes she gazes upon. For Jenny (Box 11.4), the proper way to visit sites such as Machu Picchu, the ruined Inca city in Peru, is to make your own personal effort to get there. Jenny does this by walking to the site along the former Inca route, but when she gets there she berates other travellers for visiting the site by bus, organised for them by the travel industry. Group travellers not only 'spoil' the site for her (make it an inauthentic place), but according to Jenny they mistakenly think they can properly experience the site without making their own effort to get there. The main marker of an authentic place and experience for young independent travellers is the absence of the travel industry (Plate 11.2). 'Individual' experiences of place are only available to 'travellers', as opposed to the 'collective' experience of group 'tourists' who are supported by the travel industry.

The contradiction here is that independent travellers do obviously rely on the tourism economy to some extent, but their desire to consume the Other 'authentically' by cutting down on commodification is an example of the power their 'framing' of the world holds to 'shape culture and nature to its own needs'. Travel representations 'fix' the Other, and it is only by complex reshapings of themselves into the image of the traveller that people and places can become part of the tourist economy (see for example Urry, 1990 on working under the tourist gaze). Firstly, young travellers have the power to determine which places are brought into the tourist economy (and which places are excluded), which put simply means incorporating those places which conform to notions of authenticity. Secondly young travellers determine the terms by which people and places are included in the tourist economy. Because for young travellers authenticity of experience is marked by as little commodification of the relationship between traveller and Other as possible, the main concern of the independent travellers I interviewed was often how to cut down on paying to consume the Other. At times this means attempting to create completely non-commodified relationships; for example Jo described how she would avoid using commercial guides because she sees the non-commodified information she can get from 'local people' as more authentic and trustworthy.[2] At other times it meant consuming the Other as cheaply as possible, for example Justin describes bargaining down the guide on a trek because his price was 'out of keeping with the local economy'. So by representing difference as economic underdevelopment, young travellers have the power to set the terms of their relationship with the Other as justifying as little payment as possible.

To summarise, 'collecting places' is a way of framing the 'Third World' as a place where individual knowledge and personal experience can be gained through travel. A key discourse informing the travel

Plate 11.2 'Real travelling': the search for authenticity.
Photograph: Luke Desforges

practices of 'collecting places' is authenticity. The economic power held
by young travellers means that discourses of authenticity, particularly
where non-commoditisation is used as a marker, at least in part determine the terms of the relationship between travellers and places such as
Peru, playing their part in the reshaping of cultures, employment practices and economies which are all part of tourism's role in the production
of place.

'BRINGING IT ALL BACK HOME': CULTURAL
CAPITAL AND TRAVEL

'Collecting' necessarily means bringing things home to a place where they can be put on display. The knowledge, experience, souvenirs, photographs and sun tans collected by young travellers are not part of an institutional collection, unlike Pratt's 'planetary projects', but as Ceri suggests in the opening quote, are more about the identity of individuals. For Munt this is a class identity, as travel is used as a series of social '"doors and bridges" to unite and exclude' (Featherstone, 1991 quoted in Munt, 1994: 108). By using travel as a form of cultural capital which serves as a sign of distinction, travellers gain access to a social class and its consequent privileges. In this section I show that Munt is absolutely right as far as some travellers are concerned, but for others travel as cultural capital has to be carefully and subtly negotiated in sometimes less than friendly social circumstances.

For Ceri, drawing on the cultural capital of travel is fairly unproblematic (Box 11.5). Her hopes of promotion from a secretarial to a production research position at the film company where she works were realised because of her trip to Peru. This move from what she portrays as routine administrative work to intellectual and cultural work is explicitly linked to class status, as Ceri claims 'the average production secretary does not up sticks and go to Peru for a month'. Going travelling positions Ceri in the class bracket which can do research work because she can present herself as having grown in self-confidence and competence (Box 11.5). Ceri also uses travel as cultural capital in that she uses it to claim an authority about 'the world' which she hopes will lead to the more long-term goal of producing her own ethnographic film. As she puts it 'I can't make films about the world without having seen it first'. Ceri intends to use this claim to authoritative knowledge to sell her abilities as a researcher specifically for film projects on Latin America. Munt's suggestion that the power of cultural capital can be used to gain economic power is certainly true in Ceri's case.

Travel can also be used outside the work sphere to create a sense of social solidarity through distinction, or as Ceri puts it: 'I wanted to be part of the club of people that had been away on a big trip'. Ceri (Box 11.5) talks about going out for a meal with a friend who had visited Latin America. Their trips form a mutual social bond in that both value and respect the knowledge and experiences gained through travel, which serves to distinguish them from others who Ceri sees as lacking in knowledge about the world. Travel can also be used to create social solidarity with non-travellers who share a respect for travel knowledge; for example Ceri describes showing her photographs to her grandparents, who she

Box 11.5

Ceri	' . . . there's this guy and he's one of the founder directors of the company. About a year ago he said to me, because I was talking about my trip to South America and he said "Yes, you know I certainly noticed the difference in you when you came back from travelling in Peru." It took him a little while to say but we were just at a party and he said "We think you were really changed by that experience. Made more confident from it."'
Jenny	' . . . my next door neighbour, last summer sometime [we were] talking about what I want to do, and as an employer he said he'd rather have someone who'd temped for a year and done six months travelling, worked *hard* for what they wanted, than someone who'd just got a mundane job for a year. He said that's far better. A more positively challenged, organised sort of person, is much more employable.'
Ceri	'And at Christmas we had a meal and he [another traveller] was really welcoming the opportunity to have someone lapping all the information from him. Because he was like "Yeah, and then I worked in the rainforest and then I did this and then I did that." And I was like "Oh yeah, and where did you go then?" He was really enjoying the fact that someone was interested. Because every time I say I'm going to Central America someone says "Oh yeah. I'd love to go to South America". And I'm like "No it's Central America, it's different". And he was obviously quite pleased that someone knew what he was talking about.'

says are relatively housebound but travel vicariously through her representations of South America. For Justin it is not so much that his parents want to hear about his experiences of travel, it is more that they value the new identity he takes on as someone who has grown as a person through 'collecting places'. As long as travellers can share a similar set of cultural values with their audience, a similar framing of the Other as understandable through travel, then it provides a basis for distinction.

This account of homecoming does not recognise the tensions that emerge as travellers attempt to cash in on their trip. In the work sphere, some travellers find employers contesting their narrative of cultural capital. For Justin (Box 11.6), travellers who go to countries outside the 'Third World' are not experiencing the Other but more of the same, and would be better off staying at home. The imaginative geographies of travellers, or where they have to go to 'collect places', is open to question. Sarah's employers offer a more deeply rooted challenge to travel (Box 11.6). While travel might be good for individual self-development, they do not see that as relevant to the commitment which has to made to institutions such as the accountancy office where she is looking for a job. Sarah has to renarrate her travel in a way which she does not feel fully comfortable with, if she is to get the job. Kim faced a similar

Box 11.6

Justin 'This is a bit pompous but I think it [travel] possibly shows a lack of commitment for the job, right. Now I mean I'm guilty of that myself to some extent. I tend to work because I've got to work, not because I'm in love with the job at all. But if I'm employing somebody I'd want somebody who's reasonably enthusiastic about the job, and if you've spent a year working in Sydney selling newspapers or something, which is not going to gain you a great amount of experience or great memories or anything like that, I'd feel that you've rather wasted a year doing that whereas you could have been trained to do the job you wanted to do or doing the job you want to do at home.'

Sarah 'If I go to job interviews or that sort of thing, obviously you've got to suppress all this kind of side of you because they don't want to know . . . . You could say "Oh well it [travelling] was a big mistake and I won't do it again", and they'd be quite happy but I always go "Oh it was fantastic". Like my present job they nearly didn't take me on because they thought "She's not going to stay a year. She's going to piss off." And they told me after being there a year and a half, they told me this and they said "We had a bet on that you wouldn't stay a year you know."'

---

problem as a research laboratory assistant who left half way through a project to go travelling. The narrative of travel as providing for personal self-development does not provide cultural capital in jobs where institutional expertise is important, such as accountancy, law or science. Because the knowledge and experiences of youth travel are so individual, they can only be used for cultural capital in industries which are based around an intensive design process where individual 'flair' plays an important role, such as the film or advertising industries (see Lash and Urry, 1994 on cultural capital in the culture industries).

Narratives of travel face similar challenges outside the workplace. As a form of social distinction travel can create barriers where travellers might not want them to be. For Sarah (Box 11.7) travel puts her in a

Box 11.7

Sarah	'It is true that men don't want to feel that you know more than them on anything. And I'm not an unintelligent person, I'm a graduate, I've got a good job so I generally don't sort of go on about how much I know anyway, in terms of travel. I really do try to suppress it because people in general don't want to hear about it. But I mean if I meet a girl and talk to them and it comes up and . . . they are interested and I just don't go into great details, I sort of say "Yeah I went all around". And they go "Oh that's really good" and are encouraging. Meet a guy and they think "Oh blimey" you know "Got someone here who's independent". And it undermines them in some way. I know it's an old fashioned attitude but it's true. Men, even "New Men", still feel like this you know. That makes it very difficult in my relationships with men.'
Jenny	'I think after the third time of showing a grandparent or a relative who doesn't really want you to show them [photographs] to them because they think you're hyperactive, that's the exact words, they just took one look through them and said "No thanks I don't really want you to show me". I think it just got a bit, not tedious, I think that's the wrong word, it just got tiring going through it.'

position of authority over men, challenging a feminine identity which she sometimes wants to keep. Knowing that travel gives men an inferiority complex, she has to 'suppress' her travel history. The sense of frustration Sarah feels is shared by Jenny, who also finds herself in situations which close down the possibilities of using travel as cultural capital. When she gets home from a three month trip, she feels people in her neighbourhood treat her 'as though I'd just got back from Tesco's'. Jenny is disappointed that people are simply not prepared to spend time talking to her about the trip. Photographs and presents provide one way to get around this problem, as giving away souvenirs or narrating her trip based around a photo album provides situations where she has the chance to tell travel stories. But even this has problems. Some of her family contest the worth of travel as she shows her photographs, saying that they do no want to see them as they think she is 'hyperactive' (Box 11.7). Matt faces a more aggressive contestation of travel in the hostile racism of some of his friends when he tells them he is going to India, questioning his desire to relate to the Other as worth understanding and experiencing.

The audience is central to the transformation of travel into cultural capital, setting the terms of the narrative travellers can use to recount their trip. The main way travellers convert 'collecting places' into cultural capital is through a narrative of personal development and authoritative knowledge about the world. Audiences can conspire with this narrative, even gaining cultural capital themselves through sharing a respect for travel. In a situation such as Ceri's where an employer values travel through the same narrative, then cultural capital can be converted into economic power. At other times, where the distinction created by travel is shared by the audience, travellers might not want to use travel as a resource for cultural capital as it places them too far above others, such as when Sarah decides not to talk to men about travel. There are also tensions when the audience contests the value of travel, or alternatively just simply does not leave any room for using cultural capital. The result is that narrating travel is problematic, always out of the control of the traveller because the value placed on travel has already been set by the audience.

Travellers find a more secure use of travel in a personal and individual sense of coherency of identity, where they do not have to struggle to control the symbolic value given to travel by others. One of the organising features of self-identity in late modernity is the idea of personal self-fulfilment, of living a lifestyle which provides a sense of achievement (Giddens, 1991) 'Collecting places' gives a traveller a powerful resource for telling a story about themselves to themselves; for example Justin (Box 11.1) sees travel as a time when you feel as though you are 'actually living life', and Sarah (Box 11.8) perhaps exaggeratedly says that having been travelling she can 'die happy'. This is true of other travellers I

Box 11.8

Sarah	' . . . if I died tomorrow . . . I would look back and think I did something with my life. I've got a lot of friends now who got married very young . . . and started having kids. They can't go away. I say to them "You've got a different life from me. You've got some things that I don't think I'm going to get", you know. But they never did anything, did any travelling and they wish that they had.'
Jenny	'I don't know really. The routine of it all and just [pause] all of a sudden you're living a life of a thirty something and you're only twenty-three, twenty-four . . . . People want more from their lives and something for themselves first, to go and see the world while they're young and stuff, I guess. I know I do. There's a lot to see and a lot to do . . . . I think everyone I've spoken to has wanted to go away again and is not ready to settle down into a life, because they want to see the world, and there's more important things in life than commuting up to London every day.'
Jo	'I remember being quite scared about having to go into a pub by myself, but I had to do it in the end, whereas now I wouldn't give a scrum, it wouldn't bother me in the slightest if I had to go and sit and have a drink somewhere, waiting for someone, or even going in there on my own and talking to people. Whereas before that would have freaked me out a bit.'

interviewed, but for younger travellers it is often worked into a discourse of youth identity (Jenny, Box 11.8). Youth is seen as a stage in the life course where individuals have the freedom to find out about the world and themselves as part of a transition into full adulthood of making responsibilities and commitments to others. This is very much a 'standard narrative' of youth, about personal growth before 'settling down' into adult life. But while this sense of self-fulfilment through travel is held privately, it is used strategically in the public sphere, for example Jo (Box 11.8) talks about the personal confidence which travel has given her to feel secure in herself as an individual. Travelling to become the 'person you want to be', to use Ceri's words, can be as much about private fantasies of self-identity as public shows of cultural capital.

CONCLUSION

Long-haul travel is at the centre of a largely white middle-class youth identity and its representations of the world beyond the home, drawing globalised spaces into the construction of localised identities. Travel 'fixes' the Other for consumption as knowledge and experiences, framing the world as a series of places to be 'collected'. The problem with travel is that its Othering of people and places throughout the world forms the basis for the material relationship between travellers and what are euphemistically called 'hosts' in the travel literature. Although this Othering is contested and subverted in ways which are beyond the scope of this chapter, the young traveller's representations of place still have the power to 'reshape culture and nature' to its own image. But the Othering of travellers today is different from the colonial project, and the power to be gained from consuming the Other is less clear-cut. The desire to 'collect the Other' supports a narrative of personal self-development, but using this narrative depends on situations which are beyond the control of the traveller. What travel does offer is a sense of security in oneself. The power of the young travellers I interviewed is the power to tell stories about identity which stretch in two directions: towards a fixing of the identity of the Other, and towards a fixing of their own identities. As one of the older travellers I interviewed says: 'it beats having to take pills'.

ACKNOWLEDGEMENTS

The material quoted in this chapter was produced as part of a Ph.D. thesis on identity, representation and long-haul travel, funded by the UK Economic and Social Research Council. Many thanks to the Department of Geography at University College London, especially to my supervisor Dr Jacquelin Burgess. All photographs credited to the author.

NOTES

1 There are of course other versions of youth travel, such as the leisure practices of Ibiza and Goa, or British Asians and Afro-Caribbeans visiting places where roots might be claimed. 'Collecting places' is one specific but powerful discourse in travel.

2 A good example of the conflicts this causes between travellers and people in tourist regions is in taking photographs. For independent travellers, photographs which are paid for are 'staged' and inauthentic, so some travellers went to great lengths to photograph people surreptitiously, or using telephoto lenses. Conflicts arose when the person being photographed demanded payment for being 'consumed'.

REFERENCES

Driver, F. (1992) 'Geography's empire: histories of geographical knowledge', *Environment and Planning D: Society and Space* 6: 117–26.

Frow, J. (1991) 'Tourism and the semiotics of nostalgia', *October* 57: 123–51.

Giddens, A. (1991) *Modernity and Self-identity: Self and Society in the Late Modern Age*, Cambridge: Polity Press.

Janik, A. and Toulman, S. (1973) *Wittgenstein's Vienna*, New York: Simon and Schuster.

Lash, S. and Urry, J. (1994) *Economies of Signs and Space*, London: Sage.

MacCannell D. (1993) *Empty Meeting Grounds: The Tourist Papers*, London: Routledge.

Munt, I. (1994) 'The "Other" postmodern tourism: culture, travel and the new middle classes', *Theory, Culture and Society* 11: 101–23.

Nochlin, L. (1991) 'The imaginary Orient', in L. Nochlin (ed.) *The Politics of Vision: Essays on Nineteenth-Century Art and Society*, London: Thames and Hudson.

Pratt, M.L. (1992) *Imperial Eyes: Travel Writing and Transculturation*, London: Routledge.

Said, E. (1978) *Orientalism: Western Conceptions of the Orient*, London: Penguin.

Urry, J. (1990) *The Tourist Gaze: Leisure and Travel in Contemporary Societies*, London: Sage.

—— (1995) *Consuming Places*, London: Routledge.

three

*place: geographies
of youth cultures*

Here five chapters analyse different geographical concepts of 'place'. While each different place is established as central to the production and negotiation of particular youth identities and cultures, the boundaries of each are not discrete but rather intersect and cut across each other. What clearly threads through these chapters are the complexities of choices and decisions which face young people in everyday locations and the ways in which these places are sites of conflict, contestation, as well as places of fun, adventure and security.

First we begin at home with Sara McNamee's chapter on youth and video games. She argues that the home has traditionally been the site of young women's leisure activities but that increasingly male youth are spending their spare time indoors too. By discussing the use of computers and video games, McNamee investigates the way teenage boys and girls contest each other's rights in the home. Teenage girls' attempts to resist and contest masculine authority and control are also a theme in the following chapter which focuses on the site of the school. Here Shane Blackman investigates the ways in which a school group dubbed the New Wave Girls engages in resistant practices and performances in different social settings, inside and outside the school, in a bid to present themselves as both cultural occupiers and social definers.

Sophie Bowlby, Sally Lloyd Evans and Robina Mohammad interrogate the notion of the workplace. In 'Becoming a paid worker: images and identity' the authors consider the ways in which Pakistani Muslim young women have to make employment choices which are conducive to religious and familial expectations while also negotiating the gendered and racialised stereotypes perpetuated by employers.

Paul Watt and Kevin Stenson's chapter also explores the racialised and gendered nature of a particular space – this time the public place of the street. Drawing on interview material with young people from three different ethnic groups – Afro-Caribbean, white British and South Asian – Watt and Stenson examine concepts of safety and danger in relation to the street geography of 'Thamestown'. The final place to be considered is 'The Club'. Highlighting the unique blend of pleasures – musical, tactile, sensual, sexual and chemical – in clubbing. Ben Malbon argues that 'The Club' is a place of the clubbers own creation, of which they are also the consumers.

12

THE HOME

youth, gender and video games: power and control in the home

•

Sara McNamee

The current economic climate, then, when combined with the feminist critique which, as well as stressing the need to focus on girls, has always emphasised the importance of the domestic sphere, indicates also the need for a much closer examination of the activities and regulation of boys and girls *within the home*. As a subject, the domestic lives of young people have rarely been considered of significance either in studies of the family or in studies of youth

(Nava, 1992: 73, emphasis in original)

YOUTH CULTURE AND THE INVISIBILITY OF GIRLS

It has often been stated that youth is in a process of infantilisation – that young people are economically dependent on their parents for longer than has ever been the case in the past (Jeffs and Smith, 1990; Jones and Wallace, 1992). Along with this extended period of dependency, it has been noted that the culture of childhood and youth is increasingly being controlled by parents, and that youth culture is now more often taking place in supervised and protected spaces (James, 1993; Büchner, 1990) rather than on the streets as previous commentators have found (see for example Corrigan, 1979; Hall and Jefferson, 1976). It should not be forgotten that the classic studies of youth culture and youth sub-culture were in the main studies of young working class men (McRobbie and Garber, 1976). Girls were 'invisible' in these studies of youth on the streets, and were said instead to be taking part in youth culture in their bedrooms, which has been seen as 'one way in which girls resisted boys' domination of the streets, that is using their homes as the base from which to explore aspects of teenage culture' (Griffiths, 1988: 53). In the present social and economic climate, then, it can be argued that the siting of male youth cultures has moved from the street into the home. In that case, it becomes timely to examine what effects, if any, this move into the home has on gender relations in the domestic sphere. What does the increased use of the home as a site of youth culture by boys mean for girls? This chapter addresses Nava's argument quoted above and is concerned with an examination of the activities and regulations of boys and girls within the home as becomes manifest when focusing on sibling relations around computer and video games.

COMPUTER AND VIDEO GAMES AND THE INVISIBILITY OF GIRLS

The ownership and use of home computer and video games have increased since the development of the Sega and Nintendo games consoles in the late 1980s. In 1993 the UK market was reported to be worth £420 million for Sega and £300 million for Nintendo (*Education Guardian*, 1993: 8–9). The focus for academic research in this area has tended to be psychological in perspective, and led by a concern to establish what, if any, effects use of the games might have on individuals who play them (see for example Schutte *et al.*, 1988), rather than examining what impact this large volume of consumer items aimed at young people might have on societal relations. Some researchers have examined gender differences in game play and the most frequent finding is that boys tend to play with computers and video games more than girls do (Dominick, 1984; Funk, 1993; Kaplan, 1983; Kubey and Larson, 1990). One reason often stated for this is that it is the content of the games and the predominantly masculine themes contained within them (violence, competition, etc.) which do not appeal to girls, thereby failing to capture their interest (see Provenzo, 1991 for a full discussion of the content of video games). Others have argued that the explanation lies in differential socialisation practices (Kubey and Larson, 1990: 125).

While the effect of gendered power relations in the home has not so far been examined, Haddon's (1992) study briefly touches on gender differences in the home. He contends that for male youth, a sub-culture is being created through the ownership and use of computer and video games. He states that:

producers clearly had some bearing upon the ways in which boys experienced home computing and games playing, as exemplified by the fact that so much of classroom talk was based on reading computer magazines. But ultimately this male youth also made creative use of the raw material – they made their own culture through the way they used 'talk' about micros and about games, through developing and changing products.

(Haddon, 1992: 89)

Haddon (1992) also states that this was not the case for young women. He argues that in his study girls gained their knowledge about games from their brothers, and 'just played' the games which their brothers had and which were available, but were not in the main games of their own choice. There are echoes here of earlier studies on male youth sub-cultures – we have an impression of boys as being central to the sub-culture while girls only become visible in relation to their brothers. In effect, Haddon claims, this sub-culture around computing and game playing was creating a new space for male youth culture, but it is one which is developing in what has traditionally been seen as girls' space – the home, and more especially, the bedroom.

THE PERFORMANCE OF GENDER RELATIONS IN THE HOME: GIRLS ON THE MARGINS

In contrast to Haddon, my study[1] found that there is little gender difference in liking to play with computer and video games – girls report that they like playing as often as boys do (Box 12.1). However, boys report actual playing significantly more than girls do. This chapter examines one possible reason for this discrepancy: that, for some girls, their access to computers and video games is controlled by their brothers. In the light of the increase in ownership and use of computer and video games in the home, then, this chapter positions the ownership and use of computer and video games centrally in order to allow us to see the ways in which the machine becomes a symbolic focus around which gender relations are negotiated and expressed in domestic space.

SM: You've got a sister, Phil. You said that your Nintendo was shared with her – do you ever fight about going on it?
Phil: Well, it's always kept in my room and there's a lot of arguments because I won't let her on it because it's in my room. I won't let her in my room.
SM: Did you say your sister was older or younger?
Phil: Younger.

Box 12.1

There's this game, it's about these two brothers, and it turns out they're both evil . . . there's this magazine . . . luckily it had all the cheats for it that tell you exactly what you have to do, and I tell you if we had not got that, it would *never* have been solved. I tell you, it was *so* difficult. I mean, we thought it was good when we found the spaceship and that was one of the easiest things you can do, and the last world that we got into, that was the first world you're supposed to get into, and we thought it was so difficult, but I mean it's really, sort of like, you think it's simple when you've got it all written down but it's not, because you have to look for clues and books sort of thing, and you think it's good if you've just like opened a page or something, but it's not, it's just, you have to collect sort of like pages to tell a story, and both the brothers are locked in books by their father and you have to collect the pages to see what they say, and the first clue they both say 'let me out, it's the other one's fault' but it turns out that it's both their faults, and if you go to the father and he destroys both the books, that's it! And, it's completely finished then, you can't do anything else, it just leaves you in the world, it doesn't like say you've finished the game and you've got it, it says they'll send someone, but it never happens. Once you've done it you can't do anything else. It's not like a platform game, where there's different levels and it's not really interesting enough to repeat . . . one of the worlds is called 'tree world' and it's so hard, because, I mean, we had all the cheats and you have to turn on all these water pipes, and you have to go upstairs, downstairs, upstairs, downstairs, before you can get back, and it was even hard *with* the cheats because you didn't know if you'd turned it the right way, and if you got it the wrong way you had to go back, change it, get it right. Oh! It's depressing!

(Kayleigh, aged 14)

SM: How much younger?

Phil: Ermm, she's eleven.

SM: So you won't let her in your room, you won't let her use it?

Phil: She likes different types of games to me so if I let her, she like . . . .

SM: Does she ever complain to your mum about it?

Phil: Yeah. She starts getting real annoyed and that and starts saying 'Oh well that's it now. Next time I'm gonna trash your room' and all this lot.

SM: What does your mum say?

Phil: She tells us both to pack it in. She usually blames it on me sister.

SM: Does she? Does she ever threaten to take it off you?

Phil: Yeah. She has done a couple of times.
SM: Why? For fighting?
Phil: She says it causes nothing but problems.

Phil, quoted above, is 14 years old. Like many of the young people in my study, he shares his computer game machine with another family member, in this case with his younger sister. What is also common is that, even though the machine is shared property, it is situated in his bedroom. Mitchell (1985) informs us that:

Videogame consoles were usually considered property to be shared by all members of the family. However, in families with sons, possession by the boys was considered appropriate; and sisters had to request permission for access to the games. When the games were located in a bedroom, instead of the living room or family room, it was always the boy's bedroom even though he may have been the younger of the siblings.

(*Mitchell, 1985: 129*)

Julia informed me that the computer in her house belonged to all the family, but that it was kept in her brother's bedroom. When I asked her why this was, she simply shrugged her shoulders. Wheelock (1992) has noted that the location of a computer in the home can lead to unequal usage between family members. Conflict such as that described above by Phil can arise when one sibling refuses to allow access to their space, as Raffaelli argues:

conflicts centred on property issues are regarded primarily as evidence of rivalry between siblings. It has been theorised, however, that possessions are integral to self definition (Furby 1978), and control over possessions has been linked to a sense of competence and self identity (Bettleheim 1974). Thus, property disputes may reflect not rivalry but rather age-appropriate issues of self definition and personal boundaries. This is supported by the fact that in this study, youngsters did not describe fighting over ownership but rather over unauthorised use of possessions or personal space.

(*Raffaelli, 1992: 660*)

Phil justifies excluding his sister from his room by telling me that it is because she likes different types of games to him, but they appear to be arguing not so much over the use of their machine, but over access to his personal space. Twice he tells me he won't let her in his room. His sister retaliates by threatening to 'trash' his bedroom, indicating that she is aware that it is his room, rather than his control over the machine, which is important to him. In this sense the machine becomes a focus for disputes over gendered domestic space. The conflict over the machine is often resolved by their mother physically removing it. He says 'she tells us both to pack it in', acknowledging that both of them are involved, but then qualifies this with 'she usually blames me sister'. This might be an effort on Phil's part to avoid appearing to behave at the same level as a girl – the interview with Phil was carried out as part of a group interview and his best friend Levi was also in the room while we were talking.

This issue of (threatened or actual) destruction of personal property by siblings is also demonstrated in an excerpt from an interview with Jenny, a parent of two teenagers. Her children Colin and Louise have each got a games machine in their own bedrooms. She describes a recent event in their house:

> *Jenny*: Louise never wants to borrow Colin's, but Colin would like to often use the Super Nintendo. Which can cause problems! [laughs] In fact I was thinking about this . . . it's only the other . . . well, week before last, 'cos I was in the bath, and there was all hell on earth upstairs, and I wondered what the hell was happening! He wanted to play a game on hers, and she didn't want him to, so he wanted to then take the system out of her bedroom and put it in his, and she didn't want that either. She said 'I might want to play it later'. She probably never did, but . . . she was just being awkward. So, yeah, there was fireworks then . . . .
> *SM*: So who won that time then? Did he get it in the end?
> *Jenny*: He didn't, no! She might be the youngest, but by hell!

Colin came into the lounge where Jenny and I were talking after the interview had ended. Jenny asked him to tell me about this particular fight. He made it clear that although Louise had won that argument he later went into her room and ripped up her favourite poster of a soap star in retribution. Although Louise appeared to have exercised control over her brother in terms of the defence of her belongings and space, he later ensured that this was repaid by the destruction of one of her prized possessions. Like Phil and his sister, Colin and Louise fight over and destroy (or threaten to destroy) not the games machines themselves, but other items which may be more closely concerned with personal space and identity. More pragmatically, one young woman said she would damage things belonging to her brother if he wouldn't let her use their games machine rather than damaging the machine itself because 'I wouldn't be able to play with it then, would I?'

Carl and Simon, two teenagers, also have conflicts with their siblings over the games machine. Carl has two older sisters, and Simon has an older brother.

SM: Do you ever fight with your brothers and sisters about . . . .
Carl: Yeah, my sister beats me up.
SM: Your sister beats you up?
Simon: My brother has a go at me, 'cos if he wants to go on the machine I'm on, but he wants to play a different game, like, I don't wanna get off my game so he can have a go so we end up having a fight.
SM: Who wins?
Simon: My brother, now, 'cos he's a kick boxer! [laughter] So I have to watch it!
SM: What about you, Carl, you said your sister beats you up.
Carl: Yeah. [unclear]
SM: So does she get her own way – does she get to play on it?
Carl: Yeah, probably.
Simon: [giggling at Carl] Does she? [surprised tone]
Carl: Yeah.
Simon: [to Carl] Does she win you if she wants a game?
Carl: Yeah mostly, 'cos she always punches me in the arm and dead-arms me and then mum goes [puts on a high pitched whine] 'now then stop it you two' and then I always get into trouble.
SM: Does your mum ever threaten to take it off you?
Carl: [nods]
Simon: She does with me.

Carl's older sister 'beats him up' in order to get access to the game machine. This entertains Simon, who keeps asking him about it and giggling. Simon was not surprised at the news that Carl and his sister have physical fights, but his astonishment is at the knowledge that Carl's sister wins the fights. Carl resolves his discomfort at this seeming loss of face by both stressing the extent of his injuries and mimicking his mum's reaction to the conflict. Similarly, Simon lets us know that he can't hope to win against his brother because 'he's a kick boxer'. Both of these young men attempt to save face during this exchange.

The conflicts around the machine are not simply, then, to do with wanting to use the games, but involve a whole set of power issues which in the recounting of these tales extend outside the home. Powerless against older siblings, these young people attempt to re-present themselves in a different light to their peers, and to the interviewer. Colin makes it clear that he ripped up his sister's poster for revenge, Phil says his mum usually blames his sister, Simon can't compete against a kick boxer and Carl presents his mum as whining and himself as stoically accepting injury. In this way, the stories told to me by these male youths about the events at home can be seen both as an attempt to regain control over those events, and also as re-presenting themselves as more 'masculine' – they feel that they cannot be seen as 'losing' in a conflict with their sisters – than might be inferred from their stories. A focus on computer and video games shows, then, how sibling conflicts may permit the performance of gender relations in the home and that this perform-ance can be reconstructed to suit a different audience.

However, access to video game machines is not only an issue of personal space, for, even where the machines are located in a family room, conflict arises. Amy first told me that she never had to ask permission to play video games because the family has televisions which can be used for the games console in two rooms. Whoever wanted to use the machine could set up in either room. I asked her whether she liked to play on her older brother Tom's video game machine. She told me that she did, and that she and her younger brother Paul played it the most, unless Tom was in the mood to play it himself. She went on to describe the ways in which she and Paul were at the mercy of Tom's whims:

> *SM*: What do you think of it, Amy? Do you like it, or have you gone off it since Tom first got it, or . . . .
> *Amy*: I quite like it actually, but I just . . . it's just when my brother's on it, he says 'oh you can play on it' then he just faffs about, then he goes 'oh I'll have it back now', so really, he's just messing about with it.
> *SM*: So you just get going and he pinches it back?
> *Amy*: Yes, but sometimes I play on it when he's out, me and Paul.
> *SM*: So you get a chance to play it when he's out?
> *Amy*: Yes, but when he comes back he goes 'get off it' so you get off, then he goes 'oh, play on it then' and then he goes to the other room and plays on his. He doesn't bother when I play on it, but if it's like his best game like 'Streets of Rage' or something that he's just got off someone, he won't let no-one play on it, so he said only I can play Sonic, he said.
> *SM*: So he won't let you play on anything but Sonic? Why?
> *Amy*: Yes, Sonic 1, 2 and 3. He won't let me play the other ones 'cos they're all fighting games and he says 'oh, they're too boring for you'.
> *SM*: And what do you think?
> *Amy*: I think some of them are quite good.
> *SM*: Do you play them when he's not in, then?
> *Amy*: Yeah, sometimes. He's taken most of them back to the owners, but we've still got one, but he's hidden it.

Tom had bought the games machine a year earlier, and after getting a newer computer which he used more, the games machine had been designated for family use. However, it would appear that access to the machine is still controlled by Tom. Amy and Paul use the machine when he's out of the house. If he comes in while they are playing, he reasserts his control over the machine by telling his younger siblings to 'get off it' and then after this display relenting and allowing them to play. That this control over the machine may be about sibling rivalry can be seen when Amy says 'He doesn't bother when I play on it', so perhaps it is more important to Tom to stress his dominance over access when the younger boy in the family is also present.

It is not only access to the machine which Tom attempts to control but also access to particular games. Tom only allows Amy to use Sonic, a cartoon-type game featuring a 'cute' blue hedgehog. Violent games are removed and hidden by Tom, although again Amy plays them when he is not present, and enjoys doing so. Amy's use of language is very telling here. She says 'we've still got one' (a violent game) thereby identifying that she sees the game as belonging to all the family (at least until her brother takes it back to the person he borrowed it from), but 'he's hidden it'. He has the power to remove the game and hide it so that Amy can't have access. Mica Nava has stated with regard to youth cultures:

> Girls are less of a problem on the streets because they are predominantly and more scrupulously regulated in the home. It is therefore not only through the family, but also through the interaction of girls with boys outside it, that the femininity and thus the policing of girls is assured.
>
> (Nava, 1992: 79–80)

Tom appears from Amy's description to be policing her femininity by not allowing her to play with violent games, although Amy subverts his control by playing, and enjoying a game, when he is not present. This subversion is not in any sense an act of power on Amy's part, however. Because she can only play when and at what she chooses *when her brother is not present*, she is both spatially and temporally on the margins. Girls, then, can only use domestic space on their own terms when boys are not there.

CONCLUSION: POWER AND CONTROL IN THE HOME

David Morley's analysis of families watching television offers parallel insights into the processes I have reported here. He found that the men in the family 'possess' the remote control unit, or 'zappa', for the television set, and use it to control what can and cannot be watched, often interfering in and obstructing the wife's viewing. When the father is absent, then the control of the zappa passes to the son. Morley sees this as reflecting a particular social construction of masculinity (1986: 148–9). Morley informs us that the way that men in the families he studied talk about their control over the zappa and ultimately the TV habits of the rest of the family demonstrates a fragile power. He states:

> It is noteworthy that a number of the men show some anxiety to demon-strate that they are 'the boss of the household' and their very anxiety around this issue perhaps betokens a sense that their domestic power is ultimately a fragile and somewhat insecure thing, rather than a fixed and permanent 'possession' which they can always guarantee to hold with confidence. Hence perhaps the symbolic importance to them of physical possession of the channel control device.
>
> *(Morley, 1986: 150)*

From the material I have presented in this chapter the physical own-ership of computer and video games, and the physical control of space which arises from it, are also of *symbolic* importance to teenage boys in the domestic sphere. Young men are controlling and policing their sisters' access to computer and video games in the expression of their masculine identity. As I have argued, even when their control is success-fully resisted by their sisters or their mothers, in the retelling of the stories to me and others the young men re-present themselves in a more stereotypically masculine way. Like the men in Morley's (1986) study, these teenage boys seem anxious to demonstrate their power.

In the introduction to this chapter I asked what the implications for girls would be now that boys were increasingly to be found in the home. I have argued that the increased use of the home as a site of youth culture for boys means that girls' use of domestic space as a resistance to boys' domination of the streets (Griffiths, 1988) is now being eroded. Clarke and Critcher (1985) have argued that leisure in capitalist society is not simply a matter of choice, or of role identity, but is rather constrained by class, 'race' and gender. These identities constrain leisure choice in two ways: materially (in access to resources) and culturally (which includes perceptions about what is appropriate behaviour for members of partic-ular social groups). While the material I have reported here concerns only one of these social dimensions, gender, this chapter has illustrated that playing with home computer and video games has become another leisure activity which cannot be freely chosen for girls and young women, but to which access is constrained (for studies on women and leisure see Wimbush and Talbot, 1988). Computers and video game consoles are becoming, then, objects which are invested with power, and through which gender relations are expressed in the domestic sphere.

NOTE

1 The study I refer to here is for the purpose of a Ph.D. thesis, and has involved the delivery of a questionnaire to 1,600 children and young people, which was followed up with in-depth interviews of young people and parents.

REFERENCES

Büchner, P. (1990) 'Changes in the social biography of childhood in the FRG', in L. Chisholm, P. Büchner, H. Kruger, and P. Brown, *Childhood, Youth and Social Change: A Comparative Perspective*, London: Falmer Press.

Clarke, J. and Critcher, C. (1985) *The Devil Makes Work: Leisure in Capitalist Britain*, London: Macmillan Education.

Corrigan, P. (1979) *Schooling the Smash Street Kids*, London: Macmillan.

Dominick, J.R. (1984) 'Videogames, television violence and aggression in teenagers', *Journal of Communication* 34, 2: 136–47.

Education Guardian (1993) 'Video Games', 18 April, pp. 8 9.

Funk, J. (1993) 'Re-evaluating the impact of video games', *Clinical Pediatrics*, February 1993: 86–90.

Griffiths, V. (1988) 'From "playing out" to "dossing out": young women and leisure', in E. Wimbush and M. Talbot (eds) *Relative Freedoms: Women and Leisure*, Milton Keynes: Open University Press.

Haddon, L. (1992) 'Explaining ICT consumption: the case of the home computer', in R. Silverstone and E. Hirsch (eds) *Consuming Technologies: Media and Information in Domestic Spaces*, London: Routledge.

Hall, S. and Jefferson, T. (eds) (1976) *Resistance Through Rituals: Youth Subcultures in Post-war Britain*, London: Hutchinson.

James, A. (1993) *Childhood Identities: Self and Social Relationships in the Experience of the Child*, Edinburgh: Edinburgh University Press.

Jeffs, T. and Smith, M. (eds) (1990) *Young People, Inequality and Youth Work*, Basingstoke: Macmillan.

Jones, C. and Wallace, C. (1992) *Youth, Family and Citizenship*, Milton Keynes: Open University Press.

Kaplan, S.J. (1983) 'The image of amusement arcades and differences in male and female game playing', *Journal of Popular Culture* 1, 17: 93–8.

Kubey, R. and Larson, R. (1990) 'The use and experience of the new video media among children and young adolescents', *Communication Research* 17, 1: 107–10.

McRobbie, A. and Garber, J. (1976) 'Girls and subcultures', in S. Hall and T. Jefferson (eds) *Resistance Through Rituals: Youth Subcultures in Post-war Britain*, London: Hutchinson.

Mitchell, E. (1985) 'The dynamics of family interaction around home video-games', *Marriage and Family Review* 8, 1–2: 121–35.

Morley, D. (1986) *Family Television: Cultural Power and Domestic Leisure*, London: Comedia/Routledge.

Nava, M. (1992) *Changing Cultures: Feminism, Youth and Consumerism*, London: Sage.

Provenzo, E.F. (1991) *Video Kids: Making Sense of Nintendo*, Cambridge, Mass.: Harvard University Press.

Raffaelli, M. (1992) 'Sibling conflict in early adolescence', *Journal of Marriage and the Family* 54, 652–63.

Schutte, N.S., Malouff, J.M., Post-Gorden, J.C. and Rodasta, A. (1988) 'Effects of playing videogames on children's aggressive and other behaviours', *Journal of Applied Social Psychology* 18, 5: 454–60.

Wheelock, J. (1992) 'Personal computers, gender and an institutional model of the household', in R. Silverstone and E. Hirsch (eds) *Consuming Technologies: Media and Information in Domestic Spaces*, London: Routledge.

Wimbush, E. and Talbot, M. (eds) (1988) *Relative Freedoms: Women and Leisure*, Milton Keynes: Open University Press.

●●●●●●●●●●●●●●●●●●●●●●●●●●●●●

13

THE SCHOOL: 'POXY CUPID!'

an ethnographic and feminist account of a resistant female youth culture: the New Wave Girls

•

Shane J. Blackman

INTRODUCTION TO THE NEW WAVE GIRLS

The New Wave Girls were a high profile academic and resistant youth cultural group inside a secondary school. There were ten New Wave Girls: Clare, Debbie, Sioux, Cathy, Sally, Lynne, Cat, Collen, Steff and Denise; and in addition nine marginal members of the group. The New Wave Girls' style had two features, firstly music: the girls were into new wave, punk, reggae and dub. Secondly, appearance: in a number of ways the New Wave Girls held to the idea of punk clothes, for example 'confrontational' dressing through oppositions such as wearing a skirt and Doctor Marten's (DM's) boots under a 'dirty mac'. The girls predominantly wore trousers which were usually black, with T-shirts of various types, monkey boots and large jumpers. The girls sometimes made and adapted their own clothes and shoes. A significant feature of the girls' style was omission of the iconography of sexual fetishism associated with punk. The girls questioned the dominance of sexual fetishism and inverted the meaning of its expression. Thus, the safety pin or DM's were used to establish an alternative mode to the dominant forms for teenage girls (see Widdicombe and Wooffitt, 1995).

Their appearance was a combination of challenge and alternative practice. The girls were differently disposed to the use of make-up. Most of the girls would wear little; others would use make-up in a

non-conventional manner and others banned it because of animal test-
ing. The New Wave Girls did not conform to traditional markers of
female beauty. Hebdige (1979: 123) states, 'Behind punk's favoured "cut
ups" lay hints of disorder, of breakdown and category confusion: a
desire . . . to erode racial and gender boundaries'. The New Wave Girls
seized the independence and diversity within the meaning of punk and
new wave to challenge female passivity and to reinforce their own female
solidarity.

The New Wave Girls' youth cultural style and inappropriate school
uniform made them a highly visible group. Their stylistic solidarity and
very close physical contact emphasised their confidence and strength
(Thornton, 1995). The majority of them had boyfriends who were either
in the upper sixth form or college, or in employment. The girls had
experience of working in the local labour market, on Saturdays, Sundays
or midweek and during the school holidays. The girls were were in the
upper or middle band of the streaming system, taking O level examinations: they were liked by many of the teachers because they were assertive in the classroom. Their parents were lower middle class and working
class. The New Wave Girls' shared experience of family break-up and
changes in household arrangements made a strong impact upon the girl
group as a whole. Parents gave support to the group by allowing greater
freedom of action. In exchange for independence the girls would accept
domestic responsibility: this exchange relation was a recognition of the
strength of the girl group as a whole. However, it also implicitly carried
parental expectations that the girls in the group were able to regulate
their own behaviour.

The aim of this chapter is to give an ethnographic and theoretical
account of the way in which a female youth cultural group, the New
Wave Girls, was able to resist masculine control within the space of the
school. The chapter gives initial priority to the ethnographic description
which places an insistence on lived meanings, as Willis (1982: 81) states,
'for the purpose of making theoretical argument rather than merely
raiding the evidence for appropriate illustrations of particular points'.
The strategy will be to put forward a series of ethnographic instances of
the girls' actions and performance in different social settings inside and
outside the location of school. These ethnographic descriptions will be
used as the basis for the later development of a theory of the rituals of
integrity. This theory uses as its foundation the feminist praxis of Mary
Daly (1978) who elaborated the idea of emancipatory integrity.

MEETING THE NEW WAVE GIRLS:
THE MAKING OF AN ETHNOGRAPHY

The fieldwork for this study took place inside and outside the context of a comprehensive school in the south of England called Marshlands during the 1980s; the wider findings of this comparative ethnographic study were published in Blackman (1995). I had been in the school for approximately two weeks and begun to establish useful communications with a number of male groups, one in particular was a group of 'mod boys'. On the basis that the mods knew I was undertaking a study of youth culture I asked them if they could introduce me to a group of girls and during that morning break I met the New Wave Girls. I was formally introduced and the boys left. Rather nervously I explained my broad intention and they seemed partially interested. Two days later while playing cards at lunch time with three skinhead youths I received an invitation to join the New Wave Girls outside the school gates to spend an afternoon at the café instead of attending normal lessons. This was obviously a test of trust. As we sipped cups of coffee and they smoked cigarettes the conversation embraced many topics from the subject of my research to music, style, sexism, sexual activity, and problems with teachers and parents. This conversation was directed by the youth group and although I did make certain contributions I had no intention of asking formalised or intrusive questions (Fyfe, 1992).

During the next few weeks and months I slowly began to build up closer relations with the girls, through eating school dinner with them, walking home with them after school and being asked to meet with the girls at lunch time to walk around town and go to the graveyard, where they were in the habit of gathering. Within a period of three months I was brought into a much closer relation with the girl group and individual girls within the group. It was now possible to visit the girls at their parents' houses, talk with parents, chat with the girls in their bedrooms and play music, go to discos, public houses and jumble sales or go shopping with them. The ethnographic field work also included attending live concerts and, importantly, their all night parties and all girl gatherings.

The interviewer's gender is important to the research process for feminists such as Oakley (1981) and Wolpe (1988) because it could result in the generation of different kinds of knowledge. In contrast, for male researchers of youth such as Frith (1978: 66), Jenkins (1983: 20) and Mac an Ghaill (1994: 152) the research issue concerning male exclusion from female spaces was more fundamental.

For male researchers who study women the ethics of research not only relate to the form that their contact with females takes but also to how the male interprets and writes the female within the text and how he claims ethnographic authority on the subject of women given the existing power differential within society. By control over research and writing male researchers are actively creating the 'Other' even if at face value women are the subject of the research. Such research involves putting forward their own particular interpretation using as a support data registered in a female voice (Stacey, 1988: 23).

One strategy used to reduce the formal division between researcher and researched is to enlist the help of respondents to read field notes and comment on them (Wellman, 1994: 581). Another technique employed by certain male researchers is to consult female academics about their interpretations of women (Mac an Ghaill, 1994: 173–4). In this study both strategies were used. The New Wave Girls were given access to the field diaries; they offered points of clarification which were then incorporated into the data for consideration. On the same basis collaboration was sought with female friends and academic colleagues to consider both raw data and initial drafts. Admittedly this form of research collaboration cannot solve the inequalities and exploitation endemic to social research, but it can allow male ethnographic studies to be influenced by the application of a feminist perspective.

A fundamental feature of my relationship with the girls was that it was not based on sexual relations (Warren, 1988). However, in different contexts the girls would expect different relations; for example, support in their anti-sexist arguments against certain boys or in their own counter-attack. On the one hand the girls would expect constructive advice, discussion or helpful guidance, while on the other hand, they would expect acknowledgement of their sexual attractiveness and desirability. The two roles may be contradictory but were fundamental features of my field relations with the New Wave Girls.[1]

FEMALE RESISTANCE? THE NEW WAVE GIRLS

The development of resistance theory in relation to youth culture by the Centre for Contemporary Cultural Studies (Hall and Jefferson, 1976) brought together sub-cultural and structuralist theory through an expansion of the theoretical ideas outlined by Phil Cohen (1972). The structuralist approach is to examine the order or structure underlying general meanings and in general it places less emphasis on empirical research studies (Craib, 1984: 108). Within *Resistance Through Rituals* (Hall and Jefferson, 1976), 'literary ethnography' on the three main sub-cultures, teds, mods and skinheads, was substituted for direct empirical observation of young people. The theorists took up the concept of sub-culture using secondary sources of data which predefined youth cultures as a social and moral threat. In the CCCS study, young people's experience of youth culture was not the paramount focus; priority was given to the structuralist approach to examine youth culture for a coherent system of meaning and messages related to resistance.

It is in this sense that the CCCS upheld an assumption of structuralism derived from Lévi-Strauss where the theoretical analysis is claimed as showing the underlying structure of meaning within (all) youth culture. The problem is not resistance theory per se, but rather its application as the basis for youth studies which became theory-led, with only a thin basis in empirical data collection. Thus resistance as a theoretical concept has as its major weakness a thin foundation in empirical work, upon which grand theoretical constructions have been erected (Giroux, 1992, McRobbie, 1994: 180).

Here I want to argue that the term resistance is ethnographically useful to explore the New Wave Girls' different displays because it shows their real degree of struggle and challenge with respect to family, school and masculinity. The New Wave Girls both held and occupied territory inside and outside the school. In the classroom and wider pupil spaces within the school, the girls sought to present themselves as both cultural occupiers and social definers. This form of assertion should be seen as part of their youth cultural practice to 'promenade'. However, such performances were no ordinary posture. It is suggested that this resistance is experimental and closely related to their developing feminism. It is a partially articulated female resistance, designed to establish varied spatial locations where the girls could control and assert their developing power. Here, I shall look at some of the New Wave Girls' practices and relations, in order to present the making of a female youth cultural practice. The next section will focus on a number of conversations and ethnographic examples in an attempt to explore the girls' different forms of resistance.

SKIVING

A central feature of ethnographic studies on schooling which apply the theory of resistance to youth culture is the assertion that resistant young people engage in truancy (Willis, 1977; Mac an Ghaill, 1994). Absence, truancy or skiving are terms denoting non-attendance at school for pupils up to the age of 16. There is an ambiguity within the meaning of truancy, for to 'play truant' is to stay away from school without leave, but to be a truant also means one who shirks or neglects duty. The New Wave Girls who were in the upper ability band did engage in skiving but it was always a highly selective event, for example 'We don't want to watch some boring police film about piss artists driving over people, do we', or 'We've completed this section of the course, so we might as well do something more constructive'. I attended a number of skiving sessions at the home of New Wave Girls who lived near the school. Sometimes I would set off with the girls when they made a decision to play truant, at other times I would receive a message through the 'grapevine', and make my own way to the house. During these times of absence I usually found the girls helping each other to solve problems in maths, completing history, geography or English literature homework, revising for examinations or filling in forms. Obviously, sometimes they did no work, and just drank coffee, chatted or listened to music, but even these conversations were no idle waste of time.

However, the girls' skiving was not always a happy occasion. Once they were almost caught. On Wednesday morning a group of them were gathered at Clare's house. There was a bang on the door, and the loud voice of the Head of Fifth Year Girls, Mrs Arthurs, rang out; she was with the Truancy Officer. Inside, the girls hid in silence. The banging continued for some time before the girls could get away through the back garden. Within the girl group this incident caused some initial ill feeling towards those in the group who were not in the house. However, later in the day and again early next morning, two of the girls who were in the house repeated that they were frightened and would think hard about when and where to skive next. Afterwards, the event became a tale of excitement to enhance the girls' resistant promenade. No other pupils had been able to deceive both a Head of Fifth Year and the Truancy Officer.

Overall, the New Wave Girls generally skived in one another's houses, and utilised the time in a manner not consistent with the term truancy. Their type of skiving is best described as creating an alternative frame of reference to that officially designated by the school. They developed a strategy for skiving which was based on manipulating the organisational structure of the school, to support the educational aims they had in common with the school.

It is possible to argue that the girls' challenge to school authority through the wearing of inappropriate school uniform and their acts of truancy were identifiable markers of resistance. These public acts of resistance were combined with local challenges or contestation within the classroom against teacher control and the sexism of male pupils. Here the girls would invert resistance into articulate forms of accommodation which included displays of their ability to successfully accomplish work in the classroom and achieve high grades for homework. Therefore the girls' actions such as truancy, or smoking cigarettes at breaktime in school were not single acts, but formed a part of wider intentional pattern of resistance developed to support their youth cultural discourse.

GIRLS TOGETHER: LESBIAN DISPLAYS AND PHYSICALITY

One of the boys in the school described the New Wave Girls:

'You lot of girls are fucking weird. You're always holding arms, anywhere, every time. When you say goodbye even for different lessons you kiss each other and touch one another. It's fucking crap. I'll tell you something, me and the rest of us think it's fucking funny and stupid, the way you lot act. Who do you reckon you are?'

(Kevin)

This aggressive and defensive statement by one of the 'hard nut' boys enables us to see what boys in general disliked about the New Wave Girls' behaviour. The girls would link arms when walking at school, they would also kiss goodbye, sometimes when changing classes and always at the end of the school day. This public bodily contact continued during leisure hours outside school (see Box 13.1).

Here are a couple of the girls together stories. At these gatherings the girls would sleep together, smoke cigarettes and sometimes cannabis, drink alcohol (cider, wine, martini) and engage in long conversations. When I stayed overnight at some of the girls' houses it was possible to observe this. It is a distinguishing characteristic that the girls would regularly sleep together in the same room. Sometimes they would tape record their conversations on a cassette before going to bed (Blackman, 1995). These tape recordings are an example of Skeggs' (1991) consideration of girls' 'safe places', where they can talk in private without male interruption or control. The New Wave Girls had access to a whole range of female only safe spaces from the toilet to 'all girl gatherings' (Llewellyn, 1980; Griffiths, 1995).

The New Wave Girls' level of intimacy was intense, their collective rituals and basis of group behaviour supported and strengthened solidarity. By revealing, sharing and accepting their bodily vulnerability and pleasures the girls acquired information about their own sexual responses (Pringle 1988). In this respect they are unlike the girls in the two studies by McRobbie (1991) and Lees (1986). Physical closeness was part of their internal group relations, it was also used as a promenade of resistance to promote particular messages.

Box 13.1

Cathy	It's lovely and warm in here isn't it.
Debbie	Right, last night in London, it was. This is me and this is Cathy. [practical demonstration] All of a sudden she just decided that she wanted to lie over me. Because my – because it was nice and warm on my side of the bed, and then I went like that over her. And then in the end I was over here curled up in the eiderdown, no pillows, just one blanket. She's got about six billion pillows and all the blankets, and the electric fire was over your side.
Sioux	It's like at Sally's party. We all sort of staggered in like this and we were lying there. We had the two air beds on the floor and somebody was on the sofa. Lynne and I were mucking around, pulling around, sitting around and she was thinking about spilling her coffee and she just sat there and she spilt it all down her and she goes isn't it funny how, when you think about spilling your coffee you do it. And then we all got into bed. There was Sally and Pat opposite us, like, because we had them head to head you see. Anyway, Steff was talking in her sleep 'oh, oh Gaz' – patting the sofa. And next morning Sally woke up and she goes where's Pat like this. And apparently Sally had kept rolling nearer and nearer to Pat in the night, so that Pat had got out of bed and got in the other side!
All girls	[laughter]
Sioux	You can imagine, so funny.

SJB: There'll be some life in the sixth form with you lot there.

Sioux: Yeah.

Cathy: Yeah, us two lesbians.

Sioux: Me and Sally all over the floor together. My hip hurts, right there. Is that your hip?

Clare: Yeah.

Cathy: Did you hear what Gaz [boyfriend] said?

Sioux: What did he say?

Cathy: He goes, oh I have been hearing strange things about you and I goes what? [expression of leading him on] He goes, oh it is going round the sixth form that you two are becoming lesbians, and he said, no, really, he goes I don't believe it but you know that the 'stiffs' [straight people] do.

Here they discuss how their close physical relations have been understood as lesbianism. I have decided to call the girls' open promotion of lesbian behaviour 'lesbian displays'. What is crucial is the reason for these 'lesbian' performances. In the girl group there were two apparent lesbian pairings, Sioux and Sally, and Cathy and Debbie. The girls did not identify fully with lesbianism, as each of these girls had boyfriends. Therefore it is possible to suggest that the aim of these 'lesbian displays' was to initially define personal lesbian relations but primarily to exclude others, promote group unity and to strengthen close relations between pairs of girls. Earlier we saw that boys felt threatened when observing the girls' physicality. 'Lesbian' displays frighten the boys because they render their masculine sexual bravado pointless (Nava, 1982; Batsleer, 1996).

The New Wave Girls did not see menstruation as a taboo. Inside the group a girl would say she had started her period; outside the group some of the girls would occasionally tease or try to humiliate a boy by stating 'Do you know I'm on'. A number of the girls realised the potential use of the subject of menstruation because boys feared it (Lees, 1993; Prendergast, 1989). Therefore to mention their period provoked images of defilement and consequently disrupted a boy's pattern of sexist abuse. However, they were not immune to intimidation by boys, but abuse was less of a problem for them than for other girls. The girls' lesbian displays and close physical contact was an everyday feature of their relations and bodily stories were normal events. They possessed an ability to 'shock' both boys and other girls by their forthright discussions, stories and jokes about female (and male) bodily functions normally understood as taboo.

Taken together the New Wave Girls' challenge to compulsory heterosexuality, at the school, brought forward opportunities to consider alternative notions such as bi-sexuality, celibacy and lesbianism. Some of these different forms of gender performance have also been identified by Holland *et al.* (1993) and Epstein (1994) whereby the young women can contest the normalised male authority within schooling. However, both Griffin (1993) and Cockburn (1987) have shown the severe limitations of young women's forms of resistance towards masculine defined heterosexuality. For Cockburn lesbianism operates within school culture largely as a term of abuse which ensures that girls see close female friendship as difficult or negative. The New Wave Girls' sexual identity was composed of a multiplicity of messages, including their promotion of lesbian ideals and relations, also combined with their public hetero-sexual relations with sixth form boyfriends. These different forms of gender identity were extended further by their specific youth cultural style and context for group based activities: control of physical and cultural territory presented the girls with the opportunity to promote different gender performances.

DRINKING ALCOHOL

At the ages of 15 or 16 the New Wave Girls who consumed alcohol were breaking the law. The places where the girls would regularly drink were public houses, wine bars, parties, gatherings in a girl's house, and discos or gigs. Additionally three of the girls had jobs in local public houses. The New Wave Girls identify the pub as a positive place and this was demonstrated throughout the fieldwork by many visits to pubs or wine bars. For the girls under age drinking and the telling of drinking stories were crucial aspects of their promenade as a youth cultural group. This shows their territorial movement and place in the community; the New Wave Girls were recognised. Occasionally, the girls would go to The Bear during lunch time at school, though this was not a regular activity, as the majority of their 'heavier' drinking sessions took place at parties or all girl gatherings (see Box 13.2).

Party stories and drinking stories were common currency among the New Wave Girls. This was in contrast to other girls at the school who only entered a public house with a boyfriend (see Coggans and McKellar, 1995). Amongst other pupils within the school the New Wave Girls' reputation for drinking alcohol in public houses, all girl gatherings and at their all night parties created a series of myths due to the circulation of rumours in school. The all night nature of the New Wave Girls' parties and all girl gatherings, which took place without the presence of parents, fuelled intrigue among other pupils. Due to the different sex composition of parties (male/female) and gatherings (female) the girls were able to secure heightened interest as outrageous stories and rumours would snowball while the real meaning of events remained with the party participants only.

For pupils at the school the New Wave Girls' parties held a powerful curiosity value because they suggested an 'adult world' which was not accessible. This gave the girl group a powerful resource to develop and elaborate which further secured status and expressed their disaffection, thereby demonstrating resistance. Attendance at a party or gathering was strictly controlled by the girl group and at some of the events I attended only selected marginal members of the group were invited. This

Box 13.2

Debbie	Do you realise that time we got drunk, we didn't cry?
Cathy	I BLOODY DID. Well, we didn't together, I cried after I had been sick on the dog.
Sioux	Well hold on, at that [party] Lynne was sitting there, she was rocking in the rocking chair and crying 'I want Julian', like this and I was so pissed off. I will never forgive myself. I sat there and burst out laughing. I goes 'you can't fucking have him because he's not fucking here', and then I burst out laughing.
Cathy	When you're drunk, you say some really evil things you do. At Clare's party when I was crying, you goes 'Oh for fuck's sake SHUT UP'.
All girls	[laughter]
Sioux	All right. Cathy and Debbie had being crying and Cathy came out and she was sitting on the stairs, and my mother had had a go at me, and I was depressed. So I said what's the matter Cathy, I put my arm round her and she told me why and then I thought she would stop crying, now she's confided in somebody. But she wouldn't stop crying. Anyway, she sat there still crying so I got up and said 'For fuck's sake stop bloody crying', and I walked off.
Debbie	All night long, all we did was cry.
Cathy	Oh bloody hell that was terrible!
Sioux	Cathy was going, 'Where are my cocktail cigarettes. I want something very special, I want to give one of these to' – and she was howling at me and she thought I had lost them and they were only on the side.
Debbie	She was really funny when she was pissed. Walking into the kitchen, staggered in, there are all those boys sitting round and she goes 'Hello Gaz' and fell over.
All Girls	[laughter]

SJB	Classic entry.
Debbie	Then she is – was leaning over talking to him and she just toppled over. It was really funny. Oh sorry.
Sioux	Just like Sally the other night, Sally came in, that night we had all been down to Sally's [all girl gathering]. And there was this ashtray right by the door. She comes in, she goes huuhh and she stood straight in this ashtray with all dogends in there and it went crack and it fell all over the place.
Cathy	Yeah when I walked out of Sally's place we were walking up the road and it was still really dark and we were talking and I goes Sioux I think someone has dumped an ashtray in my pocket. There was about fifty dogends in there. I got home and I was going like that, there was about three handfuls, all these matches. I goes hold on a minute Sioux. I've got the wrong coat on.
All girls	[laughter]
Sioux	So we had to go all the way back.
Clare	Whose coat was it?
Cathy	Sally's with an ashtray in her pocket.

kept the real basis to their drinking stories, sexual behaviour and use of cannabis closely tied to their own group. What other pupils within school failed to discover was the degree of negotiation which the New Wave Girls entered into with their various parents in order for parties or gatherings to occur. It was possible for the girl group to gain significant opportunities for independence on the basis that the girl group was recognised by parents as being able to police their own activities. Thus parties and gatherings can be identified at one level as a means to display resistance in terms of access to drink, drugs and sexual activity. However, at a deeper level it is possible to see the girls' events as opportunities to develop co-operation: we can identify a rule of the girls' internal group relations: do not hold your feelings back, show your real self. Here the girls' private displays of honesty by crying and consoling one another and also by being able to laugh at their own vulnerability are an essential dynamic feature of the group. Parties and gatherings were a chance to experiment, share and cope with vulnerabilities, promoting group identity.

Box 13.3

TOSS CHOPS UNITED

If we could show how much we want to hate
We'd make you live in hell
But you're just so fucking boring
We're just too scared to tell.

Your psychedelic talks on music
You really think you know
You worship all the poxy groups
And go to every show.

At parties you smoke pot
That's the 'Hip' thing to do
Tiring quickly of your old friends
Searching for someone new.

But let's just get this straight.
You're the ones 'who're' stupid
We never even tried to act
Or think like poxy Cupid.

'copyrite' Sioux and Debbie

WRITING POETRY

In Box 13.3 is a poem composed by the New Wave Girls concerning all their boyfriends. On the page for each stanza the two girls had drawn large heart shape bubbles, within which were written together two boyfriends' names, such as Slim and Peter or Stephen and Gaz, with a comment alongside stating 'true love forever', '4 ever and ever' and 'true love'. These phrases, taken from teenage girl magazines, suggest that their boyfriends are in love with each other.

The poem and love bubbles are an active response by the New Wave Girls, heavy with irony, wit and contradiction. The title is a play on words which ridicules male sexuality – toss – to masturbate, and male leisure – united – is a football team, e.g. West Ham United. Put together the negative assertion is a doubly weighted reference that defines the boyfriends as collective masturbators or 'Toss chops united' (chops = mouthy, or rude). The first verse establishes the girls' intention, they have minds of their own, refuse to be taken for granted and find boys tedious and boring. The second verse focuses upon how their boyfriends un-critically accept things, blindly assert that they know everything about today's music and discount any girl's contribution. The third verse is critical of the boys' self-centred behaviour and fantasies induced by smoking cannabis which interfere with social relations (Blackman, 1996). It makes them selfish and fickle in their friendships: reinforcing the meaning of the title, in that it implies that the boys use their relationships only as a means of self-gratification. The final stanza also suggests their autonomy in that it redefines the relationship they had with the boys, thus countering and asserting as false the boys' assumptions that they were 'in love' with them. Thus it rejects tradi-tional notions of romantic love. It explicitly reverses the patriarchal opposition of male = knowledge and female = emotion, thus countering the male symbolic order upheld by the boys (Cixous and Clement, 1991).

The poem and their celebration of it reclaim the right to speak and make definitions denied them by masculinist society. This is also affirmed as the aim of the poem in the first stanza which asserts the way in which the girls cannot show their real feelings or 'tell' what they think, i.e. it emphasises the restrictions placed on them. The boys exercise a right to speak ('Your . . . talks'), which is denied the girls. Both in the poem and their public conversations the girls effectively break a double taboo, since within patriarchy not only are girls not supposed to possess such sexual knowledge but also they are denied the power of speech on these issues (Shiach, 1991).

Thus far throughout the chapter we have seen the girls' resistant behaviour includes drinking, lesbian displays, cigarette (and cannabis) smoking and truancy all of which are examples of acts of challenge against power relations across four sites: sexuality, family, school and society. Forms of oppositional behaviour do not in themselves demon-strate resistance (Roman et al., 1988; Amit-Talai and Wulff, 1995), but taken together these examples can be described as aspects of the New Wave Girls' resistance, and could be described as an anti-patriarchal practice. It would be an over-interpretation to argue that their truancy, drinking or wearing non-regulation school uniform indicate this, although these aspects of behaviour show defiance, independence, asser-tiveness and creativity, not always associated with teenage girls (Pearce, 1996; Walkerdine, 1984). However, the girls' cultural practice did en-courage critical thinking, and was feminist in the sense of prompting them to question male dominance. Thus it is possible to argue that the concept of resistance is potentially useful to explain and interpret the

New Wave Girls' actions because the underlying principle of their anti-patriarchal practice is emancipatory.

THEORY OF THE RITUALS OF INTEGRITY: NEW WAVE GIRLS

On the foundation of the ethnographic data outlined I want to take the discussion on resistance further by offering a theoretical interpretation of the New Wave Girls' resistance in terms of feminist theory. The New Wave Girls' rituals of integrity are a concentrated marshalling of symbols and practice. The rituals amount to a kind of economy in expressing personal affection, feelings and meaning. These rituals are a carefully developed means of communication constructed within the private face and extended to public face interaction. For example, throughout the fieldwork I observed only one major incident of conflict between the two (i.e. male and female) youth cultural groups. I was standing in the corner of the fifth year area talking with a couple of boffin boys. It was half way through afternoon breaktime and the pupils were milling round, as they usually do. Five of the 'mod boys' and some of the 'criminal boys' rushed into the area through the main doors, after secretly smoking a cigarette and began to pester other pupils and 'have a laugh'. The 'criminal boys' went back outside again, while the 'mod boys' stood in front of the girls. The boys started to show off, trying to intimidate the girls by shouting and touching them. To counter the abusive insults a couple of the New Wave Girls placed tampons in their mouths and moved rapidly towards the boys, thrusting out the tampon between their lips asking the boys for a light to their 'cigars'. The boys began walking backwards, stumbling and waving their hands in disturbed motions attempting to knock the tampon away from their faces. There was an uproar in the fifth year area, girls laughing, boys shocked and pupils attempting to get out of the way. Two days later I interviewed girls from a remedial class in the fifth year area and they described what the tampon incident meant to them. They did not censure the New Wave Girls as 'slags' for putting tampons in their mouths (Shacklady, 1978). They considered that for these boys it was 'up their's' and 'really amusing'. The girls in the remedial class 'hated' these boys and others, as they were on the receiving end of much sexist terrorism. Interestingly, they said 'we couldn't have done it'. One girl concluded that the New Wave Girls were 'leaders in style, the only real girls, group of girls, in the school'.

Thus far I have identified, firstly, the girls' close physicality, that is, their intimate private and public bodily contact, and secondly, their use of the female body, its representations and its signifiers to challenge male power relationships through an anti-patriarchal practice. In order to develop the analysis further I will make use of the term 'integrity', referring to the proposition by Daly (1978: 39) that radical feminism 'is affirming our original birth, our original source, movement, surge of living. The finding of our original integrity as re-membering our Selves'.

The aim here is to use the concept of integrity on a social basis to offer an insight into understanding the New Wave Girls' conscious actions and behaviour. Integrity refers to the socially genderised female physicality; the integrity of the female body is seen not in terms of an addition or an exclusion of parts or functions, but as an integral whole, a unity from which no part may be taken. For the New Wave Girls the body is a site of power and resistance whether with reference to the sexual or the physical. Their rituals of integrity are expressions and actions which use physicality and sexuality openly as a symbolic resource; as part of their whole cultural practice and relations.

On a metaphorical level the girls operate a cultural positioning of their bodies. They celebrate the positional structure of their group, whereby the group makes it possible to speak of the body, to think with it and through it. Integrity then is the use of the body as a means to create a code independent of the dominant gender code (Griffin, 1993; MacDonald, 1980). The girls speak about their legs.

Sioux: [to Sally] Oh look at those legs.
Sally: They're classic! They'll go down in the history books.
Lynne: We ought to start a rugby team because we all have fat legs.
All girls: Yeah.
All girls: [laughter]

This is a spontaneous group joke which plays on the conventional expectation that girls should have 'slim' legs. They are highly amused by the contradictions which they perceive here and indulge in a heavy irony. They could take part in rugby, an essential male context of competitive physical violence. The New Wave Girls create a collective identity through individual experience, assessment and interpretation of bodily problems. The New Wave Girls deployed conversational choreography which means that bodily tales may come in a variety of forms: stories, jokes, puns or ordinary conversation (Douglas, 1968).

In Box 13.4 the girls are in a humorous and nostalgic mood, relating stories about ritual bodily behaviour. Collectively they celebrate the dramas of bodily functions, suggesting that the body may be out of control; hence the inability to control 'shit velocity', farting and burping. The stories of the body, whether in control or out of control, reveal

Box 13.4

Sally	When I'm in the toilet having a fag [at home] and I'm flicking the ash when I'm having a cigarette. I'm flicking the ash when it goes sssch, sssch, schsssssch. I go [cough twice] or something. I try to time my shits to land at the same time as the ash goes down. But my shit velocity sometimes is greater than the ash.
All girls	[laughter]
Sally	And I miss – close.
All girls	[burping noises]
Sioux	All you've got to do is fart now.
Cat	That'll be no problem.
Sally	Oh do you remember that night you stayed at my house, we went down to the beach that night? Do you remember with Cathy and Bloc as well as Mick? When we came back to my house do you remember? I was doing one after the other. I had a whole tin of baked beans.
Lynne	What about when we were at camp then?
All girls	[laughter]
Lynne	Sitting there burping having saveloys.
All girls	[burping and farting noises]
Sally	There's all these farts.
Sioux	You should've taped it.
Sally	We're sitting there, all got our sleeping bag flaps out, five of us are all there like drunks slurping away at this cider.

how the girls share the body as a practice; it becomes socialised within the group. The mutual appreciation of the 'body out of control' not only binds them together, it shows their capacity to promenade. Paradoxically, the ability to apply this idea actually shows their control and use of the female body to deal with sexual contradictions. All the girls accept that they have the same bodily problems and functions – there is no differentiation.

The discussion of the girls' rituals of integrity illustrates their self-conscious celebration of the female body as natural. In other conversations they spoke openly about bodily functions and were not embarrassed by their bodily fluids. The girls demonstrated their power and skill in using the gendered female body and their sexuality openly as a symbolic resource of exclusion, opposition and independence. The New Wave Girls understood farting or menstruation as a natural expression of their female body and did not assess them as unclean, separate or even 'unfeminine': rather they understood the body in terms of integrity – an original unity from which no part may be taken (Daly, 1978).

CONCLUSION

Theoretically the social basis of the girls' resistant integrity is their group relations which create a space for individuals to engage in practices which can be considered as anti-patriarchal. The girls' celebration of their own sexuality as natural becomes the communal responsibility of the group, and it is through this wholeness that they begin to collectivise the female body and thereby re-member themselves towards a position of integrity. However, outside the context of the girls' group, patriarchal culture forces them into a position where they are made to respond to and challenge masculinity. The collective action taken by the girls allows them to confront the contradictions they meet daily, although the ground rules of this confrontation may remain on male terms (Holland, 1985; Jones and Wallace, 1992).

The New Wave Girls' resistance can be understood as becoming feminist. Their capacity to exploit and use different forms of sexual expression derived from their celebration of their sexuality as 'natural'. Their rituals of integrity: close and intense group relations created a powerful social base for opposing and challenging the patriarchal stance of the school hierarchy, male pupils and boyfriends. Sexuality is an area where male control over women is strongest and is exercised at the level of language: in naming, defining and denying speech to women/girls. The New Wave Girls' resistant discourse is about the right to speak and to define, and about identifying and challenging male control of language. The poem 'Toss Chops United' is an example of where they identify the power differential in language, between female and male use of meaning. The New Wave Girls, through their spoken interactions and written communications, reverse the symbolic order of language; they demonstrate the denial of speech and then set about reclaiming the right to speak, define and know.

Within the pupil community at Marshlands the New Wave Girls' sexual identity was powerfully controversial, owing firstly to rumours of unorthodox sexual/social behaviour arising during their parties, and also from their displays of 'lesbian' or homo-erotic behaviour. The New

Wave Girls entered into heterosexual relationships with boys without necessarily desiring marriage and motherhood. This is possibly an effect of their shared experience of family break-up, and of their household responsibilities which may have given these girls a more critical understanding of marital relations, and dispelled notions of romantic love. It might also be suggested that the break-up of the family which according to feminist analysis is the first site of woman's oppression had a positive result for the New Wave Girls, by giving them a sense of autonomy and independence from the traditional domestic pattern (Barratt and McIntosh, 1982; Ramazanoglu, 1989). The New Wave Girls' group was strongly cohesive and developed its own autonomous cultural identity of female sexuality. The girls' reputation for lesbianism amongst boys in the fifth and sixth years gave them protection from heterosexual aggression and sexist harassment. In this sense the girls' resistant play undermined the boys' masculinity by posing a specific threat to patriarchal social relations.

ACKNOWLEDGEMENTS

I would like to thank Professor Basil Bernstein for his support and encouragement, the head teacher at the school for his assistance and the New Wave Girls for always welcoming me. I would also like to thank the editors of the volume for their advice, as well as Debbie Cox for her critical comments and for reading this document.

NOTE

1 The field work was influenced by the early Chicago School researchers, including Nels Anderson, Frederic Thrasher and Vivien Palmer, and more recent feminist methodology on the importance of shared understanding between the researched and the researcher (see Padfield and Procter, 1996: 364).

REFERENCES

Amit-Talai, V. and Wulff, H. (eds) (1995) *Youth Cultures: A Cross-Cultural Perspective*, London: Routledge.
Barratt, M. and McIntosh, M. (1982) *The Anti-Social Family*, London: Verso.

Batsleer, J. (1996) 'It's all right for you to talk: lesbian identification in feminist theory and youth work practice', *Youth and Policy* 52 (Spring): 12–21.

Blackman, S.J. (1995) *Youth: Positions and Oppositions: Style, Sexuality and Schooling*, Aldershot: Avebury Press.

—— (1996) 'Has drug culture become an inevitable part of youth culture? A critical assessment of drug education', *Educational Review* 48, 2: 131–42.

Campbell, A. (1984) *The Girls in the Gang*, Oxford: Basil Blackwell.

Cixous, H. and Clement, C. (1991) 'The newly born woman', in M. Eagleton (ed.) *Feminist Literary Criticism*, London: Longman.

Cockburn, C. (1987) *Two-track Training: Sex Inequality and the YTS*, London: Macmillan.

Coggans, N. and McKellar, S. (1995) *The Facts about Alcohol, Aggression and Adolescence*, London: Cassell.

Cohen, P. (1972) 'Subcultural conflict and working class community', in *Working Papers in Cultural Studies*, CCCS, University of Birmingham, Spring, pp. 5–51.

Craib, I. (1984) *Modern Social Theory*, Brighton: Harvester.

Daly, M. (1978) *Gyn-Ecology: The Metaethics of Radical Feminism*, Boston: Beacon Press.

Douglas, M. (1968) 'The social control of cognition: some factors in joke perception', *MAN*, n.s.: 361–76.

Epstein, D. (ed.) (1994) *Challenging Lesbian and Gay Inequalities in Education*, Buckingham: Open University Press.

Fornas, J. and Bolin, G. (eds) (1995) *Youth Culture in Late Modernity*, London: Sage.

Frith, S. (1978) *The Sociology of Rock*, London: Constable.

Fyfe, N. (1992) 'Observations on observations', in R. Lee (ed.) *Teaching Qualitative Geography*, Research Paper No. 6, Queen Mary and Westfield College, University of London, pp. 127–33.

Giroux, H.A. (1983) *Theory and Resistance in Education*, London: Heinemann Educational Books.

—— (1992) *Border Crossings*, London: Routledge.

Griffin, C. (1993) *Representations of Youth*, Cambridge: Polity Press.

Griffiths, V. (1995) *Adolescent Girls and their Friends*, Aldershot: Avebury Press.

Hall, S. and Jefferson, T. (eds) (1976) *Resistance Through Rituals: Youth Subcultures in Post-war Britain*, London: Hutchinson.

Hebdige, D. (1979) *Subculture: the Meaning of Style*, London: Methuen.

Holland, J. (1985) 'Gender and class: adolescent conceptions of the division of labour', unpublished Ph.D. thesis, Institute of Education, University of London.

——, Ramazanoglu, C. and Sharpe, S. (1993) 'Wimp or gladiator', *WRAP/MRAP*, Paper No. 9, London: Tufnell Press.

Jenkins, R. (1983) *Lads, Citizens and Ordinary Kids*, London: Routledge.

Jones, G. and Wallace, C. (1992) *Youth, Family and Citizenship*, Buckingham: Open University Press.

Lees, S. (1986) *Losing Out*, London: Hutchinson.

—— (1993) *Sugar and Spice*, London: Penguin.

Llewellyn, M. (1980) 'Studying girls at school: the implications of confusion', in R. Deem (ed.) *Schooling for Women's Work*, London: Routledge and Kegan Paul.

Mac an Ghaill, M. (1994) *The Making of Men: Masculinities, Sexualities and Schooling*, Buckingham: Open University Press.

MacDonald, M. (1980) 'Socio-cultural reproduction and women's education', in R. Deem (ed.) *Schooling for Women's Work*, London: Routledge and Kegan Paul.

McRobbie, A. and Garber, J. [1975] 'Girls and subcultures', in S. Hall and T. Jefferson (1976) (eds) *Resistance Through Rituals: Youth Subcultures in Post-war Britain*, London: Hutchinson.

—— (1991) *Feminism and Youth Culture*, London: Macmillan.

—— (1994) *Postmodernism and Popular Culture*, London: Routledge.

Nava, M. (1982) 'Everybody's views were just broadened: a girls' project and some responses to lesbianism', *Feminist Review* 10: 37–59.

Oakley, A. (1981) 'Interviewing women: a contradiction in terms', in H. Roberts (ed.) *Doing Feminist Research*, London: Routledge and Kegan Paul.

Padfield, M. and Procter, I. (1996) 'The effect of interviewer's gender on the interviewing process: a comparative enquiry' *Sociology* 30, 2: 355–66.

Pearce, J. (1996) 'Urban youth cultures: gender and spatial forms', *Youth and Policy* 52 (Spring): 1–11.

Prendergast, S. (1989) 'Girls' experience of menstruation in school', in L. Holly (ed.) *Girls and Sexuality*, Milton Keynes: Open University Press.

Pringle, R. (1988) *Secretaries Talk: Sexuality, Power, and Work*, London: Verso.

Ramazanoglu, C. (1989) *Feminism and the Contradictions of Oppression*, London: Routledge.

Roman, L., Christian-Smith, L., with Ellsworth, E. (eds) (1988) *Becoming Feminine: The Politics of Popular Culture*, London: Falmer Press.

Shacklady, L. (1978) 'Sexist assumptions and female delinquency', in C. Smart and B. Smart (eds) *Women, Sexuality and Social Control*, London: Routledge and Kegan Paul.

Sharpe, S. (1976) *Just Like a Girl: How Girls Learn to Be Women*, London: Penguin.

Shiach, M. (1991) *Hélène Cixous: A Politics of Writing*, London: Routledge.

Skeggs, B. (1991) 'Challenging masculinity and using sexuality', *British Journal of Sociology of Education* 9, 2: 127–40.

Stacey, J. (1988) 'Can there be a feminist ethnography?', *Women's Studies International Forum* 11, 1: 21–7.

Thornton, S. (1995) *Club Culture: Music, Media and Subcultural Capital*, Cambridge: Polity Press.

Walkerdine, V. (1984) 'Some day my prince will come', in A. McRobbie and M. Nava (eds) *Gender and Generation*, London: Macmillan.

Warren, C. (1988) *Gender Issues in Field Research*, London: Sage University Paper, Qualitative Research Methods Series, No. 9.

Wellman, D. (1994) 'Constituting ethnographic authority: the work process of field research, an ethnographic account', *Cultural Studies* 8, 3: 569–83.

Widdicombe, S. and Wooffitt, R. (1995) *The Language of Youth Subcultures*, Hemel Hempstead: Harvester.

Willis, P. (1977) *Learning to Labour*, Gower: Farnborough.

—— (1982) *Male School Counterculture*, V203 Popular Culture, Block 7, unit 30, Milton Keynes: Open University Press.

Wolpe, A. (1988) *Within School Walls*, London: Routledge and Kegan Paul.

....

14

THE WORKPLACE

becoming a paid worker: images and identity

●

Sophie Bowlby, Sally Lloyd Evans and Robina Mohammad

INTRODUCTION: 'WHAT WILL YOU DO WHEN YOU GROW UP?'

'What will you do when you grow up?' is a question that adults address to small children with somewhat amused condescension. As they grow up, however, it is a question which those children begin to ask of themselves, sometimes with anxiety, sometimes with excitement. Paid work, and different types of jobs, bring with them particular social identities which play a major part in conferring social status and public acceptability as an adult. In this chapter we examine the factors which affect whether or not a young person looks for and finds employment and the nature of the paid work that they may find.

Finding their first paid job is a step of great significance to most young people. Firstly, a paid job – even if part-time – offers an independent source of income. For many young people this gives them their first taste of independence from parental control of their finances. Alternatively, it is a means of contributing, willingly or unwillingly, to their parents' household expenses and thereby assuming an adult position in the household. For some, a paid job allows them to leave home and set up their own independent household. Secondly, and perhaps most importantly, gaining a paid job can be a crucial step in the process of creating a new, adult identity. Consequently, for those who are unable to find paid work and who are not undertaking training that is likely to lead towards such work, the prospect of continuing unemployment can be both depressing and alienating and lead to the loss of self-esteem. Finally, for young people in their late teens and early twenties the success or otherwise of their search for paid work and the nature of the work they find is often a major influence on their subsequent employment and personal history.

As Pahl (1988) argues, 'work is becoming the key personal, social and political issue of the age'. Our lives are dominated by the need to do paid and/or unpaid work. A central problem in understanding work, relates to the ambiguity over its nature and meaning. The question 'what is work?' is not easily answered. For example, the preparation of meals might be a full time occupation for a chef, an unremunerated household task for a mother, something that is undertaken for pleasure and relaxation, or undertaken as a social occasion between friends (Pahl, 1988). It is not, therefore, the type of task or whether it is paid or unpaid but the social situation in which the task is done that determines what is 'work'. At present, the social organisation of and the divisions between both paid and unpaid work are changing and young people are having to cope with the impact of many of these changes in the form of unemployment, insecure work or extended periods of training.

The timing of young people's entry to paid work and their own and society's expectations of its role in their lives are also changing. Moreover, there are major differences between young people in relation to such familiar social divisions as gender, sexuality, class, race and dis/ability which affect their chances of aspiring to and finding different types of paid work. In what follows we will draw out some of those differences. However, in order to provide a better understanding of the different influences on young people's labour market situation we will use a study of the employment opportunities and barriers facing young Pakistani Muslim women in the town of Reading in Britain to illustrate our discussion (Bowlby et al., 1995).[1]

The impetus for this study came from the concerns of women at the Pakistani Community Centre (PCC) about the lack of appropriate job opportunities for their daughters. They approached the Borough Council for help and the Borough appointed us to carry out a study of the job opportunities and barriers facing young Pakistani Muslim women. The main focus of the research was on in-depth, qualitative interviews with twenty-five young Pakistani Muslim women. Twelve were aged 15 to 16 and were still in school or just leaving school. Of these, one had a part time job, three were seeking part time work, the rest were not currently seeking work. Eleven were between the ages of 17 and 20. Nine of these were still in education and, of these, two had temporary holiday work.

One women had a part time job and another had left school and was hoping to find paid work. We also interviewed two older women, one of 27 and one of 25 about their earlier experiences in the labour market. The 27-year-old was in full time employment, the 25-year-old was not employed since she had just had a baby. Although their families' length of residence in Britain varies, all but two of the respondents were born here. These two women were born in Pakistan and have lived in the UK for under ten years. However, most of the young women's parents migrated to the UK in the early 1960s, and Reading was the first destination for fourteen of their families. In other words, the majority of the women interviewed had obtained the whole of their education and experience of work in Reading and can be described as second generation immigrants. Occasionally we will also use material from a similar study of employment opportunities we have conducted with Pakistani Muslim women aged 25 to 60 (Lloyd Evans et al., 1996) and will draw on interviews from an earlier study carried out by Robina Mohammad in which she also interviewed young Pakistani Muslim women in Reading (Mohammad, 1995).

In the next section, we will discuss the changes which have taken place in the job opportunities facing young people in Britain, including the processes of school–work transition, and illustrate their implications today for young people in Reading. We then explore the ways in which young people develop particular job aspirations and ideas about how to achieve those ambitions, once again drawing on the Reading study to illustrate our points. Finally, we draw together the discussion in a conclusion.

YOUTH LABOUR MARKETS AND WORK IDENTITIES

BECOMING AN ADULT: SCHOOL-WORK TRANSITIONS

The post Second World War period has seen major changes in the labour market for young people in industrialised countries. There has been a marked increase in the level of youth unemployment, as well as unemployment in general, since the 1970s. As a result, more young people continue in education – for example, in Europe only 30 per cent of young people between the ages of 15 and 19 were in work or seeking work in 1992 compared to 50 per cent in 1960 (European Commission, 1994). Young men can no longer expect to find stable full time paid work; young women increasingly expect to stay in paid work whether or not they have children – a step which a growing number are deferring into their thirties – but, like young men, they are finding it increasingly difficult to find stable full time employment in their late teens and early twenties (Pollock, 1996). In all industrialised countries unemployment and unstable employment are experienced most acutely by young people with few qualifications and those from ethnic minorities (Osterman, 1995).

The move from school to paid work can be seen as a process of transition. In some cases it may involve simply leaving school, finding a job and remaining in paid work thereafter but for most young people today this is becoming the exception rather than the rule. A young person may dip their toe in the paid work water by taking a part-time Saturday or holiday job while still at school. Leaving school and finding a job may be followed by re-entry to education either full or part time, spells of work may alternate with unemployment or involvement in a government training scheme. The routes into work are becoming both longer and more diverse, and these changes in the transition from school into paid work are linked to changes in the timing and organisation of other aspects of the transition from childhood to adult status, such as living independently of parents, marrying and having children, and the ways in which different life course stages are resourced and organised (Irwin, 1995).

Through the process of recruitment and hiring, different insititutions contribute to the creation of gender and racialised divisions of labour. British studies have shown class, gender and racial biases in both the careers service and government training schemes (Cross et al., 1988). Cockburn (1987) argues that it is the school rather than the careers service or the Youth Training Service (YT) which is the main source of gender and racial stereotyping. Careers officers have argued that there is little chance of breaking sex segregation in training schemes as 'it's too late at 16. Their minds are made up' (Cockburn, 1987). In all industrialised countries people's final work destinations are strongly linked to their class, gender, race and family circumstances and the labour markets in which they live. Young people with few qualifications or social skills, and those from ethnic minorities, fare particularly badly in the new service economy (Roberts et al., 1994; Osterman, 1995; Casey, 1995; Duster, 1995).

The main influences on the job aspirations of the young Pakistani Muslim women we interviewed come from family and friends and the individuals involved in the various institutions aimed at training and giving career advice to young people – in the Reading case, schools, training agencies, colleges and universities and the careers service. The jobs that are available depend on the nature of the local economy but also on employers' ideas about what kind of person will make a 'good' employee. Both employers and existing employees often have stereotyped images of the kind of person who should be doing a particular type of job – for example, 'secretaries' are expected to be women and 'engineers' to be men. Employers may also have particular expectations about young people's work capacities, or lack of them.

YOUTH LABOUR AND JOB IDENTITIES

Occupational choice is shaped by individual conceptions of self-identity and preference but at the same time, work aspirations are constrained by qualifications, skills and social obstacles related to gender, race and class. Youth work aspirations are often hampered by the specific reluctance of many employers to hire young people. What is it about young people that employers do not like? Perhaps, since there has been a rise in the skill requirements of many jobs in the new service industries and technologically advanced manufacturing, the problem is that many young people are not learning appropriate skills at school. This view would fit in with theories which suggest that the wages different people receive and their chances of unemployment are related to the supply of, and therefore investment in, their human capital – that is their training and acquisition of relevant skills – in relation to the demand for those skills.

However, other commentators regard the main problem as being the change in the nature of the skills demanded in an economy dominated by service sector occupations. Despite emphasis on the move to a more skills demanding economy voiced by those who argue that the training of young people is inadequate, the skills demanded in entry level service sector work often are the 'social' or 'transferable' skills of meshing with the culture of job supervisors and the general public. The language, attitudes and behaviour of young people from working class backgrounds and from ethnic minorities often seems distinctive from such mainstream culture (Bernstein, 1971; Willis, 1977; Jenkins, 1983; Brown, 1987; Bourgois, 1991).

In support of this argument, Duster (1995) reports on two experiments (Culp and Dunson, 1986 and Turner et al., 1992) in which researchers examined the job search experience of matched pairs of black and white job applicants in the USA. In the first (Culp and Dunson, 1986) the white and black applicants had similar success in finding jobs in the manufacturing sector but in the service sector white applicants were four times more likely to find jobs than the black applicants and this differential was greatest for retail establishments. Duster emphasises the racial differential but it is also important to emphasise different reactions to men and women. Employers have different expectations about the capabilities of men and of women.

It is also true that employers often prefer workers over 25 to the young. The competition for service jobs has increased since the 1960s as more and more married women entered the labour force, many as part time, low paid workers. For many service sector employers married women seem a more attractive cheap, part time labour supply than young workers since their greater maturity and life experience are believed to improve their dependability and women are generally believed to have better social skills than men. Young people are thus relegated to casual, part time and temporary jobs.

Consideration of employers' dislike or liking for certain categories of worker on the basis of characteristics such as age, gender and race suggest that theories of segmented labour markets may provide a better analysis of the position of young people than theories relying on notions of human capital (Stitcher and Parpart, 1990; Redclift and Sinclair, 1991). Segmented labour market theories suggest that individuals are socialised at school and home in ways that channel them into particular segments of the labour market, which in turn will affect their attitudes to and behaviour at work. The segmentation of the labour market allows employers to control workers and limit entry to the primary sector to groups with the right 'credentials' and thereby perpetuate inequality between groups. People from ethnic minorities, women and young people without further or higher education will be disproportionately represented in the secondary sector and moving from one sector to another is very difficult. This type of approach has been discussed in relation to the employment of young people (Parcel, 1987; Rosenbaum *et al.*, 1990) and found useful but incomplete. There is significant movement of young people without many educational qualifications between sectors (Parcel, 1987), and moreover the theory does not explain how or why, as these young people become older, they become more attractive employees.

Feminist critiques of segmented labour market theory have emphasised that its proponents have not paid enough attention to the socially constructed nature of skill definitions (Phillips and Taylor, 1980; Sinclair, 1991). They have shown how men have used their economic and political power to foster belief in women's inferior capabilities which then become used as justifications for claiming that women's work is 'unskilled'. Similar arguments can be advanced to explain some employer and union attitudes to people from ethnic minorities.

One of the implications of feminist work on skill definitions is to pay attention to the social processes and power relationships through which different groups of people become represented as 'good' or 'bad', 'skilled' or 'unskilled' employees. Brah (1994) suggests that processes of 'racialised gendering' are central to the way in which labour markets operate. That is, through a variety of means of social communication, ideas of the characteristics of, say, Pakistani Muslim women or Pakistani Muslim men come to be accepted both by employers and other workers and Pakistani Muslim men and women themselves. Such ideas of the 'nature' of Pakistani Muslim women or men are not fixed but are likely to have some persistent characteristics. Moreover, such ideas are formulated, negotiated and communicated through social networks in situations in which the different groups involved in articulating them – say, Pakistani Muslim men and women, white male and female employees, employers, government agencies – have varying degrees of power and conflicting interests in the outcome. Thus these ideas are likely to be related to structural characteristics of the economy and polity. Analysis of the creation and negotiation of such ideas, and ideas about people of different ages and class backgrounds, do seem to offer a useful way in to understanding the operation of sub-divided and segmented labour markets.

SPACE, PLACE AND JOB ACCESS: THE READING EXAMPLE

Clearly the nature of the local economy will affect the kinds of jobs available to young people. If general unemployment is high many young people will become marginal to the labour market and move in and out of low paid and temporary employment and training schemes (Wallace, 1987; Roberts et al., 1986). Young people tend to lack access to a car and must rely on public transport or walking or bicycling. In America where a high proportion of jobs have de-centralised this has created particular problems for young people in inner city locations, especially those from ethnic minorities living in ethnically spatially segregated residential areas (Lewin-Epstein, 1986; Ihlanfeldt and Sjoquist, 1991).

Residential segregation can also have an impact on job access via the socio-spatial extent of the social networks to which young people have access. Since word of mouth and personal contacts are very important in finding jobs, young people who live in residential areas with strongly localised social networks and/or networks confined to a small group of people may have poor information about both job opportunities and ways of gaining jobs (Hanson and Pratt, 1995; Fernandez Kelly, 1995). Furthermore, social networks play an important role in social learning and contribute to an individual's general 'social and cultural capital' – that is the set of social norms, reciprocal relationships, expectations and ways of communicating understandings of the world that an individual builds up through their social interactions. An individual's social and cultural capital is often of great importance in gaining access to work. The nature of a person's social and cultural capital is place-specific, depends to a large extent on geographical location and is often biased according to gender, race and class (Granovetter, 1988; Fernandez Kelly, 1995; Hanson and Pratt, 1995).

The nature of the social networks and contacts of the Reading Pakistani population is important in understanding their patterns of job search and access. The population is small (c. 3,000) and spatially segregated into two fairly small areas and the informal social relations of adults, and more especially of adult women, tend to be with other Pakistani Muslims. Some families try to limit interaction between their children and non-Pakistani Muslim children (Mohammad, 1995). Very few of the mothers or older female kin of the young women we interviewed are employed. Since the jobs done by their male kin were in sectors not considered appropriate to young women by their families or themselves it appears to us that these young women were disadvantaged by their lack of extensive family or community based, female social networks connecting them into the labour market. Only twelve of the young women we interviewed had either yet looked for or found paid work. Of the 8 in some form of employment, 2 had found it as a result of work experience placements organised by the school, 2 had found temporary work through family contacts, 2 had found a job through involvement in government training schemes and 2 through answering advertisements. All relied on word of mouth or newspapers for their job search rather than using employment agencies or the Job Centre. One young woman expressed the belief that the large retailers tend to take students from particular schools on 'recommendation from an employee'.

Reading has followed national trends in shifting from a situation in which there were a substantial number of skilled and unskilled jobs in manufacturing to one in which the labour market is dominated by service occupations. One symptom of this trend has been the high levels of unemployment amongst ethnic minorities (Breheny and Hart, 1994). With regard to youth employment, there is a trend to stay on in education after Year 11.[2] In 1994, 78 per cent of Year 11 and 56 per cent of Year 12 pupils stayed on in education (Berkshire County Council, n.d.). Year 11 leavers are more likely to experience unemployment (14 per cent of leavers) than those who stay on into years 12 and 13 (11 per cent and 9 per cent of leavers respectively). Year 11 leavers enter a rather small range of occupations and show a strong gender bias in their choices. The majority of women start in clerical/sales, hairdressing/beauty or health/childcare work while men enter engineering/electronics, construction and the motor trades (Berkshire County Council, n.d.). In Berkshire as a whole, of those who stay on in education into Year 13 and then enter the labour market, the majority enter jobs in banking, insurance, retailing, administration and science and engineering.

The jobs in Reading are not equally available to young people of different genders or from different ethnic groups. Occupations such as construction and the motor trades are strongly 'masculine' in image while hairdressing and beauty and childcare are 'feminine'. Men and women who enter or try to enter occupations uncommon to their gender will have to contend with prejudice expressed by work colleagues, friends and family as well as by (potential) employers. Accounts of racial harassment from work colleagues are commonplace and, as discussed above, in service sector occupations racialised gendering is likely to structure recruitment. One of our respondents described the requirements of the retail firm where she worked as follows:

'I think [my employers] believe that they [Pakistani Muslim women] don't look as professional as somebody else, I mean when they show you the uniforms and the pictures that they have in their magazines . . . it's like their way of doing it, she has nice little earrings, normally a bobbed hairstyle, perfect make-up, it's always the same sort of colours, they've got their ideas of the colours of shoes and tights they prefer too. They've got their idea of the image they are looking for and Asian girls don't fit it.'
(*Young woman, working for a high street retailer*)

Another respondent gave this account of being interviewed for a job as a shop assistant at a supermarket (which she got):

'She [the person interviewing her for the job] said "you seem quite Westernised to me", I was going "thank you" [ironically], she goes "you don't mind wearing the overalls that come to the knee" and I said "I don't mind" . . . it's just a natural assumption that you will mind.'

Many young Pakistani Muslims marry spouses from Pakistan and in order for a spouse to come and live in Britain they have to show that they are able to support him or her. This means they must have paid work. One respondent recounted how employers view Asian women and also comments on the young women's naivety about the labour market:

'I know about 4 who . . . are older than me, who having left school at 16, they haven't worked all this time and now are looking for jobs because they've got married and they need jobs to bring their husbands over and a lot of them are so dumb that they actually go to these places [applying for jobs] and actually say that "I'm looking for a job so I can get my husband over . . ." and then they [the employers] can say "Oh yes and then you'd be wanting more leave blah blah blah and then you'll be having kids, we know what you Asians are like".'

> *(Young woman, in employment with a high street retailer)*

JOB ASPIRATIONS AND IMAGES OF FUTURE WORK IDENTITIES

Young people develop ideas about the jobs they might do and those they wish to do both formally and informally. Informal information is picked up in social interactions with family, friends and teachers and their own observation of people doing jobs in everyday situations. Formal information can be gained from careers advisors, information leaflets and teachers on training courses. Information gained from 'the media' is also very important (Jackson, 1994). The media provide both direct and indirect information about the existence of different types of jobs, images of the kinds of jobs open to different types of people, and ideas of the kinds of jobs which seem socially valued. There are also other, more subtle, gendered expectations about the kind of work that men and women are 'suited' to and the kind of characteristics they will display as workers. In our study, there was a consensus amongst our informants (both the young women and those we interviewed from local educational and careers institutions) that many people in local educational and employment organisations have a stereotypical image of the abilities and aspirations of 'young South Asian Muslim women' as submissive, uninterested in work and prevented by parental or community pressures from pursuing academic or professional careers. Such attitudes can affect young women's aspirations while at school:

> throughout the whole school time I was under the impression that I'm not going to become anything, I ain't going to be able to do anything [paid work] so why bother, so it was a big doss.
>
> *(Young woman, who retrained after leaving school and is now in paid work)*

These racialised and gendered expectations are also influenced by gendered and racialised expectations amongst parents as well as young people themselves. In what follows we will show how these influences affected the job aspirations of the young Pakistani Muslim women we interviewed.

PARENTAL ASPIRATIONS

In the case of the young Pakistani Muslim women in Reading the attitudes of their parents and of the local Pakistani Muslim community were central to the formation of their ideas about their future. The Pakistani population in Reading is poor (male unemployment was 20 per cent in 1991), predominantly working class and socially and spatially segregated. Many Pakistanis feel themselves to be the object of racial discrimination. In such a situation a high premium is put on maintaining cultural solidarity and pride in what are felt to be the superior moral values of and social behaviour consistent with Islam. The lax sexual morals of white society are seen as a major threat to the maintenance of a distinctive Pakistani religio-culture. Women's behaviour and conformity with cultural norms are crucial to the maintenance of cultural separation from white society. Women are expected to bear and socialise the next generation and hence it is vital that they adhere to and transmit appropriate values and maintain a monogamous marital relationship with a Pakistani Muslim husband. The major emphasis in a women's life is on making a 'home' for husband and children, and paid work is expected to remain secondary to this obligation. For example, one of the older married women we interviewed said that unsuitable work for a Pakistani Muslim woman was *any work which undermines women's ability to run the home*. For individuals and their families – especially those without many resources which might allow achievement outside local Pakistani society – it is important to maintain acceptance within the Pakistani group since the outside, white world is seen as one of hostility and difference. Surveillance of the behaviour of young women is particularly important since failure to maintain approved standards of behaviour will limit their marriage ability. In particular, mixing with men who are not close kin in a situation that is not seen as adequately controlled can compromise the young woman's reputation and also, very importantly, the 'honour' of the family.

These attitudes lead to parental concern about the dress and movement of young women in public space – concerns which are not in evidence for young men (Mohammad, 1996). One young 17-year-old school girl interviewed by Mohammad (1995: 27) said:

> 'We can't go to the cinema or even just shopping. It's different for the boys, though they aren't allowed to go to the cinema either, but because they are allowed to wander around outside, at all hours, their visits to the cinema go unnoticed.'

Such attitudes also affect the nature of the working environment that is considered suitable for women. When questioned about suitable work for Pakistani Muslim women an older, professionally qualified woman, who did a paid job herself said that *'office environments where there is more freer mixing of the sexes, it is not so suitable . . . other than that I don't see any specific requirements for Muslims'*. Another recommended working in schools because *'there is not so much involvement with males in the school environment as there is in the office world'*. Thus, although parents may wish their daughters to find paid work, it must be in an environment that allows the daughter to maintain her reputation as sexually pure and respectable.

It must be emphasised however, that there are very considerable differences between families and young women over these concerns. Some young women are allowed to wear short skirts and Western dress and their parents are happy for them to stay away at university or on training courses while others are obliged to wear the *shalwaar kameez* and *hijab* and are not allowed to go outside the house alone. The behaviour of the majority is expected to lie somewhere between these two extremes. Although the fear of community disapproval acts as a brake on change, parents who have been educated in Britain or who have been able to establish a middle class status are more likely to feel able to allow their daughters greater freedoms.

Many young women shared their parents' views about what constituted acceptable paid employment and behaviour in public. But others did not. Some negotiated greater freedoms outside the home – to travel alone, go to university or college, or take paid work – by accepting the wearing of Pakistani Muslim dress or restrictions on the times at which they could be away from home. As one put it, *'playing the perfect daughter in the home, by dressing in Punjabi dress and covering my hair* [buys me trust and freedom] *to be the imperfect daughter outside it'* (young women, working for a high street retailer, interviewed by Mohammad (1995)). Other young women wanted to adopt different ways of life but felt unable to challenge parental and community authority. Some young women emphasise the need to maintain Islamic standards of dress but often in a Western mode (wearing long skirts and a scarf) and do not see paid work and freedom to move within public space as incompatible with their religion (Dwyer, forthcoming).

Parental concerns can affect young women's ability to aspire to and undertake college or university education since this may involve living away from home and mixing freely with young men.

'I think the people in Reading see . . . going off to university and college is not so good because you are not under any supervision from your parents or uncles and aunties.'

(*Young woman, not in paid work*)

These concerns can also limit aspirations to better paid jobs since these often involve undertaking training away from home or travelling. One respondent felt that her parents considered secretarial work the ideal career for females, and management positions for the boys. When pressed as to why her parents would oppose management positions offering higher pay for their daughters she said:

> 'Put it this way, if the rise in pay from £150 to £250 [as the result of a rise to management status] meant travelling, then they would say that £150 was fine [i.e., enough].'

WHY WORK?

Although paid work is desired by most young people in high income industrialised societies, there are different expectations about the importance of paid work to men and to women. Women are thought to have the alternative of marrying and having a family without doing paid work and it is sometimes suggested that young unmarried women will find unemployment less painful than young men because they are able to occupy themselves with domestic chores. However, this effect is short-lived. It is increasingly expected that married or cohabiting women with children as well as men will need to and, indeed, *ought* to work in order to contribute to the family finances.

For the young Pakistani Muslim women in Reading there were strong parental expectations that they would marry and have children. Moreover, only three of their mothers and few older female kin did paid work and for most there were many younger brothers and sisters to help care for. Thus it would not have been surprising if many of the young women we interviewed had not intended to do paid work. However, this was not the case. All except three of the twenty-three expressed a clear intention to do paid work in the future. Sixteen said they wanted to work to give them an interest, fifteen said that work was 'something to do'. For the majority money is not the primary motivation to do paid work, money was mentioned as 'fairly important' by only eleven of the respondents. In the main, these young women are and expect to be fully supported financially (by parents and later husbands). One girl, for example, told us that she was reminded by her father that she should work only at a job that she enjoyed because if she needed money he could give it to her. Only two of the young women living at home said that they were expected to contribute to the household. The young women interested in pursuing particular professional careers seemed more concerned for their status and need to maintain an interest outside the home than in money. The majority felt that bringing in the money was the male responsibility, whilst theirs, now and in the future, was and would be the home. Most showed a lack of awareness of and/or interest in current wage levels.

In the main, these young women rated suitability of the working environment and colleagues above career progression and salary levels despite the high levels of unemployment in Pakistani families in Reading. It seems that young Pakistani Muslim women's motivation to work is based mainly on desires for the social experience of work and, for some, status, and this is unlikely to be very different from the motivations of their white peers. Their expectations about their future roles as paid workers run counter both to the increasing tendency for most families in Britain to view the woman's wage as a necessity and the probable future pressure on these young women to earn money for their families. These attitudes indicate a gap between their ideas about paid work and the likely realities which, for many, will include poorly paid and uninteresting employment.

'GOOD' JOBS AND 'BAD' JOBS

Through formal and informal social contacts young people will start to build up an idea of the kinds of jobs that are desirable and those that are not. We asked our respondents to tell us what they saw as 'good' and 'bad' jobs and what they thought 'women's' and 'men's' jobs were.

Many of the 15/16-year-olds and some of the 17+ respondents initially found it difficult to conceptualise 'good' or 'bad' jobs and needed a lot of explanation before they could give their opinions. This in itself reveals a lack of knowledge of the labour market. However, we were able to identify images of what our respondents felt 'good' and 'bad' jobs to be.

Firstly, amongst all age groups, being a 'doctor' or a 'solicitor' was the most frequently cited 'good job'. Job characteristics such as good pay, career prospects, job satisfaction and working with friendly people were also mentioned. 'Cleaning', 'menial' and 'low paid work' were the most frequently cited 'bad jobs', particularly amongst the youngest respondents. Many respondents were unable to identify specific occupations, as their knowledge was limited, and concentrated on their characteristics instead. For the young Pakistani women in Reading conceptions of jobs are derived either from previous experience and/or family and media. The jobs mentioned as 'bad' jobs are often those they have seen in the home such as cleaning or catering which are known to be hard physical work or 'messy' jobs such as nursing.

Secondly many bad jobs were also identified as 'women's jobs'.

Secretarial work, working in a factory and cleaning were all considered to be 'bad jobs'. Most respondents were aware that part time jobs are both 'female' jobs and often in low paid service occupations. Most of the older respondents were aware of and accepted the current gender divisions in paid work because, *'women's homelife affects the type of job you have'* as one respondent put it. Many hold traditional views of what is appropriate work for women and men. One 16-year-old commented that her perception of a man's job was *'being a boss in a suit'*! But another respondent commented:

> 'Women are stereotyped into working in gentler, 'caring' professions or jobs or just not working. In reality, I think women are able to do anything they set their mind to. Men have more options – but particularly the professions, engineering, science etc. and, of course, more manual work are seen as more masculine fields.'
>
> *(Young woman, not in paid work)*

Thirdly, when respondents were asked about what they 'wouldn't consider doing', cleaning and factory work (jobs which may have been undertaken by other family members at some time) were high on the list. Another high priority was working near home due to parental concerns over well-being and safety.

If parental awareness of career possibilities is fairly limited this in turn limits the awareness of children. As the majority of respondents had fathers who were either unemployed or in semi-skilled occupations and mothers who had never worked, exposure to different occupations was often very limited, particularly as the mothers are the ones primarily involved with the socialising of the daughters. Therefore what are seen as 'good' jobs are restricted to professions that are highly 'visible' to parents – that is they are professions that people come in contact with in their daily lives – visiting the doctor, carrying out legal business, going to the chemist and running businesses – and that clearly have respect from the community. Persuading parents and daughters that there are many other good and respectable occupations available could greatly enhance young Pakistani Muslim women's position in the labour market. For the majority at present, to be successful in parental terms, as well as their own, is to work in medicine, law or accountancy. The construction of this yardstick of success leaves those (the majority) who do not reach it with a sense of failure.

CONCLUSION: IMAGE, IDENTITY AND SOCIAL CAPITAL

The transition from school to paid work is part of a young person's development of a new relationship to the wider world – the creation of a new adult identity. For every young person the process of transition takes place in specific social and physical 'spaces' which strongly influence the experience and its outcome. For most it is the spaces of home, school and the local labour market which are crucial. It is through locally based experiences that young people develop their job aspirations, attitudes to and experiences of education, training and employment and gain ideas about the opportunities available in the locality for people 'like them'. We have suggested that the social spaces and places which young people are able to access and experience as they seek to define their adult identity in the world of paid employment are shaped by the social networks in which the young person is enmeshed.

In line with the findings of other researchers, we have illustrated the importance of social networks to the exchange of information between employers and prospective employees. We have also argued that social networks have a broader significance as the means through which social and cultural capital are created and maintained. It is through their social networks that groups of people learn socially acceptable ways of behaving and communicating, acquire understandings of the expectations of others within the group and develop relations of trust and reciprocity with them. A variety of studies are now emphasising that having adequate social and cultural capital can be vital to the economic and social success of poor and disadvantaged groups (Portes and Sensenbrenner, 1993). However, as we have illustrated, a problem for some of the young women that we interviewed was that some of the distinctive behaviours and relationships used to maintain the social capital of Reading's Pakistani Muslims and provide support against racism also created barriers against engaging successfully with prospective employers and establishing individual identities which did not 'fit' group norms.

We have also suggested that the various social networks through which employers and young people build up conceptions of the 'ideal employee' and the 'ideal job' circulate racialised and gendered representations of workers and paid work which also interact with images of age and stage in the life course. These representations hinder the employment of young Pakistani Muslim women in Britain, as they hinder the employment of young women and young people of both sexes from ethnic minorities more generally. These representations and understandings are not fixed, however. They are changed as the participants in exchanging information through social networks try to make sense of their experiences, accommodate to larger social changes and structures, and negotiate a more personally satisfactory interpretation of their situation. Many of the young Pakistani Muslim women we interviewed were actively engaged in challenging existing representations through negotiating a variety of new identities as employees, as women and as British Muslims, which were distinct from those of their parents and their peers at school.

NOTES

1 Reading is a town of approximately 200,000 people, in the county of Berkshire, located about 40 miles west of London. This study was funded by Reading Borough Council and the Government Office for the South East under their 'Fair Play for Women' initiative.
2 In England the first year of instruction at school is called 'Year 1', the second 'Year 2' and so on. Children must start school once they are 5, so most of the children in Year 1 classes are 5-year-olds and most in Year 2 are 6-year-olds. Most students in Year 11, therefore, will be about 15 years old and this is also the year when the majority of young people will take their GCSE exams – national exams which are intended to be taken by the majority of school students. At 16 young people can leave school or they can stay on and study for two more years. It is also possible to continue to study Year 12 and 13 'school' subjects at sixth form colleges and further education colleges.

REFERENCES

Bernstein, B. (1971, 1973, 1975) *Class, Codes and Control*, 3 vols, London: Routledge and Kegan Paul.
Bourgois, P. (1991) *In Search of Respect: The New Service Economy and the Crack Alternative in Spanish Harlem*, Working paper No. 21, New York: Russell Sage Foundation.
Bowlby S.R., Lloyd Evans, S. and Mohammad, R., in association with the Pakistani Community Centre Reading (1995) *Pakistani Women's Experience of the Labour Market in Reading: Barriers, Aspirations and Opportunities*, report for Reading Borough Council.
Brah, A. (1994) '"Race" and "culture" in the gendering of labour markets', in H. Afshar and M. Maynard (eds) *The Dynamics of Race and Gender: Some Feminist Interventions*, London: Taylor and Francis.
Breheny, M. and Hart, D. (1994) *Reading in Profile*, Reading Borough Council.
Broadbridge, A. (1991) 'Images and goods: women in retailing', in N. Redclift

and T. Sinclair (eds) *Working Women: International Perspectives on Labour and Gender Ideology*, London: Routledge.

Brown, P. (1987) *Schooling Ordinary Kids*, London: Tavistock.

Casey, B. (1995) 'Apprentice training in Germany: the experience of the 1980s', in K. McFate, R. Lawson and W.J. Wilson (eds) *Poverty, Inequality and the Future of Social Policy: Western States in the New World Order*, New York: Russell Sage Foundation.

Cockburn, C. (1987) *Two-Track Training: Sex Inequalities and the YTS*, London: Macmillan.

Cross, M., Wrench, J. and Barnett, S. (1988) *Ethnic Minorities and the Careers Service: An Investigation into Processes of Assessment and Placement*, London: Department of Employment Research Paper Series, No. 73.

Culp, J. and Dunson, B.H. (1986) 'Brothers of a different color: a preliminary look at employer treatment of white and black youth', in R.B. Freeman and H.J. Holzer (eds) *The Black Youth Employment Crisis*, Chicago: University of Chicago Press.

Donovan, A. and Oddy, M. (1982) 'Psychological aspects of unemployment: an investigation into the emotional and social adjustment of school leavers', *Journal of Adolescence* 5: 15–30.

Duster, T. (1995) 'Postindustrialisation and youth employment: African Americans as harbingers', in K. McFate, R. Lawson and W.J. Wilson (eds) *Poverty, Inequality and the Future of Social Policy: Western States in the New World Order*, New York: Russell Sage Foundation.

Dwyer, C. (forthcoming) 'Contradictions of community: questions of identity for young British Muslim women', forthcoming in *Environment and Planning A*, theme issue on 'Community'.

European Commission (1994) *Employment in Europe*, Luxembourg: Office for Official Publications of the European Communities.

Fernandez Kelly (1995) 'Social and cultural capital in the urban ghetto: implications for the economic sociology of immigration', in A. Portes (ed.) *The Economic Sociology of Immigration: Essays on Networks, Ethnicity and Entrepreneurship*, New York: Russell Sage Foundation.

Granovetter, M. (1988) 'The sociological and economic approaches to labor market analysis: a social structural view', in G. Farkas and P. England (eds) *Industries, Firms and Jobs: Sociological and Economic Approaches*, New York: Plenum, pp. 187–216.

Grieco, M. (1987) 'Family networks and the closure of employment', in G. Lee and R. Loveridge (eds) *The Manufacture of Disadvantage*, Milton Keynes: Open University Press.

Hanson, S. and Pratt, G. (1995) *Gender, Work and Space*, New York: Routledge.

Ihlanfeldt, K.R. and Sjoquist, D. (1991) 'The effect of job access on black and white youth employment: a cross sectional analysis', *Urban Studies* 28, 2: 255–65.

Irwin, S. (1995) 'Social reproduction and change in the transition from youth to adulthood', *Sociology* 29, 2: 293–315.

Jackson, P. (1994) 'Black male: advertising and the cultural politics of masculinity', *Gender, Place and Culture* 1, 1: 49–60.

Jenkins, R. (1983) *Lads, Citizens and Ordinary Kids*, London: Routlege and Kegan Paul.

Lewin-Epstein, N. (1986) 'Effects of residential segregation and neighbourhood opportunity structure on the employment of black and white youth', *Sociological Quarterly* 27, 4: 559–70.

Lloyd Evans, S. (1994) 'Ethnicity and gender in the informal sector in Trinidad: with particular reference to petty commodity trading', unpublished Ph.D. thesis, University of London.

—— , Bowlby, S.R. and Mohammad, R., in association with the Pakistani Community Centre Reading (1996) *Pakistani Women's Experience of the Labour Market in Reading· Barriers, Aspirations and Opportunities*, Report II for Reading Borough Council.

Mohammad, R. (1995) 'Birds, bees, birds, bees . . . an enquiry into the cultural logic of identity and "othering"', unpublished undergraduate dissertation, University of Reading.

—— (1996) 'An exploration of the reproduction of the "other" patriarchy: the case of an urban-based working-class Pakistani Muslim community', unpublished Master's Dissertation, University of Bristol.

Osterman, P. (1995) 'Is there a problem with the youth labour market, and if so, how should we fix it?', in K. McFate, R. Lawson and W.J. Wilson (eds) *Poverty, Inequality and the Future of Social Policy: Western States in the New World Order*, New York: Russell Sage Foundation.

Pahl, R. (1988) *On Work: Historical, Comparative and Theoretical Approaches*, Oxford: Blackwell.

Parcel, T.L. (1987) 'Theories of the labor market and the employment of youth', *Research in the Sociology of Education and Socialisation* 7: 29–55.

Phillips, A. and Taylor, B. (1980) 'Sex and skill: notes towards a feminist economics', *Feminist Review* 6: 79–88.

Pollock, G. (1996) 'Uncertain futures: young people in and out of employment since 1940', paper presented at the British Sociological Conference, 1–4 April.

Portes, A. and Sensenbrenner, J. (1993) 'Embeddedness and immigration: notes on the social determinants of economic action', *American Journal of Sociology* 98: 1320–50.

Redclift, N. and Sinclair, T. (1991) *Working Women: International Perspectives on Labour and Gender Ideology*, London: Routledge.

Roberts, K., Dench, S. and Richardson, D. (1986) 'The changing structure of youth labour markets', London: Department of Employment, Research Paper No. 59.

——, Clark, S.C. and Wallace, C. (1994) 'Flexibility and individualism: a comparison of transitions into employment in England and Germany', *Sociology* 28, 1: 31–54.

Rosenbaum, J.E., Kariya, T., Sttersten, R. and Maier, T. (1990) 'Market and network theories of the transition from high school to work: their application to industrial societies', *Annual Review of Sociology* 16: 263–99.

Sinclair, T. (1991) 'Women, work and skill: economic theories and feminist perspectives', in N. Redclift and T. Sinclair (eds) *Working Women*, London: Routledge.

Stitcher, S. and Parpart, J. (1990) *Women, Employment and the Family in the International Division of Labour*, Philadelphia: Temple University Press.

Turner, M.A., Fix, M. and Stuyk, R.J. (1992) *Opportunities Denied, Opportunities Diminished: Discrimination in Hiring*, Washington, D.C.: Urban Institute Press.

Wallace, C. (1987) 'From generation to generation: the effects of employment and unemployment upon the domestic life cycle of young adults', in P. Brown and D.N. Ashton (eds) *Education, Unemployment and Labour Markets*, Lewes: Falmer Press.

Willis, P. (1977) *Learning to Labour*, Farnborough: Saxon House.

....

15

THE STREET: 'IT'S A BIT DODGY AROUND THERE'

safety, danger, ethnicity and young people's use of public space

•

Paul Watt and Kevin Stenson

'Streetville . . . it's where all the black and Asian people live. I don't think I'd feel safe there at night.'[1]

'Well I know a lot of people around the area, around Thamestown, so I'd be able so go somewhere and say see somebody I know and say "what's happening", so I feel safe . . . I don't really need safety, I know everybody, I know all the Asians and the blacks and all the whites, so I know I'm alright.'

This chapter is concerned to investigate young people's uses and perceptions of public space and issues of safety and danger. It is based upon a study of South Asian, Afro-Caribbean and white youth who live in a medium-sized town in the south-east of England which we have called 'Thamestown'.[2] The chapter explores the question of how an ethnically mixed group of young people negotiate public space as part and parcel of their everyday lives when moving about the town. This is set against a backcloth of continuing racial, class and gender divisions within an area of the country which is stereotypically characterised as the 'affluent south-east', but which has been under-researched in terms of youth studies. As the above quotations from our interviews with these young people suggest, certain people and places can gain local reputations based upon racialised[3] perceptions of 'danger' and 'trouble', whilst on the other hand personal familiarity and 'knowing people' from a diverse range of backgrounds can facilitate feelings of safety and movement across the invisible borders of youth territory.

YOUTH IN THE SOUTH-EAST OF ENGLAND

Many of the most influential sociological studies on British youth have been based in London (Cohen, 1972; Hall and Jefferson, 1976; Hewitt, 1986), or alternatively the cities and towns of the north and Midlands (Coffield *et al.*, 1985; Griffin, 1985; Jones, 1988; Willis, 1977). As a result, young people in London or the north and Midlands are assumed to represent the lifestyles and social conditions of English or even British youth in general; this is a questionable assumption. Young people living in the home counties of the south-east of England are among the groups whose lives have been relatively under-explored (although see Ashton *et al.*, 1990; Banks *et al.*, 1992; Wallace, 1987 for notable exceptions), and this is particularly true for ethnic minority youth who are most frequently portrayed as living in inner city areas. The relative neglect of youth in the south-east is all the more surprising given the demographic significance of the region; in 1991 nearly one in five of all people in Britain aged between 16 and 24 lived in the south-east region outside of London (OPCS, 1993).

The socio-economic position of the south-east altered dramatically from the 1980s to the 1990s. During the 1980s, the south-east was a booming region within the UK economy, based upon dynamic high-tech industries, a rapidly expanding service sector and escalating house prices (Breheny and Congdon, 1989). The region has been seen as an area of high social and geographical mobility and the major site for the growth of the middle classes in Britain (Fielding, 1995). However, the differential impact of the recession in the early 1990s, which was felt most strongly in the south, has led to a growing recognition of 'poverty amidst plenty' with unemployment, deprivation and homelessness scarring many towns and cities in the south-east (Mohan, 1995). A study carried out in Oxford, for example, found a growing polarisation between wealthy and poor areas within the city (Noble *et al.*, 1994). The changing circumstances of the region were dramatically highlighted by the public order disturbances involving young people on a housing estate in Luton in July 1995 which further indicated the way in which issues often thought to be largely confined to the inner cities also exist in the supposedly affluent and 'trouble-free' towns of the south-east.

THAMESTOWN

The local authority district which contains Thamestown largely consists of small towns and commuter villages, and it is, by most indicators, one of the wealthiest in Britain. The suburban and rural areas around Thamestown are dominated by a large and generally affluent middle class, but the town itself contains areas of marked social deprivation, especially amongst the older terraced housing near the centre of town and the council housing estates.[4] Physically these council estates appear very different from the high rise and deck access flatted estates found in London or other major cities, in the sense that in Thamestown they primarily consist of low rise flats and semi-detached houses. Only a fifth of households in the town are council tenants, whilst over two-thirds are owner–occupiers.

Smith (1989) has argued that residential segregation in Britain has a marked racial dimension to it, and this is certainly the case in this part of the south-east. In contrast to the surrounding suburbs and villages, which are virtually all white enclaves, Thamestown has a sizeable ethnic minority population largely consisting of South Asians, mainly Muslims of Pakistani origin, and a smaller black Afro-Caribbean population. Whilst Afro-Caribbeans and South Asians can be found in all of the town wards, the latter make up over 20 per cent of the population in one particular ward, according to the 1991 Census. Within this ward, the Asian population is particularly concentrated in a deprived area of older, owner occupied terraced housing near the town centre which also has a growing number of students; we have called this area 'Streetville' (Figure 15.1). The Afro-Caribbean population is more evenly distributed throughout the town, both in owner occupied and rented housing. Thamestown has experienced a significant decline in manufacturing over the last twenty years and a more recent increase in unemployment during the recession of the early 1990s which has impacted particularly on the white working class living in the council estates in the town, but even more heavily on the Asian and black population. At the time of the 1991 Census, one in three 16–24-year-old Asian and Afro-Caribbean males and Asian females were unemployed, as were over one in five Afro-Caribbean females.

It is important to note that one atypical feature of the Thamestown area is the fact that it does not have a comprehensive system of secondary school education, unlike most of the rest of Britain. Instead, pupils are selected by examination for entry into either one of the grammar schools if they 'pass', or they go to one of the secondary modern schools if they 'fail'.[5] In relation to our interviews, about half of the white respondents were either attending or had just left one of the grammar

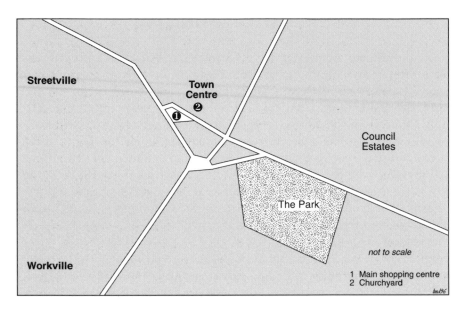

Figure 15.1 Schematic plan of 'Thamestown'.
Source: Linda Dawes.

schools, and they were mainly from middle-class families living in the suburbs and commuter villages a few miles from the town. The rest of the white interviewees were from working-class backgrounds and they lived nearer to the town centre; they either attended one of the local secondary modern schools, had a job or were unemployed. The majority of the South Asian respondents attended a local further education college and nearly half lived in the Streetville area of the town. A large proportion of the young Afro-Caribbeans were unemployed at the time of the interview, with the rest either at secondary modern schools or college. Most of our non-white[6] respondents came from working-class backgrounds and resided in the town itself as opposed to the suburban or rural areas.

THE RESILIENCE OF PLACE: BUT 'IS IT SAFE'?

Despite the postmodernist claims about the declining significance of local neighbourhoods as important for young people's identities and lifestyles (Featherstone, 1991), empirical studies of youth have demonstrated the ways in which young people continue to identify with local places (Hendry *et al.*, 1993; Pearce, 1996; Taylor *et al.*, 1996). By 'place' we mean a space which people in a given locality understand as having a particular history and as arousing emotional identifications, and which is associated with particular groups and activities. Sometimes identification with places can be somewhat confining and 'localist' in the sense of people having restricted spatial horizons, and studies in Belfast (Jenkins, 1983) and Sunderland (Callaghan, 1992) have shown the importance of localism for understanding young white working-class people's spatial orientations; this can include a strong identification and pride in place.

In relation to non-white youth, studies in an inner city Midlands area (Westwood, 1990) and in Manchester and Sheffield (Taylor *et al.*, 1996) have found that young Afro-Caribbean and Asian people tend to lead a localist existence in the sense that they rarely leave the part of the city where they live, partly because of local ties in the area and restricted transport access, but also because of feelings of danger moving outside the 'safe space' of their own neighbourhood. In this sense, the 'localism' of ethnic minority youth partly arises from the fear of racial harassment, as well as from positive attachments to local areas. This fear of racial harassment can be regarded as resulting from 'white territorialism' (Hesse *et al.*, 1992: 171) which refers to the way in which some white people, mainly males, use various forms of violence to defend 'their' areas against non-white people. 'Territoriality' can be 'best understood as a spatial strategy to effect, influence, or control resources and people, by controlling area' (Sack, quoted in Hesse *et al.*, 1992: 172). Control over area can take place on the basis of residence, religion, or other social markers; it can also take a racist form, and racist white territorialism is most visible in particular working-class locales such as parts of the East End of London (Cohen, 1988; Toon and Qureshi, 1995). Some researchers have argued that territoriality is no longer just the province of white males, but has also developed as a form of defensive street masculinity by Asian and Afro-Caribbean males against racist attacks. For example, a study of Asian and white youths in a town in the north of England suggests that the development of such territoriality on the part of Asian males occurred over time because of racist attacks on them by whites (Webster, 1996). This defensive version of street territoriality by young Asian and black males has been found in other inner city areas (Westwood, 1990) and also amongst Bengalis in east London as a result of the ever present threat of racial violence there (Keith, 1995).

An alternative scenario to territoriality based upon exclusive racial or ethnic groups is the vision articulated in several ethnographic studies that working-class youths living in inner city areas are developing multi-ethnic cultural practices which transcend racism and ethnic stereotypes (Back, 1996; Hewitt, 1986; Jones, 1988; Wulff, 1995). In these studies, inter-ethnic friendships were recorded in which shared residence and locality played a key role. Instead of racially exclusive spatial demarcations, local neighbourhoods were actually the arena and also the basis of multi-ethnic harmony. One element in such inter-ethnic friendships is the respect shown by white youths for aspects of black street culture, including music (Jones, 1988). Back (1996: 71) argues that in areas of south London 'nationalisms of the neighbourhood' have developed in which territorial ties and allegiances amongst non-white and white young people are being created so as 'shared locality offers an alternative identity option to divisive and exclusive notions of "race"'. However, Back is also cautious about such neighbourhood nationalisms since inter-ethnic mixing can actually co-exist with racist practices and the espousal of racist views.

TENSIONS AND FRIENDSHIPS IN THAMESTOWN

During the 1970s and early 1980s, there had been a history of racial harassment by groups of white males, some of them skinheads with far Right political affiliations, against Asian and black youths in the centre of Thamestown. By the 1990s this had subsided, partly as a result of a major clash at a town centre pub which local folklore portrays as a symbolic victory by black and Asian youth against the racists, and partly as a result of the subsidence of the skinhead phenomenon. Nevertheless, public displays of racial conflict still occur, as indicated by a Saturday night fight early in 1996 between groups of around twenty Asian and white youths in the town centre. There has also been increasing concern expressed in reports in the local press, and to us in interviews with local authority officials and the police, about what is regarded as growing public disorder in the town shopping malls centring on conflict between groups of young Asian and Afro-Caribbean males (Stenson and Watt, 1995). However, these anxieties, coded within a racialised and criminalised[7] view of black and South Asian youth (Keith, 1993, 1995), distort a complex reality.

We found that there was some evidence of inter-ethnic tension between some of the young Afro-Caribbean and Asian males in the town centre (cf. Stenson and Factor, 1994 on youth in north London) which had resulted in a much discussed and arguably mythologised fight in the main shopping mall (see Figure 15.1):

'There was a fight here a few months ago, yeah, between blacks and Asians in the [shopping mall] . . . it was a massive fight, the blacks were outnumbered in it, and they got into a fight in the [shopping mall], it was in the newspapers, innit?'

(Asian male, aged 16)

Young black males sometimes did hang out with other young blacks, and white youths similarly could be seen in groups containing only whites. Certainly, the dominant tendency was for Asian males in the town centre to go around in exclusively Asian, single-sex groups, partly for defensive purposes. These groups usually numbered three or four, but they could be larger: 'if we're fighting against the blacks then it would be a big crowd . . . we've got to stick up for each other', one young Asian male said to us.

Despite such tensions between the various ethnic groups, the patterns of conflict and co-operation between the young people in our study were far more nuanced than those expressed in the local press and by local officials. Observation in the town's main shopping mall, a place with a local reputation as an assembly point for young people in general, and ethnic minority youth in particular, revealed that the groups of young men and women milling around, white and non-white, were not consistently divided by ethnicity or 'race', with evidence of contacts and acquaintanceships across ethnic and racial lines for both males and females. Although several of our respondents mentioned being in 'posses', a term used by the mass media to refer to allegedly violent black criminal gangs, in Thamestown it was simply a phrase used to describe friendship groupings. The groups of young men did not constitute fully fledged, coherent 'gangs' based upon ethnic allegiances. Instead, the groupings observed in the shopping mall, for example, often had a haphazard nature: groups would form, break up, regroup with different members, all propelled along by the attempts of the security guards to keep the young people on the move and to at least look like genuine bona fide shoppers as opposed to young people just 'checking each other out' on a Saturday afternoon. Open conflict in the town centre between the various ethnic groups, white and non-white, was sporadic, despite being the source of much anxiety amongst the social control agencies, and was ameliorated by personal knowledge of 'others' across ethnic boundaries.

Our interviews revealed evidence of friendship between some of our black, Asian and white respondents:

'I mean I sometimes go around with white people, black people, Indians sometimes and Pakistanis, so it's all mixed . . . it's like some people just sort of stick to one race, because they feel comfortable with them, but I don't take people for colour really or what race they come from, I just take them for what they are. So I could be with white people, black people, whatever.'

(*Asian male, aged 18*)

Sometimes these friendships occurred on the basis of shared musical tastes, as one Afro-Caribbean male said, 'I hang out with black and white, Asians, just everyone who's into hip-hop music.' However, inter-ethnic friendships were more frequent amongst youths who came from the ethnically mixed council estates and other predominantly working-class areas around the town where young people often shared the same schools and leisure facilities. More than half of the white working-class young people said they had black and Asian friends, but there were far fewer non-white friendships amongst the grammar school respondents living in the affluent suburbs.

SAFETY AND DANGER FOR ETHNIC MINORITY YOUTH

Whilst we uncovered evidence for inter-ethnic friendships amongst some of our respondents living in ethnically mixed neighbourhoods, this must nevertheless be set against views which indicated that those same places could also be sites of racial harassment against other black and Asian young people if they didn't live there or didn't personally 'know anyone' there. Safety was by no means guaranteed for black and Asian youth and a survey commissioned by the Thamestown local authority showed that 14 per cent of a sample of Asians and Afro-Caribbeans had been victims of racial attacks in a single year. Given such levels of racial attacks on blacks and Asians in Thamestown, one important issue is whether or not non-white youth restricted their movements to their local neighbour-hoods and adopted survival strategies in order to traverse public space, as other research has found in large cities (Taylor *et al.*, 1996; Westwood, 1990).

Some of the young Afro-Caribbean people in Thamestown did admit to feeling unsafe in particular areas. One young man mentioned being threatened by white youths in one particular housing estate on the outskirts of the town, also mentioned by several Asians as a dangerous place, and one 18-year-old black female indicated that her feelings of safety were in fact predicated upon not venturing very far out of home territory. Nevertheless, surprisingly the majority of Afro-Caribbean males said that they did not feel unsafe in the town and as individuals felt they could go anywhere. The second quotation at the beginning of the chapter illustrates how the spatial confidence of these black males was based upon their personal knowledge of people in different parts of the town. This in turn arguably reflects life in a town small enough so that many people know each other on the basis of shared schooling, neighbours, use of sports and entertainment facilities, and so on, a situation likely to differ from that found in the large cities in which people are more familiar with particular locales, but not others.[8]

On the other hand, more of our Asian respondents, both males and females, identified areas of the town where they personally, or their friends, had been the victims of racist abuse or physical attacks, and which consequently they avoided. For example, one Asian male respondent, aged 16, said of 'Workville' (Figure 15.1):

'We used to go up Workville. There are all white people up there, you're just going to get into a fight, you're going to get yourself beaten up bad. No-one's going to go up Workville, especially, you know, Asians. They are racist there.'

Such views might well imply that the Asians in the town led a localist existence, corralled by the fear of racial attacks into particular safe 'Asian neighbourhoods'. To some extent the Streetville area of the town did in fact have this function; it was referred to by many of the Asian respondents as an 'Asian area'. For example, one young Asian male said of Streetville: 'it's always filled with Asians there, it's like no-one messes around with you, we just hang around with each other'. The reputation of Streetville as an Asian area also had benefits for some of the young Asian women:

'Where I feel really at home is the Streetville area 'cos there's a lot of Asians in Streetville . . . I feel safer when there's more Asians and when I know them.'

(Asian female, aged 18)

Streetville was an area which in many ways represented the clearest expression of local place identity and pride amongst all of the young people we spoke to. This aspect of localism, however, co-existed with a group oriented masculine street strategy which allowed the young Asian males to move into other, potentially hostile, parts of the town. Around half of the Asian males said that they felt safe anywhere in the town in relation to perceived threats from both white and black males, and much of this seemed connected with safety in numbers and going around in large all male groups, as witnessed in the town's main shopping mall; one young Asian man said he felt safe anywhere, 'because I've got too many people to back me up'.

Several of the Asians who lived outside of the Streetville area in the ethnically mixed council estates had a less negative view of safety and danger in these places, in comparison with the Streetville Asians, reflecting the kinds of spatial confidence expressed by the young Afro-Caribbean males based upon knowledge of individuals. The theme of 'knowing people' as individuals as a way of avoiding 'trouble' was a recurrent one amongst many of our male respondents, both white and non-white. As one Asian male, aged 16, who expressed confidence about his ability to move between areas, put it: 'Thamestown is a small town . . . and you get to know a lot of people if you have been living here for quite a while . . . 'cos I went to two schools as well.'

A few of the Asian males and females, and particularly those who lived outside the Streetville area, saw it in a less sanguine light, and several mentioned a local café as a place to avoid because of 'trouble' connected with drugs. There was also some evidence of rivalry between young Asian youths from different neighbourhoods; one young Asian man who lived in one of the areas away from the town centre suggested that the Streetville Asians had 'all got attitudes . . . they just think they are it and all that, they come around our area and give you the evil eye so we don't even bother looking at them'.

Views on Streetville by the Afro-Caribbean males were mixed. A couple who lived in another part of town said they felt unsafe in Streetville with one, for example, reporting that he had been beaten up by a group of Asian youths after he had been verbally abusive to them. Other Afro-Caribbeans had very different views of the area based upon what they saw as a shared 'black' identity spanning both Afro-Caribbean and Asian people in opposition to white racism:

> 'I feel at home where there's black people to be honest, y'know, 'cos you know there's an area like Streetville, y'know, and when I walk down there I feel safe, y'know, 'cos I know all the people, but if I go to some of these areas where it's just mainly white people, the way they look at me is like I'm an alien or I don't belong here so I try and avoid those kinds of areas.'
> *(Afro-Caribbean male, aged 17)*

This particular young man hung around with other Afro-Caribbeans, but also with Asians, again pointing to the part played by inter-ethnic friendships in the maintenance of street confidence.

Despite clear attachments to their local neighbourhoods expressed by many black, and especially Asian youths, coupled with the very real threats of racial attacks in certain parts of the town if they didn't personally 'know anyone' there, surprisingly we did not find ethnic minority youth to live a *solely* localist existence, contrary to research in inner city areas (Taylor *et al.*, 1996; Westwood, 1990). Many of the non-white respondents, although by no means all, seemed to frequently visit a number of out of town places for leisure activities, with London as a popular destination (Watt, 1998).

IN FROM THE STICKS: SUBURBAN YOUTH

The young white middle-class people who lived in the suburban owner occupied housing estates or the up-market commuter villages several miles from the town centre constituted a distinctive group in several ways. Unlike many of our working-class respondents, both white and non-white, there was little allegiance to their areas of residence; most of the suburban youths found the places where they lived 'boring'. Instead, when asked about where they felt at home, many of them mentioned various places in the town centre, including the park, certain pubs and fast food outlets, and the main churchyard. The latter (the churchyard – Figure 15.1) was regarded as a common hang out for the 'Goths' and 'Indie' young people, both males and females, identifiable by their long hair, dark clothes, big boots and a preference for independent, alternative bands, such as The Levellers:

'the churchyard . . . is quite a good place because you've got lots of people there, sort of into my kind of music and everyone knows each other, 'cos it's like a small little, almost kind of community. . . people sit and talk.'
(*White male, aged 17*)

However, by no means all of the young people from the suburbs liked either 'Indie' music or the 'churchyard scene', and some of the females felt wary about the latter:

'I wouldn't go there . . . 'cos people are a bit intimidating to be honest. I'd kind of go through there as quickly as humanly possible . . . there's like groups of leather jacketed types. I mean they're not all, but just the general impression, you just think I don't want to be here. And there's people taking drugs quite a lot'.
(*White female, aged 17*)

The key problem for suburban youth was that in order to get to the town centre meeting places they had to cross what many of them saw as potentially hostile space, and both males and females reported having to take security measures. Whilst the vast majority of young women from all ethnic groups reported fears of several areas of the town at night, it was very noticeable that the middle-class males we interviewed also reported that they had to take security measures in relation to fears about intimidation from working-class 'Kevs', as well as racialised anxieties about black youth. The Kevs were seen as being into rave or hip-hop music and were likely to wear baggy jeans and bomber jackets. One of the suburban males described Kevs as follows:

'I mean I'm not saying Kevs are thick, because I do have some Kev friends who are pretty good blokes, they're quite intelligent. . . . I think if you live in a poor area you're better off being a Kev, because there's going to be more Kevs around. . . . It's like the Detroit thing, y'know, wearing the right trainers or you get shot for it. It's that kind of thing.'

(White male, aged 17)

Unlike the white working-class youths, several of whom were involved in inter-ethnic friendship patterns, the young people from the suburbs did not have the personal contacts across the ethnic divides which were so important for moving confidently about the town. Not only were these young people living in geographically different parts of the town, but they were also less likely to meet black and Asian youngsters at school, given the selective schooling system which seemed to operate in such a manner as to disproportionately filter ethnic minority youth into the secondary modern schools. All of the suburban youths in our sample were either attending, or had attended, one of the grammar schools. It could also be that the anxieties expressed by these young people about those from the secondary modern schools are an anachronistic reminder of the grammar/secondary modern split which so divided secondary school age children nationally during the 1950s and 1960s prior to the introduction of comprehensive schools.

Several of the female and male young middle-class people told 'cautionary tales' (Anderson *et al.*, 1994) about the Streetville area as a place of 'danger':

'The Streetville area . . . it's a bit dodgy around there to be honest . . . I mean some of my friends go round there and well occasionally because they mix with other groups and they say this happened and this happened and you think "I'm not going there" then.'

(White male, aged 18)

The reliance on racialised stereotypes about the Streetville area as a dangerous 'black ghetto' echoes some of the findings of Taylor *et al.* (1996) on Moss Side in Manchester. As in Moss Side, such stereotypes seemed to be most common amongst those with little personal knowledge of the area; most of the young white people in Thamestown had very little personal contact with the Streetville neighbourhood.

Given the relatively separate social and spatial worlds in which the young middle-class people from the suburbs lived, it is unsurprising that they often felt wary about moving around the town. Such anxieties, ultimately arising from class and racial differentiation, were overlaid in the case of the young women with more gender specific concerns about the town at night.

YOUNG WOMEN AND PUBLIC SPACE

In relation to perceptions of safety and danger, there were certain continuities between the young women, irrespective of ethnicity or class background.[9] There was, as other researchers have noted (Stanko 1990; Valentine 1989), a common emphasis on the need to take precautions to avoid danger, especially when moving around public space in the evening. For some, this constituted a generalised apprehension about the town at night:

> 'I don't think it's specific places [where I don't feel safe], but I wouldn't go down, sort of, late at night into Thamestown at all, I mean I just wouldn't be there, especially if I was just with my girlfriends. I mean if there was a couple of blokes with us, or something, then maybe, but the chance is I wouldn't go down. It's nowhere in particular, it's just the whole idea of going down to town in the evening'.
>
> *(White female, aged 17)*

Young women from all ethnic groups frequented the town centre shopping malls and burger and pizza restaurants during the daytime, whilst the fast food outlets were also popular in the evenings, as were the town centre pubs for some of the young white females from the suburbs. In comparison with the streets, especially in the evenings, these places offered a more secure environment to meet other young people, minimising the risk of unwanted or threatening male attention.

In the context of research on young people in the very different urban environment of the East End of London, Pearce (1996: 7) says that, as well as window shopping, young women go to shops in order 'to see friends, to hang around and, in their words, to have somewhere safe to be'. The same can be said for the young women of Thamestown in relation to the popularity of the town centre shopping malls as daytime meeting places.

CONCLUSION

We have argued that there are some similarities in the uses of public space among young people in Thamestown and in larger urban areas elsewhere, not least in relation to the ways in which young women take precautions in pursuing their leisure activities when moving about the town. There is also evidence, through white racist attacks, of the attempt to exclude ethnic minority groups from certain parts of the town, and of defensive organisation by young Asian males in the Streetville neighbourhood. The town centre is itself a contested space for many of the young people, and tensions can be said to exist between the different ethnic groups, notably the males, at various times.

However, our research in Thamestown did not reveal the same degree of territorialism, amounting to what Webster (1996: 26) refers to as 'ethnic apartheid' in one town in the north of England, which also exists in certain parts of the East End of London (Keith, 1995; Toon and Qureshi, 1995). The tendency of young working class males to defend 'their' areas on the basis of racial or ethnic exclusivity was, at least to some extent, mitigated in both the town centre and the ethnically mixed council estates by evidence of shared neighbourhood and school friendships which crossed racial or ethnic divides within what was frequently referred to as a 'small town'. Personal knowledge of 'others' could counteract group or place based stereotypes and enhance feelings of safety when using public space. The young middle-class people, on the other hand, who lived in the predominantly white suburbs and commuter villages, had less recourse to such personal funds of knowledge about people from different ethnic or class backgrounds, not least because of racialised patterns of residential segregation and the particularities of the schooling system in the area.

For youth in this superficially affluent town, the rigidities of socio-spatial inequality co-exist uneasily with postmodern fluidity. We found a complex pattern of ethnic, racial and class rivalries, as well as the crossing of social and spatial borderlines. That the borders were crossed does not mean that they were removed.

ACKNOWLEDGEMENTS

Many thanks to the following: Claire Doherty and Karen Eden for conducting some of the interviews and assisting with initial data analysis; Jenice Layne, Sheila Patel, Janet Sarju, Danny Spencer and Mavis Thomas for carrying out the interviews; Martyn Hudson, Shirley Koster, Max Travers and Colin Webster for their comments on previous versions of this chapter; Buckinghamshire College for funding the project; and not least our respondents.

NOTES

1 We use the term 'black' in this chapter to refer to people with either an African or Caribbean ('Afro-Caribbean') ethnic origin, and 'Asian' to refer to people with a South Asian ethnic origin based in the Indian sub-continent. This was the dominant nomenclature used by the young informants in our research, although we recognise that the term 'Asian' is itself a generalisation for a number of minority groups differentiated by religion and place of origin (see Modood et al., 1994).

2 The research used a variety of methods, including semi-structured interviews, carried out between June 1994 and July 1995, with 70 young men and women (35 of each) aged between 15 and 21 about their leisure time and use of public space. The respondents were drawn fairly evenly from the three main ethnic groups in the town and included South Asians (mainly Muslims of Pakistani origin), Afro-Caribbeans and whites. As far as possible, we tried to match interviewers and interviewees by ethnic group, although not by gender since only one of our interviewers was male. Our respondents were drawn from those who tended to regularly 'go out' and use public space, so they cannot be taken to be representative of all young people in the town. As well as the interviews with the young people, we also carried out observations of patterns of youth group organisation in the shopping malls and on the streets, interviewed local government, commercial and voluntary agency officials, and analysed official documents.

3 By 'racialised' we mean the attribution of certain types of behaviour and characteristics to particular groups of people defined by surface physical features such as skin colour (Miles, 1989).

4 Council estates consist of social housing for rent in which the local authority acts as landlord.

5 It is undoubtedly the case that the grammar schools in the area have a higher status than the secondary modern schools. The majority of the town's pupils attended one of the latter, and although we do not have conclusive evidence, further research carried out by the authors suggests that children from ethnic minority backgrounds disproportionately attend the secondary modern schools rather than the more prestigious grammar schools.

6 'Non-white' refers to both Asians and Afro-Caribbeans.

7 By 'criminalised' we mean the tendency to highlight criminal behaviour by particular groups of people, hence reinforcing the likelihood of such people coming into contact with the criminal justice system.

8 It is possible that some of the male respondents exaggerated various aspects of their streetwise confidence in the interviews with the female interviewers.

9 We intend to analyse the issues of space, safety and identity for the young women in greater detail in another paper.

REFERENCES

Anderson, S., Kinsey, R., Loader, I. and Smith, C. (1994) *Cautionary Tales: Young People, Crime and Policing in Edinburgh*, Aldershot: Avebury.

Ashton, D., Malcolm, M. and Spilsbury, M. (1990) *Restructuring the Labour Market: The Implications for Youth*, Basingstoke: Macmillan.

Back, L. (1996) *New Ethnicities and Urban Culture: Racisms and Multiculture in Young Lives*, London: UCL Press.

Banks, M., Bates, I., Breakwell, G., Bynner, J., Emler, N., Jamieson, L. and Roberts, K. (1992) *Careers and Identities*, Milton Keynes: Open University Press.

Breheny, M. and Congdon, P. (eds) (1989) *Growth and Change in a Core Region: The Case of South East England*, London: Pion.

Callaghan, G. (1992) 'Locality and localism: the spatial orientation of young adults in Sunderland', *Youth & Policy* 39: 23–33.

Coffield, F., Borrill, S. and Marshall, S. (1985) *Growing up at the Margins*, Milton Keynes: Open University Press.

Cohen, P. (1972) 'Sub-cultural conflict and working class community', in Centre for Contemporary Cultural Studies (eds), *Working Papers in Cultural Studies*, No. 2, Birmingham: CCCS, University of Birmingham.

—— (1988) 'The perversions of inheritance: studies in the making of multi-racist Britain', in P. Cohen and H.S. Bains (eds), *Multi-Racist Britain*, Basingstoke: Macmillan.

Featherstone, M. (1991) *Consumer Culture and Postmodernism*, London: Sage.

Fielding, T. (1995) 'Migration and middle-class formation in England and Wales, 1981–91', in T. Butler and M. Savage (eds) *Social Change and the Middle Classes*, London: UCL Press.

Griffin, C. (1985) *Typical Girls? Young Women from School to the Job Market*, London: Routledge and Kegan Paul.

Hall, S. and Jefferson, T. (eds) (1976) *Resistance Through Rituals: Youth Subcultures in Post-war Britain*, London: Hutchinson.

Hendry, L., Shucksmith, J., Love, J.G. and Glendinning, A. (1993) *Young People's Leisure and Lifestyles*, London: Routledge.

Hesse, B., Rai, D.K., Bennett, C. and McGilchrist, P. (1992) *Beneath the Surface: Racial Harassment*, Aldershot: Avebury.

Hewitt, R. (1986) *White Talk, Black Talk: Inter-racial Friendship and Communication Amongst Adolescents*, Cambridge: Cambridge University Press.

Jenkins, R. (1983) *Lads, Citizens and Ordinary Kids: Working-class Youth Lifestyles in Belfast*, London: Routledge and Kegan Paul.

Jones, S. (1988) *Black Culture, White Youth: The Reggae Tradition from JA to UK*, Basingstoke: Macmillan.

Keith, M. (1993) *Race, Riots and Policing: Law and Disorder in a Multi-racist Society*, London: UCL Press.

—— (1995) 'Making the street visible: placing racial violence in context', *New Community* 21: 551–65.

Miles, R. (1989) *Racism*, London: Routledge.

Modood, T., Beishon, S. and Virdee, S. (1994) *Changing Ethnic Identities*, London: Policy Studies Institute.

Mohan, J. (1995) 'Missing the boat: poverty, debt and unemployment in the South-East', in C. Philo (ed.) *Off the Map: The Social Geography of Poverty in the UK*, London: Child Poverty Action Group.

Noble, M., Smith, G., Avenell, D., Smith, T. and Sharland, E. (1994) *Changing Patterns of Income and Wealth in Oxford and Oldham*, Oxford: Department of Applied Social Studies and Social Research, University of Oxford.

Office of Population Censuses and Surveys (1993) *1991 Census: Sex, Age and Marital Status, Great Britain*, London: HMSO.

Pearce, J. (1996) 'Urban youth cultures: gender and spatial forms', *Youth & Policy* 52. 1 11.

Smith, S.J. (1989) *The Politics of 'Race' and Residence*, Cambridge: Polity Press.

Stanko, E. (1990) *Everyday Violence: How Women and Men Experience Everyday Sexual and Physical Danger*, London: Pandora.

Stenson, K. and Factor, F. (1994) 'Youth work, risk and crime prevention', *Youth & Policy* 45: 1–15.

—— and Watt, P. (1995) 'Young people, risk and public space', paper presented at the Youth 2000 International Conference, University of Teeside, 19–23 July.

Taylor, I., Evans, K. and Fraser, P. (1996) *A Tale of Two Cities: A Study in Manchester and Sheffield*, London: Routledge.

Toon, I. and Qureshi, T. (1995) 'The contestation over residential and urban space in the Isle of Dogs', paper presented at the British Sociological Association Annual Conference, University of Leicester, 10–13 April.

Valentine, G. (1989) 'The geography of women's fear', *Area* 21: 385–90.

Wallace, C. (1987) *For Richer, For Poorer: Growing Up in and out of Work*, London: Tavistock Publications.

Watt, P. (1998) ' "Going out-of-town": youth, "race" and place in the South East of England', *Environment and Planning: Society and Space* (forthcoming).

Webster, C. (1996) 'Local heroes: violent racism, localism and spacism among Asian and white young people', *Youth & Policy* 53: 15–27.

Westwood, S. (1990) 'Racism, black masculinity and the politics of space', in J. Hearn and D. Morgan (eds) *Men, Masculinities and Social Theory*, London: Unwin Hyman.

Willis, P. (1977) *Learning to Labour: How Working Class Kids Get Working Class Jobs*, London: Gower.

Wulff, H. (1995) 'Inter-racial friendship: consuming youth styles, ethnicity and teenage femininity in South London', in V. Amit-Talai and H. Wulff (eds) *Youth Cultures: A Cross-Cultural Perspective*, London: Routledge.

....

16

THE CLUB

clubbing: consumption, identity and the spatial practices of every-night life

•

Ben Malbon

> All I need is a little space to express myself – it's not a lot to ask but you'd
> be surprised.
>
> *(Nicolette, 'Beautiful Day', 1996)*

Clubbing[1] is now one of the major forms of experiential consumption[2] for young people in Britain today, particularly so in cities and towns. In narrowly economic terms, the 'nightclub industry' is worth an estimated £2 billion a year, with over one million young (and plenty of not so young) people spending an average of £35 a week partaking in its practices (Hyder, 1995).

However, this chapter is not about the economic consequences of clubbing, nor is it about the potential regenerative possibilities that clubs and their associated cultural industries (bars, fashion outlets and so on) offer to decaying inner city areas (for example, Manchester, Leeds and more lately Sheffield). My focus is the actual *experience* of clubbing – its relationships with the overall experience of living in the city, what it involves (what clubbing is actually like, what it is about) and what it offers those who partake in its practices – the clubbers – in the form of spaces in which they can express themselves and feel an affiliation or affection with others, forging and re-forging their self- and group identities[3].

Firstly, then, as the practices of clubbing as I am conceptualising them are very much rooted in the social life of the modern city, I examine how this city life can impact upon our senses of identity and belonging. As the site of a multitude of 'social spaces' of varying levels of sociality the city intensely stimulates our emotions and senses, at times to the brink of sensory overload, but also offers sites and spaces of relief from this intensity, if only through bounding these experiences in certain spaces and at certain times (a process which often, but not always, involves commodification).

Secondly, I try to unpack 'experiential consumption' through a focus on the club as experienced by the mind and the body. I look first at the remarkably emotional and sensual nature of clubbing before examining it from a 'dramaturgical perspective' – as a performance – in which the clubbers are simultaneously actors and audience.

Finally, I link the places and practices of clubbing with the formation and maintenance of identities and identifications. I offer a reconceptualisation of the way in which youth has been theorised, building partly on the work of French sociologist, Michel Maffesoli. His notion of the 'tribe', which I rework slightly, provides a useful guide to comprehending the ways in which spaces and identities interact.

Illustrations of some of the conceptualisations and phenomena I will be touching on can be found in the accompanying vignettes (eight boxes throughout the chapter), in which I trace a 'night in the life' of the club experience of three clubbers, chosen from my early empirical work. Each clubber goes to a different club and experiences their night out with friends in their own unique way.

CLUBBING AND THE SOCIAL SPACES OF THE CITY

The speed and intensity of modern city living, structured through processes of time keeping, punctuality, spatial ordering and organisation on the one hand (Simmel, 1903), and yet appearing more as a 'maniac's scrap-book, filled with colorful entries which have no relation to each other' on the other (Raban, 1974: 129), creates a sensory and stimulating environment of such complexity and density that its citizens can find themselves feeling overwhelmed, isolated and, perhaps most paradoxically, lonely.

Many of the *physical* social spaces of the modern city (as opposed to the aesthetic or moral social spaces)[4] are organised so 'that meetings which are not actively sought may be avoided' (Bauman, 1993: 157). Public spaces in the city often seem designed more for travelling through than for socialising within – more fleeting spaces than meeting places. Potential meetings become 'mismeetings' (Goffman, 1971: 331) in which the point is to see whilst pretending that one is not looking – scrutiny disguised as indifference (Lofland, 1973).

A NIGHT IN THE LIFE OF THE CLUB EXPERIENCE

In this short vignette, I trace a route through a night out at a club for three people. Each goes to a different club and experiences their night out with friends in their own unique way. Each is presented in a different type-style to ease reading.

Decisions, decisions

I always try and look for who's playing – we're totally influenced by what DJs are playing and I like to go out and check out the better ones, the ones you're stoked on. So we went to the *Drum Club* and that was spot on. We usually check the flyers or *Time Out* or something.

WE STARTED WITH A FEW DRINKS AT HOME – MY PLACE AS IT'S CLOSEST. A LITRE OR TWO OF WHITE CIDER IS ENOUGH TO GET ANYONE LOADED AND BY THE TIME WE LEFT WE WERE MESSING AROUND AND HAVING A FEW LAUGHS. WE GOT PRETTY STONED TOO – IT'S AN EXCUSE TO GET SOME IN AND HAVE A FEW SMOKES.

Basically the more people that you know, the more friends that are into it, the more you hear about what's going on. I mean, I went to a place on Saturday, a sort of hardcore jungle do – I found out about that from surfing the Web, as well as a tip off from UK-Dance [a discussion list on the Internet].

Coming together

SOMEHOW, SOMEONE USUALLY TAKES US ALL IN HAND AND GETS US THERE. THINGS JUST SEEM TO SLOT INTO PLACE AND EVERYONE IS PRETTY CHILLED BY THEN ANYHOW. YOU CAN ALWAYS HEAR THE CLUB BEFORE YOU GET THERE, AND WE USUALLY SEE A FEW LIKELY LOOKING MEGATRIPOLANS ON THE TUBE AT EMBANKMENT. KNOWING. EXPECTING. IT'S CALLED 'THE FESTIVAL IN A CLUB' ON THE FLYERS, BUT I'VE NEVER BEEN TO A FESTIVAL LIKE IT. FESTIVALS ARE ALL ABOUT OUTSIDE AND THE AIR, FOR ME ANYWAY – THIS IS INSIDE, IT'S IN LONDON AND WHEN I'M THERE I REALLY DON'T FEEL LIKE I'M ANYWHERE IN PARTICULAR.

We drove in from just outside London and the car drive always puts us in the mood. Sort of have a spliff at six, get ready, get togged up and start going down in the car, have a few smokes and a laugh. Clothes . . . well, we kind of dress up to go out because it's like . . . the big night of the week, but I never go overboard and wear kind of poncey stuff – we like to look well-turned out, nice shirts and smart jeans.

> I went with this mate of mine that I occasionally see from back home – he calls me Friday and says 'do you fancy going clubbing later?' So I say of course, and then he says that there's one slight snag – it's in Bedford, so I say 'this better be fucking good then!' So, we went out and had a few beers and then got on the train – to Bedford. The place was called Milwaukee's – it's half way up the M5. We got there and it was mostly white, a few black people. I was just thinking 'what the fuck am I doing here?'

The presence of so many strangers can act to increase the sense of unease that city dwellers often experience as they move through various social spaces. We employ a variety of tactics in our efforts to manage this exposure to alterity – our contact with people we do not know and will not know. One of these tactics is to project an indifference to the distinctions between objects and people, and even an indifference to the presence of these objects and people – what Simmel (1903) calls the 'blasé outlook'. This response can be seen in what we commonly call 'reserve' – a lack of response which paradoxically constitutes a response[5] so that what initially appears as dissociation is actually yet another form of sociality.[6]

There does, however, appear to be a limit to the extent to which tactics such as this can provide us with the relief we search for. The quest for a release from these civilising influences or 'ceremonial codes' (Goffman, 1967: 55), a temporary escape from the obligations we impose upon ourselves and the expectations of others in public space, becomes a focal imperative for many in the city, as we search for times and spaces in which we can enjoy the experience of close proximity to others in a more positive and rewarding form than we experience in, say, the streets (Sennett, 1990, 1994).

While for some people (and for most at certain times), the conflict and juxtapositions present in these more exposed social spaces are attractive, others are driven by the urge to feel at ease in space, even if only fleetingly. This desire to be in accord with the group rather than to be exposed amongst strangers, to feel an affinity with a place and the people within it, though not necessarily a sense of unity – to feel an *identification* with them and the social situation – can be seen as integral to the experience of other social situations (Hannerz, 1980). Here one can think of numerous spaces and contexts where this togetherness and tactility might be experienced – sporting occasions, festivals, political events, even military celebrations – and the club can be constructed as one of these spaces.

PART OF THE 'IN' CROWD

We tried to get in there to see Sasha – I'd only probably seen him like once or twice before, and Sasha is kind of wicked. We were totally up for it and it's like meaty queues and then we got to the door and it was sort of like they were trying to totally balance the like sex sort of deal, so they wanted like one bloke and one girl in, sort of thing and, I mean, it's a total excuse to kind of turn you away if they don't like you – if you're not wearing the right leather waistcoat or something, or some dumb-ass shit! The thing is, if it's heavy music, if it's not going to be posey, then it's going to be like 75 per cent male pretty much all the time anyway so it's like you've got to go to the soft music if you want loads of girls really.

THERE'S THIS WEIRD METAL-DETECTOR THING AT THE DOOR WHICH I THINK IS THERE FOR SHOW BUT ANYWAY – I USUALLY TAKE A FEW READY-ROLLED AND MAYBE A LITTLE BOTTLE OF WHISKY DOWN MY PANTS. WHAT ARE THEY LOOKING FOR? GUNS? WELL WHISKY DOESN'T SHOW ON A METAL DETECTOR. HAVING SAID THAT THE BOUNCERS ARE REALLY CHILLED OUT – I DON'T THINK SOME OF THEM ACTUALLY REALISE THAT THEY ARE SUPPOSED TO BE ON THE DOOR.

It was quite weird cos jungle's roots came from these sort of nowhere towns – Stevenage, Bedford, Luton – where they had these extremely bored white and black kids in the same place. I was a bit wary at first because they were mostly white working class Bedfordshire housing estate types who aren't known for their loving kindness towards sub-continental descendants like myself, y'know? But people are always a bit sort of edgy and that's one of the things that I like about jungle – that it's got that old sort of hip-hop vibe, it's not this sort of 'Hey, we all love each other' thing.

CLUBBING: A SENSATIONAL PERFORMANCE

The clubbing experience is a total experience – very much an encounter of mind *and* body. Not only do the practices of clubbing involve specific ways of doing things, skills, customs and competence in respect of certain bodily practices (dancing, adornment, poise and so on) – what Crossley (1995, citing Mauss) calls 'body techniques' – they also involve (and result in) massive stimulation of the senses and emotions. The consumption of the club experience is socially performed and both exceptionally sensuous and sensual. Although in certain ways similar to the total stimulation of other less planned and more loosely bounded social situations, the club experience is also constituted through some other quite different practices and qualities.

Plate 16.1 Light, darkness and the body combine in the intense sensual
experience of the club.
Photograph: Alan Lodge.

As a sensuous experience, we might construct clubbing as a performance, where the lights (or darkness), the sounds, the possible use of
drugs, the practices (and rituals) of dancing and the proximity of the
'audience' all add to its intensity (Plate 16.1).

Clubbing is heavily dependent on musical forms (mediated through
technology) and the ability of music to transform (and 'create' certain
types of) space. The music may not necessarily be intensely loud –
indeed in some cases it may be no louder than one would experience
in the home.' Yet in each case it is the ability of music (and sound more
generally) to create an *atmosphere* (an emotionally charged space) which
is of crucial importance, for it is largely this atmosphere that the
clubbers consume. But it is not the music alone – the crowd is extremely
important in its response to the music.

Dancing is the most visible response to music, perhaps the most
overtly bodily practice which clubbing involves. Dancing can provide a
release from many of the accepted social norms and customs of the
'civilised' social spaces of everyday life, such as social distance, conformity and reserve or disattention. Dancing might be seen as an embodied
statement by the clubber that they will not be dragged down by the
pressures of work, the speed and isolation of the city, the chilly inter-
personal relations one finds in many of the city's social spaces.

CASING THE JOINT

Soon as we got in we had a walk around the entire place – see what's where and what's on. Then we got a beer and headed for one of the 'harder' rooms where we loafed for a bit, had a bit of a groove when it starts kicking off. If you just stand at the side of the room waiting for the perfect track to come on, you'll be waiting all night so you have to try and get into the groove pretty quick.

Eventually my mate says, 'fancy scoring a pill?' and I go 'sure, go ahead'. So we sort of score a pill off these really large raggas there with big Moschino coats on. Anyway, we score these pills, expecting them to be sort of 'London-shite' and about an hour later we're sort of staring at each other and coming up SO massively we had to go and sit down and it's always a bad mistake doing Ecstasy getting into long conversations about how we feel about each other and with me that always veers off into formal logic which is another thing that you should never talk about on E.

Once we're in we usually spend a bit of time wandering around all the rooms and little corridors just checking it, seeing what is happening, we'll get a drink maybe, and have a smoke somewhere quiet. The main dance floor was totally chaotic. They call it 'The Cathedral' and it's in this enormous railway arch buried deep under Charing Cross.

So we're sitting there and suddenly this massive black geezer runs in, slams his fist down on the table that we're on and just screams 'YES!' and runs off again. My mate turns to me and says that that's the guy we've just scored off – he must have taken one himself and has worked out how good they are. From then on it was just mental, we went straight back to the dance floor.

In terms of the pill, I guess we totally work out where the night is going to be at it's peak – you want to be up when that happens – it usually coincides with the top DJ of the night hitting the peak of his set too. If it finishes at 2 or 3 then the peak will be 12 to 1, so you'll drop between 11 and 12. I mean you can get fucked all night if you like but we usually don't do that. We dropped at about 11.30 I think and I was getting funny feelings by 12. We all usually take the same so we're kind of on the same plane out there.

Now he was going with it, his body bubbling and flowing in all ways to the roaring bass-lines and tearing dub plates. All the joy of love for everything good was in him, though he could see all the bad things in Britain; in fact this twentieth-century urban blues music defined and illustrated them more sharply than ever. Yet he wasn't scared and he wasn't down about it: he could see what needed to be done to get away from them. It was the party: he felt that you had to party. You had to party harder than ever. It was the only way. It was your duty to show that you were still alive.

(Welsh, 1996: 27–8)

Further, the tactility experienced through this being together (whether this be as a result of dancing or of less intense forms of conviviality such as chatting or literally just 'being' together) is central. This communal ethic, the pleasure of being with others, is born both from the sharing of space (a territory) and the proximity of that act of sharing, and from the establishment and maintenance of some sense of unity or membership. These feelings of membership do not necessarily have to be founded upon exclusivity (with respect to the 'outside') or conformity (with respect to the 'inside'), though both these structuring effects do impinge upon the club experience, manifested respectively as (for example) door and entry restrictions and stylisation. Another possibility, and one which is perhaps more exciting, is of clubs thriving upon the *differences* present within the club space; this phenomenon can be seen in the increasing trend for musical (and cultural) cross-fertilisation and hybridity – something which is posing problems for those involved in classifying both musical and clubbing forms (see Nowicka, 1995, for example).

Many clubbers' experiences are enhanced through the use of recreational drugs, used by some to heighten the sensuous nature of the experience – to intensify the sense of euphoria, to boost self-confidence, to give pleasure and on a more mundane level to increase stamina. *Ecstasy*, in particular, is widely used in certain strands of clubbing, although by no means all clubbing experiences. Whilst an understanding of the effects and experience of Ecstasy remains scanty, studies so far suggest that some of its most notable effects lie in its ability to increase empathy between users in a group situation (Saunders, 1993, 1995). Crucially, however, this empathy is obviously also present in a whole range of group social interactions which have no connection at all with drug use (for example shopping, theatre, spectator sports, even the reading of newspapers). This suggests that in the case of clubbing the intensity, tactility and the temporarily shared ethos of the stimulatory experience may accentuate the effect of the drug rather than the other way around.

The darkness present in some club situations is as important as any of the lighting effects used. Individuals' use of space changes as they lose sight of others (Goffman, 1971) and while the visual is vital in the club experience, the 'illegibility' of darkness functions as a kind of inverse

---

FINDING A SPACE

THE OTHERS ARE OFTEN KEEN TO HIT THE DANCE FLOOR PRETTY SOON AFTER WE GET IN, BUT I USUALLY PERSUADE SOMEONE TO COME UP TO 'THE TELLING CAVE' WITH ME, WHICH IS A SMALLER BAR THAT HAS CUSHIONS AND BEANBAGS ALL OVER THE PLACE WITH LOW LIGHTING, AND EARLY ON IN THE EVENING THERE IS USUALLY A LECTURE OF SOME SORT ON AN 'ALTERNATIVE' TOPIC. THIS WEEK THERE WAS A LIVE LINK-UP WITH TIM LEARY IN THE STATES. IT WAS TOTALLY ODD WATCHING LEARY IN LA TALKING ABOUT ECSTASY AS THE NEW PARADIGM IN SOCIAL RELATIONS AND ALL SORTS OF STUFF LIKE THAT WHILE WE WERE ALL STUFFED INTO THIS WEIRD BAR UNDER CHARING CROSS – TRULY UNDERGROUND I GUESS!

It's the music that is the driving force – music now is more exciting than at any time since '88, and that's jungle, pure and simple jungle. I though that acid house was the dog's bollocks and everything else was, well, that we were seeing the after effects of that, the after shock. Then I went to a jungle club that I heard about through the media and just thought this is really different. Because the beat's so fast you can actually slot in a reggae bass beat. Jungle has got like two beats to it, and there were like the white guys who were all going fucking crazy, the black guys were all bopping and the Asians like me were all doing something in between. It's putting the British black community on the map, worldwide . . . I mean, London is like a media clearing house, but it never really comes up with anything of it's own – this is a first, I'm a Londoner and this is MY FUCKING MUSIC!

We made a place to meet as soon as we got there as we always get split up and lose each other. The music was like – hard House I guess, slow ploddy stuff with a few bits of acidy turns on the top. The DJs do their best to provide us with a bit of variety, so they'll throw in the odd different tune that kind of sets the crowd off in a different direction. You get that big rush when a classic tune is played that everyone goes for. You can't beat a bit of pumping House to get a crowd going!

---

form of visuality – facilitating escape. So although in some ways clubbing resembles other less convivial social spaces, associated with notions of the crowd, regimes of bodily practice and rules of interaction, it can also offer respite from these same structurings – 'an-other' space seemingly in between the ordering of the outside and that of the club itself.

Similarly, the total nature of the sound present in some club experiences can effect the obliteration of the aural *outside* of the music.[8] This notion of 'losing it' or 'losing yourself' appears to be at the heart of many clubbing experiences, whether they be in 'hard house', 'salsa' or

---

LOSING YOUR SELF

The music was just on fire – it was just so good and it was just a powerful clean E. Just the unexpectedness of rushing in the middle of Bedford after fucking working all day. Y'know, suddenly, he says, 'I know let's go to Bedford', and suddenly BANG there we were – there are times when I was dancing that I just completely lost it. Music for your body, for your mind, for your soul, and that's the real thing about jungle – there's a state of complete awareness, complete self-awareness. Whenever I'm thinking, saying or doing anything, there's always a bit of me that's like monitoring everything.

THE CLUB GOT VERY HECTIC BY ABOUT 11.30/MIDNIGHT AND THE BASS IN THE CATHEDRAL JUST FILLED THE ROOM. I JUST LOVE STANDING ON THAT LITTLE STAGE AT THE FRONT AND LETTING THAT LITTLE TRANCEY STUFF SEAR STRAIGHT THROUGH ME. MY EARS WERE ALREADY COMPLETELY SHOT THROUGH. I WAS INCOHERENT BY THIS STAGE, I THINK, BUT THE DJ DIDN'T GIVE US ANY LET-UP. HE STARTED BRINGING IN A WHOLE LOAD OF REALLY MINIMAL STUFF – JUST THE MOST BASIC OF TRANCE MIXED IN WITH A FEW WEIRD TECHNO TRACKS. IT WAS FROM THAT POINT THAT I KNEW THAT I WAS GOING TO BE KILLED BY THE MUSIC THAT NIGHT AS HE STARTED BUILDING, BUILDING – WE WERE GOING FRENZIED DOWN THERE, I MEAN YOU CAN'T IMAGINE IT UNLESS YOU WERE THERE! MAD FREQUENCIES WERE BUILDING IN MY HEAD, FREQUENCIES PIERCING MY EARDRUMS, I LOST IT . . . LIGHTS, LASERS, ACHING SCREAMING RHYTHMS. I REALLY WAS IN ON ANOTHER PLANE ALTOGETHER.

By about oneish I was fairly fucked and I think everyone else was too actually. That's cool although I was next to some really screwed up nutters for a while and that kind of brought me down a bit as they weren't with it and looked a bit unpredictable. But having said that everyone was on the same buzz and I love it when it's like that – the drugs worked better, the music was better, the DJs were better – you get much more of the up vibe if you feel better naturally. The crowd made the night totally.

'easy listening' clubs. The key point seems to be that the clubbers temporarily (though semi-consciously, intentionally) forget (lose sight/ site of) aspects of their lives which they find arduous (for example work), and experience a state of 'inward emigration' or 'away' (Goffman, 1963: 69). Their individual senses of identity become (temporarily) less significant than the nature of the social situation of which they are a part.

On the other hand, many clubbers seem to enjoy the intensive socialising that can occur while clubbing and go more explicitly to enjoy others' company rather than to let themselves go on the dance floor. There are as many ways of experiencing a club as there are clubbers clubbing (Plate 16.2).

In addition to being experienced sensually, and inextricably linked to

Plate 16.2 Music, dance and conversation create the social experience of clubbing.
Photograph: Alan Lodge

this sensuous experience, clubbing is also physically performed.[9] That is, the clubbers fashion the situations in which they participate through the ways in which they conduct themselves and read the conduct of others. These practices of performance are often unconscious, yet they consti-tute the glue which acts to bind the disparate personae together encap-sulating the customs, traditions and norms that go to make clubbing a distinctive form of social interaction (and each 'strand' of clubbing within the overall scheme of things a distinctive form of social interac-tion in turn). For example the rituals of the queue and the door become situations in which the clubbers act out certain roles in order to gain entry to the club – they behave as they believe they are expected to behave. Of course, the performance of clubbing involves much more than merely a fear induced conformity. Acting out certain roles, dressing in a similar manner, dancing in a certain way, even drinking similar beers are all ways in which the affinity of the group can be reinforced, the territory of the club experience claimed.

A shared definition of the situation – an ethos – can be seen to have developed when a group atmosphere exists, which, as Goffman explains, is constructed out of a 'standardised set of emotional attitudes . . . a temporarily adopted set of sentiments towards the rest of the world' (Goffman, 1963: 97) – a set of more or less shared feelings. The emergence of this group ethos in and through the experience of clubbing highlights the significance of the concept of *identification* (Scheler, 1954; Maffesoli, 1988)[10]– the notion that when we come together in a group situation we often feel as though we have temporarily lost our individual identities and are instead part of a collective subject. The emotional community generated through the club experience is characterised by 'an ephemeral quality, a fluid composition, a lack of organisation and a rather fixed routine' (Maffesoli, 1988: 146) and can be infectious, as Simmel suggests:

the individual feels himself [sic] pulled along by the 'quivering ambiance' of the mass as if by a force to which he is exterior, a force indifferent to his individual being and will, even though this mass is constituted exclusively of such individuals.

(*Simmel, 1903: 116, cited in Maffesoli, 1993: xiv*)[11]

The clubbers consume each other – the clubbing crowd contains both the producers and the consumers of the experience and the clubbers are consuming a crowd of which they are a part; the club space comes to resemble a scene 'in which everyone is at once both actor and spectator' (Maffesoli, 1988: 148).

SPACES OF YOUTH

The spaces within and around which these performative and sensuous practices occur become sites where the clubbers' identities and identifications are formed and re-formed (in both senses of the word). While the sounds of youth (Chambers, 1986; Halfacree and Kitchen 1996) and young people's ways of dressing (Finkelstein, 1991; McRobbie, 1994) have been studied, albeit fleetingly, little attention has been given to the actual spaces or contexts of their interaction. If we accept that we all (to a lesser or greater extent) have the ability to present ourselves in different ways in different places and at different times, then the spaces or contexts of social interactions become key factors in terms of our opportunities to refashion ourselves and identify with others. This is not in a crudely deterministic manner but in a more subtle and open fashion, where the clubbers share more a sense of *unicity* than a unity of purpose (Maffesoli, 1993: 19)[12] Relatively diverse elements (groups) and individual identities are subsumed within the wider and much more fragile identification present within that space. Uniformity and unity are still apparent in certain strands of clubbing. But unity of identity, and in particular an identification with a specific sub-cultural grouping, appear

---

ON REFLECTION

They got the lights and visuals just right again. Not too light is the key for me. I mean you don't want to see yourself dancing really. Although you look at other people dancing you'll never be critical of them. It was neat that I could see the DJ at *Drum Club* as he was totally losing it back there and getting us all going too. Plenty of smoke – yeah.

I EVENTUALLY MADE MY WAY TO THE AMBIENT LOFT PLACE, 'THE WELL' I THINK IT'S CALLED – WHERE THEY PLAY THE SLOWEST AND MOST RELAXING TUNES I'VE EVER HEARD. I FOUND MOST OF THE OTHERS UP THERE ALREADY KIND OF ALTERNATING BETWEEN CHATTING WILDLY ABOUT SILLY THINGS AND LYING ON THEIR BACKS ON THE MATTRESSES JUST STARING AT THE ROOF AND SWIMMING THROUGH THE SOUNDS. THIS IS THE PLACE WHERE WE REALLY FIND WHAT WE COME LOOKING FOR – I DON'T EVEN KNOW WHERE WE ARE IN THAT WEIRD LOAD OF TUNNELS AND ROOMS. ALL I KNOW IS IT WAS BLACK, COOL, COMFORTABLE AND THICK WITH THE MOST RELAXING MUSIC I'VE EVER HEARD – LITTLE GENTLE GUITARS, SLOW DRUMS FOR HALF AN HOUR AT A TIME, LOTS OF PEOPLE SMOKING, LYING BACK, CHATTING, LAUGHING. YEAH – THIS IS IT. I HAD NO IDEA WHAT TIME IT WAS AT THAT STAGE. YOU FEEL ALMOST INVISIBLE.

By about three in the morning when everyone was up it was great and I was wandering around chatting to people. Not as much as in a house club where it seems compulsory to go 'Oh, what's your name?', but, you know, you'd ask for a light and then ask what they think of the music. I've never had too much fear of crowds like that, as long as you're confident but not too cocky. If you try posing you're going to get fucked off but you know, it's pretty obvious where I'm coming from – middle class Asian, London – can't make any bones about that. I mean what they'd be like when they were pissed up is another matter. I'd like to think that the club culture would all come together.

By the time the E really started cutting I was well into the dancing thing. I objected a little when a few of the boys started taking their tops off and they're all sweaty and wet – that kind of sucks. It's like the biggest night of the week so we were all wearing our favourite clothes so we weren't into going totally overboard like that. Although it's dark you can see people who have made the effort and taken time over their appearance. I really appreciate that as it's a big night for me, and I want to feel like I'm there among like minded people.

---

to be far less significant in contemporary youth culture than has been recognised by theorists of youth culture up to now, at least for certain people and in certain contexts. Some of the more recent work on youth has highlighted this 'stylistic non-conformity' of certain strands of youth (see especially Muggleton, 1995[13]), a point which reinforces the notion of *identification* upon which I briefly touched earlier, and the contention that so-called 'authentic sub-cultures' were 'produced by sub-cultural theories, not the other way around' (Redhead, 1990: 25).

This is not to say that certain structuring aspects of our identities (our ethnicity, gender, sexuality, age and social status, for example) disintegrate in the club experience. Instead, in specific situations and at particular times, and surely for some people more than others, there occurs a 'going beyond' of their sense of identity as the immediate time and space of the situation that they are in at that moment takes on an increased significance. The stable identity of the individual is superseded by the much more fluid and ephemeral identifications of the persona. Thus it might be possible to distinguish between certain spaces in certain contexts as being more or less 'egocentric' (where the individual's identity and her/his actions take on a primary significance) or 'lococentric' (where the immediate surroundings or context – the environment – are paramount) (Maffesoli, 1995).

In place of theorising youth cultures and identities in an aspatial and acontextual fashion then, as relatively fixed sub-cultures, we might, building on Maffesoli (1995), begin to see them as 'tribes'. These tribes are quite unlike the fixed forms of (group) sociality found in, for example

---

HOMEWARD BOUND?

I STUMBLED HOME RELUCTANTLY, BUT EXHILARATED, REFRESHED, REVITALISED. I'D BEEN 'THERE' AGAIN, FELT THAT ENERGY, THAT RELEASE, THAT SURGE OF LIFE. ANONYMITY YET COMPLETE SECURITY. THAT MOMENT OF FREEDOM.

We sort of came down, straight to earth, it was really clean. We suddenly realised how knackered we were, sort of got on the train back home and we were just saying what a fucking incredible night it was, a really nice club too. I think I'll always remember that 'cos it was just me and him just shooting off into the middle of nowhere, on a whim basically, and having a fucking brilliant night without even trying, and you know – I really liked that.

We stayed to the very end – the moment they turned the lights on was a bit of a shock as usual – you see each other how you 'really' are – sweaty, dirty but still very smiley. People didn't usually stop dancing just like that either – you physically can't for a start. We shuffled our way out of the door and into the cold blackness – headed for the car and the tension of the 'who's going to drive' decision. None of us really felt safe so we crashed in the car park for a few hours, although I didn't sleep much.

the 'classic tribes' of the 1970s (the Californian counter-culture or the European student communes, for example), which were characteristically stable in form (if not function). Contemporary tribal formations are typified by fluidity, by punctuated gathering and scattering, and to distinguish them from the groupings of the 1970s, Maffesoli labels the more recent groupings 'neo-tribes' (Maffesoli, 1988: 148), or as I prefer to conceptualise them 'transitory tribes' to reflect the highly spatialised nature of these constantly in-transit, continually re-forming and temporarily bounded social interactions.

With the emphasis more upon the flitting *between* groups than of membership *per se* of a group or community, the critical aspect of these tribal identifications becomes the *spaces* of the identifications. For young people especially, there are certain spaces and certain contexts within those spaces which appeal more to them than do other spaces and contexts. They identify more with those spaces (and the forms of sociality within them) than with others, however fleeting this identification might be.

As one of these spaces of identification, with its rituals and customs, its intensive sociality, its overwhelmingly sensuous and emotional ethos, its re-citing and re-fusing of musical histories, boundaries and cultures and its seeming dislocation from the binds of a life which appears increasingly uncertain, the club situation offers clubbers opportunities to inscribe their own creativities upon a shared space, to create a space of their own making of which they are also the consumers.

In clubbing they find a unique blend of pleasures (musical, tactile, sensual, emotional, sexual, chemical, bodily), a potential for illicit activities, the stimulation experienced through proximity to difference (and in particular of being so close to what in other contexts they would regard as 'strangers' and thus potentially a 'danger') and an escape route (albeit an ephemeral one) from the rigours and stresses of an ever quicker society which offers no guarantees and provides few opportunities for the release of deep seated emotions and desires in close proximity to others.

It is here, in this creation of a space of their own, that we find resistance, not as a struggle with a dominant, hegemonic culture (as theorised for example by Hall and Jefferson, 1976) or even as the fact of clubbing itself, but resistance as located in the most minute subtleties of clubbing, the ways of clubbing – its *arts de faire* (de Certeau, 1984). This is resistance on a micro-level, on the level of everyday life, where the unspoken is that which binds the group together, where the desire to be with others is manifested, and differences are addressed (Maffesoli, 1995). This is resistance as found in the dynamic and exciting combination of musics, in the fleeting pause before the DJ drops the bass in, in

---

CLUBBING

I SUPPOSE I JUST LIKE THE SHEER ESCAPISM OF IT ALL, THE FEELING THAT WHEN I'M THERE I CAN DO EXACTLY WHAT I LIKE, PRETTY MUCH, AND THERE'S NOTHING THAT CAN STOP ME. Y'KNOW, THE FEELING, THE ATMOSPHERE THAT FILLS THE PLACE WHEN IT'S FULL AND THE MAIN DANCE FLOOR IS TOTALLY ROCKING, BUT THEN YOU ALSO KNOW THAT THERE'S ALWAYS THE OTHER ROOMS WITH THEIR INDIVIDUAL ATMOSPHERES, EACH OFFER-ING SOMETHING DIFFERENT, A DIFFERENT MOOD, A NEW VIBE. YEAH, IT'S THE MUSIC THAT DOES IT – TOTALLY. YOU CANNOT BEAT LOUD MUSIC OF WHATEVER SORT MIXED WITH A LOAD OF LIKE MINDED PEOPLE.

I've always grown up thinking life is pretty grim, and you've got to fucking work your bollocks off 'cos otherwise you're not going to get anywhere – I've still got a very strong work ethic inherited from my Asian background . . . but what I really like about clubbing is just the trash Western, hedonistic non-sense of it all . . . and this is happening now and it's my culture, this is the only thing that I have that I can call home. I'm not English in the normal sense of the word, and I'm not Indian by any means . . . I've just been dumped into this urban nightmare scenario with a blank sheet of paper and I just, I just love it – it's my culture.

I doubt . . . you couldn't really do pills for ever. I can't imagine doing it in my thirties or anything like that but, I dunno, House music's only been around since 1989, so you're going to tell me that it's still going to be the main party music when I'm 31 or something. It's established now and you will have electronic music for ever, but clubs aren't just House music anyway. I mean there's all the ragga clubs and soul clubs and whatever other sort of club you want to go to so there will *always* be clubs and there will *always* be a club that I go to. Maybe if I go to a club when I'm 40 or 50 I'll go to jazz clubs or something.

---

the semi-visibility of a darkened dance floor, in taking Ecstasy (in not taking Ecstasy), in dressing in a certain way, in the emotional and empathetic effects of close proximity to hundreds of others, not neces-sarily like yourself, but sharing, at the very least, a desire to be right there, right now. This is the resistance found through losing your self, paradoxically to find your self.

It is . . . inscribed in the languages of appearances, on the multiple surfaces of the present-day world. There, re-working the relationships of power, resistance hides and dwells in the rites of religion, in the particular inflection of a musical cadence. It is where the familiar, the taken for granted, is turned around, acquires an unsuspected twist, and, in becoming temporarily unfamiliar, produces an unexpected, sometimes magical, space . . . . It is dwelling in this mutable space, inhabiting its languages, cultivating and building on them and thereby transforming them into particular places, that engenders our very sense of existence and discloses its possibilities.

(Chambers, 1994: 16)

ACKNOWLEDGEMENTS

I'd particularly like to thank Phil Crang for his unswerving encouragement. I'd also like to thank Peter Wood and Martin Berger for covering earlier drafts with red and green ink respectively, and my occasional research companion, Miffi. Of course, all the rash statements are my own. Most of all I'd like to thank all the clubbers who have let me impose myself on their fun times.

My Ph.D. research is funded by an ESRC Studentship – Award Number R00429434210, and aspects of the field work are funded by the University College London Graduate School Research Fund.

NOTES

1 The terms 'clubbing' and 'clubber' are used throughout the chapter in preference to 'night-clubbing' and 'night-clubber' – terms which clubbers very rarely use.
2 I go on to explore this term in further detail later. Crudely, it is used to distinguish activities where expenditure involves the purchase of 'physical' goods from those activities where no such 'physical' exchange is involved.
3 This chapter and the accompanying vignettes are based upon Ph.D. research that I am undertaking on clubbing in London. At the time of writing I am approximately one-third of the way through my field work which is based upon a blend of in-depth one on one interviews with clubbers and participant observation in the clubs themselves. It is, therefore, very much 'work in progress' and should be read as such.
4 Bauman proposes that while physical space is constructed through the acquisition and distribution of knowledge, aesthetic space exists effectively through processes of curiosity and the search for experiential intensity, and moral space is 'constructed through an uneven distribution of felt/assumed responsibility' (Bauman, 1993: 146).

5 We all both 'give' and 'give off' information in any social interaction (Goffman, 1963). Information which is 'given' is provided voluntarily, and the provider of the information is held responsible for what s/he provides – talk and dress are two examples. Information which is 'given off' is taken to be provided regardless of whether the provider intends to do so or not – for example facial expressions or poise (Kendon, 1988).

6 Sociality can be defined as 'the basic everyday ways in which people relate to one another and maintain an atmosphere of normality, even in the midst of antagonisms based on gender, race, class or other social fractures' (Glennie and Thrift, 1996: 225). Sociality can also refer to the power of the collective (the masses), the urge to get along together, to be together, to identify with others.

7 I should emphasise at this point that by clubbing I mean more than the 'raving' which has provided the focus of much of the work on clubbing to date (see Redhead, 1993; Saunders, 1995; Thornton, 1995, for example). I include within the remit of my work, for example, 'new age' or ambient clubs, jazz and funk nights, 'jungle' or 'drum and bass' nights, and more eclectic easy listening and 'new beats' type nights, as well as house and garage club nights. I don't include nights where live music forms the basis of the evening, or clubs where the *average* age of the clubbers seems much over 25. Furthermore my focus is restricted to central London.

8 The music often makes conversation between *more* than two people difficult (shouting into two people's ears at once is tricky), resulting in the creation of micro-aural social spaces in which the conversation cannot be overheard, even by other people very close by.

9 In referring to performance here I am trying to evoke some of the sense of the ways of doing (or ways of clubbing) and constant monitoring of self and others which characterises (to a lesser or greater extent unconsciously) most clubbing experiences. This notion of performance is heavily based on the work of Goffman (1959, 1963, 1967, 1971 for example) in which he develops what he calls his 'dramaturgical perspective'. This is a framework which Goffman uses for unravelling the procedures employed by people in their 'face-to-face dealings with each other at a micro-sociological level' (Drew and Wootton, 1988: 1).

10 The concept of identification is fundamentally different to that of identity in that, theoretically, it takes account of the fact that we wear many different hats at different times and in different surroundings. We present ourselves differently and identify with different things in various contexts. The logic of identification emphasises the sympathy (in the purest etymological sense) that we feel with others when we share situations with them.

11 Interestingly Simmel writes here not of the club experience (though there was a healthy club scene in London in the early part of the century – see

Wyndham and St. J. George [1926]) but of the experience of the city more generally.

12 Maffesoli uses the notion of *unicity* here to convey a much more open and heterogeneous condition than that of unity: 'As examples of the differences between the notions of *unity* and *unicity*, one might compare the following, identity–identification; individual–person; nation-state–poly-culturaliza-tion' (Maffesoli, 1991: 19).

13 Muggleton undertook in-depth ethnographic work with a sample of 57 'spectacular stylists' whom he regarded as non-conventional in appearance. He found that many respondents explicitly rejected any named sub-cultural identity, instead preferring to talk about their ability to identify with what they wished, when they wished, how they wished and where they wished. So instead of a set of highly distinctive groupings he found a movement 'towards the emergence of a more of less diffuse collection of people who are expressing their non-conformity in more varied ways' (Muggleton, 1995: 8).

REFERENCES

Bauman, Z. (1993) *Postmodern Ethics*, Oxford: Blackwell.

Chambers, I. (1986) *Popular Culture: The Metropolitan Experience*, London: Routledge.

—— (1994) *Migrancy, Culture, Identity*, London: Routledge.

Crossley, N. (1995) 'Body techniques, agency and intercorporeality: on Goff-man's *Relations in Public*', *Sociology* 29, 1: 133–49.

de Certeau, M. (1984) *The Practice of Everyday Life*, Berkeley: University of California Press.

Drew, P. and Wootton, A. (eds) (1988) *Erving Goffman: Exploring the Interac-tion Order*, Cambridge: Polity Press.

Finkelstein, J. (1991) *The Fashioned Self*, London: Polity Press.

Frith, S. (1988) *Music for Pleasure*, London: Polity Press.

—— (1992) 'The cultural study of popular music', in L. Grossberg *et al.* (eds) *Cultural Studies*, London: Routledge.

Glennie, P. and Thrift, N. (1996) 'Consumption, shopping and gender', in N. Wrigley and M. Lowe (eds) *Retailing, Consumption and Capital: Towards the New Retail Geography*, London: Longman.

Goffman, E. (1959) *The Presentation of Self in Everyday Life*, London: Penguin.

—— (1963) *Behavior in Public Places: Notes on the Social Organization of Gatherings*, New York: Macmillan.

—— (1967) *Interaction Ritual: Essays on Face-to-Face Behavior*, London: Allen Lane.

—— (1971) *Relations in Public: Microstudies of the Public Order*, London: Penguin.

Halfacree, K.H. and Kitchen, R.M. (1996) ' "Madchester Rave On": placing the fragments of popular music', *Area* 28, 1: 47–55.

Hall, S. and Jefferson, T. (1976) (eds) *Resistance Through Rituals: Youth Subcultures in Post-war Britain*, Hutchinson: London.

Hannerz, U. (1980) *Exploring the City: Inquiries toward an Urban Anthropology*, New York: Columbia University Press.

Hebdige, D. (1979) *Subculture: The Meaning of Style*, Methuen: London.

Hyder, K. (1995) 'Ecstasy's deadly cocktails', *Observer*, 13 August: 10.

Kendon, A. (1988) 'Goffman's approach to face to-face interaction', in P. Drew and A. Wootton (eds) *Erving Goffman: Exploring the Interaction Order*, Cambridge: Polity Press.

Lofland, L. (1973) *A World of Strangers: Order and Action in Urban Public Space*, New York: Basic Books.

McRobbie, A. (1994) *Postmodernism and Popular Culture*, London: Routledge.

Maffesoli, M. (1988) 'Jeux de masques: postmodern tribalism', *Design Issues*, 4 1–2: 141–51 (special issue).

—— (1991) 'The ethics of aesthetics', *Theory, Culture and Society* 8: 7–20.

—— (1993) *The Shadow of Dionysus: A Contribution toward the Sociology of the Orgy*, trans. C. Linse and M. Palmquist, New York: State University of New York Press.

—— (1995) *The Time of the Tribes: The Decline of Individualism in Mass Society*, London: Sage.

Morley, D. (1995) 'Theories of consumption in media studies', in D. Miller (ed.) *Acknowledging Consumption: A Review of New Studies*, London: Routledge.

Muggleton, D. (1995) 'From "subculture" to "neo-tribe": identity, paradox and postmodernism in "alternative style"', paper presented at Youth 2000 International Conference, University of Teesside, 19–23 July.

Nicolette (1996) *Beautiful Day*, mixed by Mark Broom, Talkin' Loud Records.

Nowicka, H. (1995) ' "Borderline" classics split music industry', *Guardian*, 11 November: 5.

Raban, J. (1974) *Soft City*, London: Harvill.

Redhead, S. (1990) *The End of the Century Party: Youth and Pop Towards 2000*, Manchester: Manchester University Press.

—— (1993) (ed.) *Rave Off: Politics and Deviance in Contemporary Youth Culture*, Aldershot: Avebury.

Saunders, N, (1993) *E for Ecstasy*, London: Neal's Yard Press.

—— (1995) *Ecstasy and the Dance Culture*, London: Neal's Yard Press.

Scheler, M. (1954) *The Nature of Sympathy*, London: Routledge and Kegan Paul.

Sennett, R. (1990) *The Conscience of the Eye: The Design and Social Life of Cities*, London: Faber and Faber.

—— (1994) *Flesh and Stone: The Body and the City in Western Civilization*, London: Faber and Faber.

Simmel, G. (1903) 'The metropolis and mental life', reprinted in D. Levine (1971) (ed.) *Georg Simmel: On Individuality and Social Forms*, Chicago: University of Chicago Press.

Thornton, S. (1994) 'Moral panic, the media and British rave culture', in A. Ross and T. Rose (eds) *Microphone Fiends: Youth Music and Youth Culture*, London: Routledge.

—— (1995) *Club Cultures: Music, Media and Subcultural Capital*, London: Polity Press.

Welsh, I. (1996) *Ecstasy: Three Tales of Chemical Romance*, London: Jonathan Cape.

Wyndham, H. and St. J. George, D. (1926) *Nights in London: Where Mayfair Makes Merry*, London: The Bodley Head.

....

four

sites of resistance

While young people participate in and contribute to the production of every-day spaces (as previous sections have outlined), they remain a relatively marginalised group in society. Rather than passively accepting dominant cultural practices and exclusionary productions of urban and rural space, young people often, as the final four chapters of this book demonstrate, actively resist them in different ways and at different scales.

In her chapter on Germany, 'Between East and West: sites of resistance in east German youth cultures' Fiona Smith clearly demonstrates the transformations that have been taking place in former East Germany since 1989 in relation to youth cultures and youth movements. While often dominated by political rhetoric youth movements have been used by young people as a site of resistance. She shows that this resistance was taking place before 1989 and has continued after reunification.

Myrna Breitbart also emphasises the amount of commitment, energy and passion young people are willing to put into changing and redefining their own environments. In this chapter she shows (echoing some of the themes in Cindi Katz's chapter) how economic decline impacts directly on young American lives through the built environment, but she then goes on to demonstrate how young people have worked collectively through a range of imaginative projects – including public art – to redefine their urban landscape.

Most work on youth cultures has an urban focus. But in 'Vanloads of uproarious humanity: New Age Travellers and the utopics of the country-side', Kevin Hetherington shows how the very lifestyles and cultures of New Age Travellers disrupt meanings and understandings of the English countryside. He debates the concept of New Age Travellers and argues how influential they are in contemporary British youth cultures.

In the final chapter of this section and indeed the book, Susan Ruddick discusses the ways in which 'homeless' youth cultures in Hollywood are oppositional and yet at the same time contribute to a larger process of place production. In 'Modernism and resistance: how "homeless" youth sub-cultures make a difference' she considers the ways in which dominant images of young people have been formulated over time. She charts the changes in youth culture through a consideration of differing experiences of youth homelessness in a particular place, Hollywood, and shows how each form has been part of a movement of youth resistance and self-definition.

17

BETWEEN EAST AND WEST

sites of resistance in east German youth cultures

●

Fiona M. Smith

In the GDR a young generation is growing up which knows all the groups and cliques which have developed in other industrialised nations, even if they have different emphases and partly 'limp behind' the developments in the West.

(Friedrich-Ebert-Stiftung 1988· 6)

The primary task in shaping the developed socialist society is to raise all young people as citizens of the state, truly devoted to the ideas of socialism, who think and act as patriots and internationalists, who strengthen socialism and protect it safely against all enemies. Young people themselves carry heavy responsibilities for their own development into socialist personalities.

(Youth Law of the GDR 1974: para. 1.1)[1]

The first quotation comes from a publication giving west German young people an insight into life for their counterparts in the German Democratic Republic (GDR). It argues that virtually all the youth cultures in the West were present in the East, even if they did seem to 'lag behind' in some respects. However, in the first comparative study of east and west German youth in 1990, Lenz (1991) showed that young people had the same priorities for their lives, attached the same importance to their peers, and that rather than being more conservative, east Germans were even more in sympathy with new social movements than western youth. The approach here echoes work by Hilary Pilkington (1994) on Russian youth cultures in arguing that simply analysing youth cultures in eastern Europe as a poor relation of western youth cultures is to miss their significance in reflecting the forces that shaped them and in shaping their societies.

It is too easy to suggest youth cultures in eastern Europe simply reflected directly state and Party ideologies, such as the Youth Law of the GDR quoted above. Instead I want to explore how young people in the GDR were actively engaged in the processes which constituted their lives and to follow this through to understand something of the complexities of how German reunification is shaped by the actions of those affected by it. This is not a comprehensive view of German or even east German youth cultures.[2] Rather, the chapter addresses geographies of resistance among east German youth cultures, in the GDR before 1989, during the revolution of 1989–90 and during German reunification, at a variety of sites and scales. To place this in context, I begin by considering the position of youth in relation to official GDR youth policy.

YOUTH AND GDR YOUTH POLICY

In the communist period, youth cultures always stood in tension between western inspiration, primarily through music and style, and the lived experiences of the period for young people. The GDR state essentially sought to control young people's spaces through the *einheitliches Bildungssystem* (unified educational system) (Waterkamp, 1989) and through the only officially sanctioned youth organisation, the Freie Deutsche Jugend (FDJ: Free German Youth). Both were built around conceptualisations of youth, common across communist countries, of youth as 'constructors of communism' (Pilkington, 1994: 4). The task of school and FDJ was to educate young people in Marxism–Leninism and in their duties to work and to defend their country (Freiburg and Mahrad, 1982). The assumed energy and romanticism of youth was to be mobilised and controlled, particularly as youth were also 'victims of western influence' from which they had to be protected (Mahrad, 1977: 198). Official GDR youth policies did not change their Stalinist forms even after the *perestroika*-inspired changes in Soviet youth policies of the late 1980s (Frisby, 1989; Pilkington, 1994).

The extent of involvement in these official policies in school, university, vocational training and in leisure was considerable. FDJ membership accounted for 75 per cent of the 14 to 25 age group, including nearly all school pupils, students and apprentices. In addition, nearly all 6- to 13-year-olds were members of the junior section, the 'Ernst Thälmann Pioneers' (Freiburg and Mahrad, 1982). According to its statute, the FDJ worked 'under the leadership of the Socialist Unity Party of Germany [the GDR communist party] and sees itself as its active helper and reserve in times of struggle' (Freiburg and Mahrad, 1982: 11). The FDJ organised young people in schools, universities, workplaces, youth clubs, summer camps, and paramilitary training and in the armed forces where its tasks included political education, encouragement to achieve and steering leisure (Freiburg and Mahrad, 1982: 12). Formally these aims were met by high membership figures. However, since almost all organised leisure activities required FDJ membership, and advancement to senior school and higher education required good grades in political education, it was by and large not membership which made a political statement but non-membership (ibid.: 19). Most young people lived with the tension between their knowledge of the west from western television and their own life experiences on the one hand, and on the other the required behaviour necessary to get on in the GDR system. Many also adopted strategies and actions to circumvent or challenge official spaces and state policies. They also wanted to challenge the dominant silent majority of the population.

SITES OF RESISTANCE

Young people sought to utilise space and place as sites of resistance, often finding other spaces, using the margins of or subverting the use and significance of official spaces. This covered a range of scales and sites which are examined in turn: the body as a site of resistance in style sub cultures; state defined youth spaces such as clubs or work; new places in the neglected spaces of old buildings and neighbourhoods; private spaces in homes and among friends; state controlled public spaces; and the free spaces offered by GDR churches.

BODY AND STYLE

There was in the GDR little of a 'drop-out' culture of those who chose not to work, since there was no official unemployment and anyone who refused to work lost all social security and income and, if under 18, could be sent to a correctional institution. The low incidence of unemployment also meant that unlike many western youth sub-cultures which grew out of concentrations of high unemployment, most 'sub-cultures' in the GDR had different origins. Skinhead groups consisted mostly of young men in education or in full time employment, as did punks or goths, or any of the other groups which were found in particular versions in the GDR. Such style groups used a range of sites to question dominant and official GDR cultures. Clubs and pubs often became associated with particular groups, and private spaces were used to meet and socialise. While skins, punks, goths and others began as a relatively undifferentiated style sub-culture movement in the late 1970s, groups later sub-divided into different styles.

In common with many other youth cultures, young people in the GDR used the body and particular styles as a site of difference, just as wearing the FDJ scarf at school signalled at least passive conformity. Punks in the GDR were largely concerned with shocking and rejecting the bourgeois dominant culture but despite their concern often with left wing issues, they were seen by the state as an attack on order and discipline. In a collection of accounts of life for young people in the GDR in the early 1980s, Büscher and Wensierski (1984) recount the experiences of East Berlin punks. They were regularly stopped on the streets by the police. One of the punks, Sid, responded:

> These wankers! Have nothing better to do than to attack me because of my clothing, my difference on the outside. My overseer and the other ones up there who really [steal materials at the factory] aren't even under surveillance, but us they lay in to.
>
> (Büscher and Wensierski, 1984: 112)

While punks were often arrested and even fined for their 'unaesthetic appearance' (Freiburg and Mahrad, 1982: 47), skins and neo-Nazi group members prided themselves on an ultra-neat appearance which at once conformed to state promoted ideas of order and discipline, and subverted this communist order and discipline to a fascist meaning by wearing certain items of clothing or certain hairstyles. Contemporary studies found a surprisingly large number of skins and neo-Nazis were skilled workers from families with a strong Party or state employment background. Moreover, many were involved in the FDJ's youth policing groups, the *Ordnungsgruppen*, or in the *Gesellschaft für Sport und Technik* (Society for Sport and Technology) which offered military style sports, such as shooting. It was therefore precisely the sites which the state designated as showing commitment to the communist order (work, order, Party, military readiness) which right wing young people used to subvert the meaning of the state. Paradoxically, the GDR denied the possibility of the existence of any neo-fascist tendencies up to the late 1980s since the state was constitutionally anti-fascist (Hockenos, 1993). Only after a particularly violent attack by skins on a punk concert in a Berlin church in 1987 did the state finally admit their existence.

CLUBS AND NEIGHBOURHOODS

A second way in which young people used space was by subverting places specifically defined by the state as youth spaces, either by ignoring the state's claims to control it or by sidelining the state's intentions. A ruling that only 40 per cent of music played at discos could be from non-socialist countries was often ignored by DJs, or the music was played in the breaks (Freiburg and Mahrad, 1982). However, young people (and others) also actively created their own spaces, including spaces for non-standard styles of living in older neighbourhoods in towns and cities across the GDR, and especially in Berlin, and there were also rural attempts at alternative lifestyles.

GDR housing policy strongly favoured construction of pre-fabricated blocks in large estates while most inner city areas were left to run down. In a situation of housing shortage, particularly for young single people who were not given a high priority on waiting lists, and of the desire for an alternative cultural scene, many moved to these areas, often squatting and creating spaces for independent action:

> In small shops and workshops in older neighbourhoods, which encour-
> aged such uses just by being available, . . . small galleries, shops for
> artists' materials, cafés, rooms for lectures, music, theatre and other
> functions were all renovated, mostly by DIY in a combination of self-
> help, state support and the market. The politics of this movement
> depended on one's own quarter and on an informal public space in the
> quarter, not on organisations. In a symbolic connection to home in the
> old quarters the distance not only to the state but also to the modernised,
> industrialised, working and functional model of civilisation is expressed.
> *(Göschel and Mittag, 1991: 1–2)*

People used these areas and actions rejected by the GDR state as bourgeois or pre-socialist and subverted them away from the dominant consumption oriented culture. However, the squatting had little in common with the often anarchistic squatter movements of many west German cities. For most the issue was primarily one of housing and they aimed to repair the flats to a level of acceptable accommodation. In recognition of this, in East Berlin at least, those who squatted could have their tenancy recognised if they could prove the flat had been empty for at least three months and paid a fine for having circumvented the allocation system.

PRIVATE SPACE: PUBLIC SPACE

While creating these alternative neighbourhoods was important, even in the areas with many squatters and an alternative culture, the predominant use of space for leisure was in the private sphere. Family, friends and home were of great importance to most young people (Lemke, 1991) and the majority of leisure time was spent listening to music and watching television, as well as playing sport and being with friends. This 'retreat to the private niche' (Gaus, 1983: 115) was seen as typical for all generations and represented a major strategy for dealing with the claims of state and Party control since in addition to youth policies the state influenced the population as a whole through strategies of organisation and involvement in everyday life and of surveillance by the Stasi, the state Security Service, whose large network of informants kept much of the GDR population in fear of being reported. In this situation, personally controlled spaces became significant. For example, the East Berlin punk, Sid, and his friends had a rehearsal room for their band:

> What is a cult in the West – the hidden and inaccessible practice room in the concrete pillar of a motorway – is for us a simple necessity, being screened off from the outside world, our space which is the way we have decided.
>
> *(Büscher and Wensierski, 1984: 112)*

However, there were also significant challenges to state control of public space, often transgressing the limits of acceptable action. Not surprisingly, punks used public spaces for particular ends. The authorities often viewed punks as potential neo-Nazis because they occasionally shouted 'Heil Hitler!' on the streets as the greatest possible provocation in the GDR, but this was often disrupted by the punks themselves. One group tried to lay a wreath at the concentration camp at Sachsenhausen on 8 May 1983 but, like a group of gay men the same day, they were prevented. They therefore went to the main military memorial in the centre of East Berlin, the Watch on Unter den Linden:

> They went past the soldiers standing to attention into the holy of holies of GDR memorials. There they laid their wreath with a ribbon bearing the words 'The punks of Berlin'. [One explained that] even if they don't think much of politics otherwise, they won't allow themselves to be pigeon-holed in the neo-fascists' corner by the state.
>
> *(Büscher and Wensierski, 1984: 113)*

It would be wrong, however, to suggest that the only elements of youth cultures in the GDR involved style groups. From the late 1970s onwards, a series of independent groups developed concerned particularly with environmental and peace issues (Rytlewski, 1989). Such groups, while often operating from a base in private homes or churches, also exploited the margins of participation in public space. For example, public gatherings were controlled by the state and were usually not permitted, but cycling was not illegal. A collection of people cycling was legally part of normal traffic, so environmental groups used bicycle demonstrations to protest at damage caused by the GDR's economic intensification policies. Bicycles themselves became a symbolic site for resistance to the logic of the state which promoted increased material wealth as its legitimating characteristic in the 1980s. Some peace groups went a step further and asked to participate in the FDJ sections of official Whitsun marches, causing consternation among local functionaries who were surprised by young people actually wanting to march, rather than the merely ritual participation of most young people. Such actions were not unchallenged by the state and participants were often kept under surveillance or were stopped by police 'checking papers'. Nor was such involvement unproblematic for the participants:

If we don't have an effect outside our own circle then our involvement will have no success in the long run.

If we march with the FDJ we will have lost our credibility with the population and they'll just lump us in with it.

(Members of a Leipzig peace group, 1983:
Büscher and Wensierski, 1984: 134)

Reactions were often particularly strong when the state was challenged on its own terms. A demonstration in January 1988 marking the anniversary of the deaths of Rosa Luxemburg and Karl Liebknecht by around one hundred of those who had applied to leave the country was met with force and many arrests, even though Luxemburg and Liebknecht were key early communist figures (Rytlewski, 1989). The state's ultimate threat was the order to shoot those attempting to cross the border to the West illegally.

FREE SPACE: FREIRAUM

The only state sanctioned public 'free space' (Freiraum) in the GDR was in the churches, where there was an uneasy 'truce' between the state and the established churches. It was here that many environmental and peace groups developed. Churches offered spaces for rock concerts, for discussions not controlled by a need to give the rote learned answers of school, and, more controversially, for those who refused to serve in any way in the army, for those who had applied to leave the GDR, or for lesbians and gay men who were subjected to discrimination (Büscher and Wensierski, 1984). However, many young people who were not interested in political action of any form found that churches simply offered a space to meet which was unstructured by state and FDJ:

> I go down to the basement [of the church where the youth group meets] especially when it's cold outside. You can meet your mates there . . . . It's good with your mates, especially driving around in a car, parties and drinking. I don't give a damn about the church.
> *(19-year-old, cited from a church study of youth work:*
> *Lemke, 1991: 179)*

Churches combined their role with other concerns (such as environmental or peace issues) in ways particularly designed for young people, such as the Blues Masses of churches in Dresden, which several thousand young people attended. These created not only a space and particular time for communal actions which questioned the state and the dominant silent society, but also offered an opportunity for contact between people and a sense of common cause, acting to undermine the effect of the Stasi.

Young people were therefore not merely 'victims' of the West or of the East, nor were they uniformly constructors of communism, although this account probably underestimates the extent to which many young people were committed to the ideals of Marxism–Leninism, if not in the precise form taken in the GDR. Many young people actively or passively resisted the ideology and actions of the state by creating other geographies which challenged and subverted state intentions at a series of scales from the body to the home, the club, the street, the churches and other niches created in the margins of the official geographies of the GDR. Such resistance grew and by the late 1980s there was a 'rapid and general collapse' of support for the principles of Marxism–Leninism among young people in the GDR, documented in various studies which could not be published at the time.[3] This unpublished material was followed by the spectacular images of thousands of young people leaving the GDR for the West via Hungary and Czechoslovakia and then of young people among the hundreds of thousands who took to the streets of the GDR in the autumn and winter of 1989.

(RE)UNIFICATION AND (RE)CONTESTED SPACES

The GDR revolution in 1989 and German reunification in 1990 changed most aspects of life for young people. It brought about the collapse of east German youth organisations and the existing education system. State youth clubs were either closed or, increasingly, privatised. The majority of young people embraced the opening to the west, particularly in relation to leisure, entertainment, new consumption patterns and the extent of foreign travel, which all show rapid convergence with standard patterns elsewhere in the West. Dominant discourses on youth moved from 'constructors of communism' and 'victims of western influence' to youth as a prime agent of consumption and, in particular, as a 'problem', in relation to crime, drugs, political extremism and the use of urban space (see Pilkington, 1994 for the comparable situation in Russia). The 'problem' is often discussed in its absence, as any formal involvement of young people in legislation and administration was lost with the GDR. Here I concentrate on experiences in the city of Leipzig in the south of eastern Germany where I was engaged in field work between 1991 and 1993.[4] I address the geographies of reactions to the change of the nation-state and to the introduction of capitalism, highlighting continuities and discontinuities between GDR and united Germany.

A CHANGE OF STATE

Perhaps the most obvious use of space by young people was the continuation of the street-based protests of the revolution. In winter 1990–1 students began a series of protests about changes to east German universities which were seen as imposing western personnel and ideas with no differentiation of any positive aspects of the GDR system. A march through the GDR, occupation of the Leipzig University chancellors' office, and even hunger strikes sought to draw attention to how reunification was affecting young people. At the same time, in common with reactions across Germany to the Gulf War, a peace camp was established in the square which had held the mass demonstrations of the revolution in Leipzig, protesting at western involvement and at German financial support for the war. Both were system critical actions using often symbolic public and official spaces to make political points.

Criticism of the nature, and even the fact, of reunification continued. On 3 October 1991 a celebration was staged in Leipzig's market square (notably not on the square where the revolution had been played out) to mark the first anniversary of reunification. A group of largely left wing young people marched through the inner city streets whistling and shouting in protest at the nature of reunification and the new state's (lack of) response to right wing extremism. In contrast to the main event where policing was unobtrusive, this small demonstration was heavily controlled with riot police on stand-by. Participants were photographed and filmed. While the intention of some of those taking part was to provoke confrontation, the march was heavily policed internally by the participants so that there was no violence. The apparent continuity of the relation between state/police and protests was reinforced when later that night the café at the centre of the city's alternative district was fire-bombed and destroyed by right wing extremists, despite warnings to the police that violence was planned. The café had become a source of irritation to the police as a place where drugs were available and those who used it argued that the police thought the right wing groups had done them a service.

Right wing youth developed different perspectives on the new state, seeing opportunities for change, having escaped from the GDR, but also seeing generational conflict between parents who had 'betrayed' them in the past and were doing so now. Paul Hockenos (1993) gives an account of a meeting with some young people in Dresden:

> One streetcar stop beyond the Square of the Construction Workers begins Dresden's largest housing project, Gorbitz . . . . Up a muddy incline sits a flat, single storey building – the uniform GDR youth club. The graffiti covering the Espe Club's walls shows that Gorbitz has earned its reputation as the stronghold of Dresden's hardcore neo-Nazi scene . . . . 'Yeah, we're all right-wing here,' says Jörg, eighteen, in a rough, working-class Saxon accent . . . . Jörg also says that he's a member of the Ku Klux Klan. Although he's never met another Klan member, he has their translated literature . . . . Gerd says that he's a member of the FAP [*Freie Arbeiter Partei*, another German right-wing extremist party] 'because they got the best program.' That program he summarizes in two words: 'Stop foreigners.' . . . Ute says that she will join the Aryan-supremacist *Wiking Jugend* (Viking Youth) as soon as she turns eighteen next year. 'The WJ just wants to restore traditional German values and Germany's real borders,' she says . . . . 'In principle, my parents agree with a lot that the WJ stands for,' she says, 'but they're your typical *Mitläufer*, the followers, the worst of the lot.' The others nod with disgust. 'First they were communists, now they vote CDU [Chancellor Kohl's party],' she says, her high voice rising higher. 'And we're the ones who pay.'
>
> (Hockenos, 1993: 58–60)

Protest against the new state therefore often used urban spaces to challenge the definition of the new state, but it also involved organisation across space, whether in relation to international right wing organisations, as shown above, or as was the case when the reform of public broadcasting after reunification threatened closure of the ex-GDR's youth radio station. The radio station, DT64, had developed post-revolution into a station with a strong identity around the experiences of east German youth, dealing with a range of social issues as well as cultural and music issues. Responses were seen in demonstrations, graffiti, petitions and benefit concerts across eastern Germany. The medium for communication was DT64 itself, but listeners ran local committees and implemented actions which even crossed into western Germany. Despite this campaign of organisation across the space of eastern Germany, success was only partial, as DT64 moved from FM to MW for the southern part of eastern Germany and was lost further north. The fight to maintain DT64 radio station which had evolved from being an FDJ mouthpiece to a critical and particularly east German radio station showed the efforts of the state to erase any traces of the GDR past. It also showed an ability among young people to organise across the territory of the ex-GDR and beyond in the cause of saving something which they believe represents them. It represents both resistance to the commercialisation of youth culture and the loss of voices expressing east German experiences, but it is also part of welcoming the musical 'imports' of Western culture. There are, therefore, contradictory elements in this aspect of youth resistance.

CAPITAL AND URBAN SPACES

Reunification was marked not only by the change of the identity of the nation-state. In common with the rest of eastern Europe, processes of transformation involved the (re)introduction of capitalism. Between the GDR revolution and reunification a period of relative openness led, in many areas which had been centres of squatting and alternative scenes, to the establishment of groupings of mostly young people who sought to change the dominant urban planning concepts and to establish alternative housing and culture projects. One such was the Connewitz Alternative in Leipzig, located in a small area of poor quality nineteenth century workers' housing:

> We want to save this unmistakable quarter . . . and we want to transform it at the same time and give it an alternative-cultural character.
>
> *(Initiator of project, quoted in*
> Leipziger Volkszeitung, *24 March 1990: 3)*

The project was described by the local newspaper as aiming to create a Leipzig 'Montmartre', with 'young workers, students, teachers, artists and traders' (ibid.). The group sought to negotiate with the city housing department for occupancy rights for some properties designated for demolition. After reunification and the concomitant (re)introduction of market values and private ownership of property (Smith, 1996), such projects found a much greater difficulty in establishing their rights to use such buildings. Connewitz Alternative were clear that their intentions were that 'Connewitz should remain Connewitz and not become a noble quarter for capitalists', and should provide space for 'other ways of living and other mentalities than those of normal citizens.'[5] They wanted to support young people with housing, sport, art and training. However, the city council designated the area as a renewal area and appointed a western developer. Thus, while the group sought a certain continuity of GDR-type housing and alternative cultural area-based movements, the structures of western planning, private ownership and market values meant their plans were undermined.

Several other changes helped to problematise the whole issue of Connewitz as a space where young people could shape their own form of living. As mentioned above, the main café of the area was seen by the police as a source of irritation. Street festivals were blocked by other local residents objecting to the noise, and the semi-legal Zoro venue was subject to protests and threats of vigilante actions by local residents (Holterdorf, 1993). The city council was divided in its reactions. The youth department supported independent youth venues because unlike most state supported forms, they were actually used by young people. The housing control office and the local licensing department, as well as the regional government's law and order department wanted strong action against the whole 'squatter scene'. The unacceptability of such non-capitalist use of land was heightened by local administrators' concerns to see the new system of planning for urban space correctly implemented.

The language used to characterise these youth geographies was one of threat and disorder – Connewitz was regularly described as 'the Bermuda Triangle'. This was reinforced as those involved stressed the area and the culture was deliberately 'other' from the dominant culture. Those involved were well aware of the level of illegal activities but were often more concerned at the extent to which the area was becoming home to many young people who either chose to drop out (something now possible) or who were homeless, leading to the city employing two street workers. In contrast, the security discourse dominated most official circles, in turn reinforced by the way Connewitz had become more open to crime and particularly to drugs which had been largely non-existent in the GDR. While areas such as Connewitz became centres for largely left wing youth cultures and for those with few political convictions, the right wing scene developed its own geographical strongholds, most often in the modern estates of the GDR period, often focusing on former FDJ youth clubs (Göschel and Mittag, 1991). One element in the geography of youth cultures then became, very obviously, the divisions and violence between left wing and right wing groups, further increasing the 'youth as problem' discourse and creating particular sites as areas of conflict.

It is important not to overplay the extent of action and resistance among young people post-reunification. Most welcomed the new order, although it also represented an increased level of uncertainty (financial, finding a traineeship, finding housing, finding employment) (Göschel and Mittag, 1991). This contrasted with a key experience of the GDR for young people which was the stability of life (Meier, 1984). Divisions also increased as the younger, lower status people tended to hold to western models while older, higher status 'youth' (often those in their twenties) were still seeking some mix and continuity from the GDR.

Many simply want to forget something that they now see as irrelevant. The new spatiality of the 'dance capital of the East' in Leipzig shows something of this. The interrogation cellars in the Stasi headquarters in Leipzig were converted into the commercially run 'U-2' club, while raves ('the biggest Techno Party in East Germany') are held in numerous post-industrial sites, monuments of GDR technology left by the collapse of east German employment (factories, breweries, trade fair halls). In the new socio-economic system, more commercial dance venues are generally less controversial than the semi-legal attempts to create alternative spaces. However, when the local council closed down the 'Destillery' club because of fire regulations and licensing problems, young people had not forgotten the power of public protest and took their party to the square outside the town hall until the council helped to find an alternative site in an unused railway building (Busse, 1995).

CONCLUSION

There are continuities between the geographies of youth cultures in the GDR and in eastern Germany after German reunification, such as using urban spaces as sites for articulating criticism of the state and of systemic issues. Young people acted at a variety of scales, creating local spaces and organising across space. Divisions between youth cultures also remain, particularly between left and right wing youth, and more generally between the minority seeking alternative cultures and the majority who adapt(ed) to the dominant culture. There are changes in how urban space can be used and in the ways the state sought/seeks to regulate youth cultural uses of space, though sometimes these forms show more similarity between the GDR and the reunited Germany than the state is willing to accept. East German youth cultures continually present(ed) a challenge to the pure spatiality of the nation-state both in the GDR (rejecting the state's claims to control) and in the new Germany which refuses to recognise the possibility of continuities from the GDR. At the same time they are rooted in particular experiences of place, are divided and contested, and are negotiated within hegemonic relations of power and control.

NOTES

1 Translation from the law on the participation of youth in the formation of developed socialist society and its comprehensive promotion in the German Democratic Republic – Youth Law of the GDR, published in Gesetzblatt der DDR (1974). All translations from German sources are the author's.

2 There is more written in German than in English about youth cultures in the GDR or in eastern Germany after reunification. However, two English texts are useful here. Paul Hockenos (1993) writes about the rise of right wing extremism in eastern Europe and has two chapters dealing with the situation in east Germany before and after 1989. Hilary Pilkington (1994), while dealing with Russian rather than GDR youth, gives a detailed and insightful account of ethnographic work in Moscow on the development of youth cultures both in the period of *perestroika* and after the collapse of the Soviet Union.

3 Förster and Roski (1990) reported a series of surveys of apprentices which indicate falling support for the GDR. While 33 per cent identified strongly with the GDR in 1979, the number had fallen to 9 per cent by May 1989. Similarly agreement with a statement that 'the goals of the FDJ are my goals' fell from 37 per cent to 4 per cent over the same period. However, even in 1970 only 24 per cent of apprentices surveyed agreed strongly that 'the Party has my complete trust' (53 per cent agreed with limitations) while by May 1989 53 per cent agreed 'hardly' or 'not at all'. Such survey work was strictly controlled by the GDR Academy of Sciences both in what could be asked and in what could be published. Results such as these which were deemed detrimental to the government and the Socialist Unity Party were censored and could only be published after the collapse of the GDR government.

4 Material about youth geographies was collected as part of a larger project which focused on the geographies of urban and neighbourhood action in the

period of transformation. The work was supported by the Carnegie Trust for the Universities of Scotland.
5 Author's notes from meeting of Connewitz Alternative, Leipzig, 28 January 1992.

REFERENCES

Büscher, W. and Wensierski, P. (1984) *Null Bock auf DDR: Aussteigerjugend im anderen Deutschland*, Hamburg: Rowohlt.

Buuuu, A. (1995) 'Die Party-Hauptstadt Ost heißt Leipzig', *Die Welt* 26 April 1995: WR14.

Förster, P. and Roski, G. (1990) *DDR zwischen Wende und Wahl: Meinungsforscher analysieren den Umbruch*, Berlin: LinksDruck.

Freiburg, A. and Mahrad, C. (1982) *FDJ: Der Sozialistische Jugendverband der DDR*, Opladen: Westdeutscher Verlag.

Friedrich-Ebert-Stiftung (1988) *Jugend in der DDR: Zu Jugendalltag und Jugendproblemen im 'Sozialismus'*, Bonn–Bad Godesberg: Friedrich-Ebert-Stiftung.

Frisby, T. (1989) 'Soviet youth culture', in J. Riordan (ed.) *Soviet Youth Culture*, Bloomington and Indianapolis: Indiana University Press.

Gaus, G. (1983) *Wo Deutschland liegt*, Hamburg: Hoffmann und Campe Verlag.

Gesetzblatt der DDR (1974) 'Gesetz über die Teilnahme der Jugend an der Gestaltung der entwickelten sozialistischen Gesellschaft und über ihre allseitige Förderung in der Deutschen Demokratischen Republik – Jugendgesetz der DDR', *Gesetzblatt der DDR* 1: 45–59.

Göschel, A. and Mittag, K. (1991) *Stadtteilkultureinrichtungen in Ost und West*, Berlin: DIfU.

Hockenos, P. (1993) *Free to Hate*, New York: Routledge.

Holterdorf, B. (1993) 'Zukunft des Zoro ungewiß – Anwohner fordern Schließung', *Leipziger Volkszeitung*, 24 August: 11.

Leipziger Volkszeitung (1990) 'Leipzigs Montmartre', *Leipziger Volkszeitung*, 24 March: 3.

Lemke, C. (1991) *Die Ursachen des Umbruchs 1989*, Opladen: Westdeutscher Verlag.

Lenz, K. (1991) 'Kulturformen von Jugendlichen: Von der Sub- und Jugendkultur zu Formen der Jugendbiographie', *Aus Politik und Zeitgeschichte*, B27/91: 11–19.

Mahrad, C. (1977) 'Jugendpolitik in der DDR', in W. Jaide and B. Hille (eds) *Jugend im doppelten Deutschland*, Opladen: Westdeutscher Verlag.

Meier, A. (1984) 'Soziale Sicherheit und Zukunftsbewußtsein der lernenden Jugend in der DDR', *Informationen zur soziologischen Forschung in der Deutschen Demokratischen Republik* 20: 15–18.

Pilkington, H. (1994) *Russia's Youth and its Culture: A Nation's Constructors and its Constructed*, London: Routledge.

Rytlewksi, R. (1989) 'Politische Kultur und Generationswechsel in der DDR: Tendenzen zu einer alternativen politischen Kultur', in B. Claußen (ed.) *Politische Sozialisation Jugendlicher in Ost und West*, Bonn: Bundeszentrale für politische Bildung.

Smith, F.M. (1996) 'Housing tenures in transformation: questioning geographies of ownership in eastern Germany', *Scottish Geographical Magazine* 112, 1: 3–10.

Waterkamp, D. (1989) 'Erziehung zur Identifikation mit dem Staat in der DDR', in B. Claußen (ed.) *Politische Sozialisation Jugendlicher in Ost und West*, Bonn: Bundeszentrale für politische Bildung.

••••••••••••••••••••••••••

18

'DANA'S MYSTICAL TUNNEL'

young people's designs for survival and change in the city

●

Myrna Margulies Breithart

In the last two decades and current post-industrial economy, US youth have experienced 'the most rapid deterioration in economic and social conditions since the Depression' (Males, 1993: 18). Since 1973, youth poverty has increased by 51 per cent, one in four young people now live in poverty, and violent crime and drug deaths among youth have doubled. These worsening conditions and a deepening youth unemployment crisis have generally been met by denial and a 'blaming of the victims'. Negative images of young people perpetuated by the media disproportionately impact upon non-white low income youth and severely compromise the quantity and quality of social programmes available to urban youth in general (Gooding-Williams, 1993). These images are often used to justify punitive rather than socially supportive policies, resulting in reductions in physical recreation amenities, the de-funding of youth organisations, curfews and other restrictions on young people's free access to public space (Rose, 1994; Medoff and Sklar, 1994; Jennings, 1992; Dawsey, 1995; Nauer, 1995).

Some have argued that a refusal on the part of policy makers to acknowledge the severity of problems affecting youth, and the promotion, instead, of negative youth stereotypes, is motivated by a desire to divert attention away from the desperate economic conditions and social problems currently perpetuated by adults (Males, 1996). Others have taken a more historical view, pointing to repeated 'moral panics' that have arisen since the late nineteenth century whenever the young urban working class was thought to be operating outside of adult authority. These panics are said to have generated representations of 'delinquent youth' as young people with personal characteristics, cultural practices, and family forms that diverge from 'normal' youth (Griffin, 1993).

One of the clearest demarcations of power, wealth and influence in the urban landscape has always been the ability to invest one's living space with meaning – to literally occupy, define and decorate one's surroundings. Graffiti writing and hip-hop, as they originated in the 1970s among youth in the devastated landscape of the South Bronx, attracted attention, in part, because they appeared to invert this relationship. Observers described the rise of these cultural forms as an effort on the part of marginalised and powerless young people to inscribe their living space with meaning against high odds, and in response to the brutal and destructive processes of neighbourhood decline, unemployment, crime, drug use and violence (Kelley, 1997). Rap music, for example, was seen to create a place from which young people could identify and target the negative social impacts of these problems and challenge their source (Rose, 1994). Graffiti writing, on the other hand, was seen to provide a symbolic means of escape for young people who might otherwise feel trapped within the geographical and social confines of their neighbourhoods (Cresswell, 1992). Taken together, these various forms of cultural expression and politics were thought to provide a means for youth to establish their unique identities in an urban setting while also drawing attention to, and, at times, resisting publicly assigned meanings to their lives.

What appears to have led many adults to interpret these urban cultural practices of youth in the extreme, as either positive expressions of freedom and opposition, or criminal activity, is their important locus in public space. When graffiti were confined to low income neighbourhoods, they did not engender multi-billion dollar clean-up campaigns. Only when they appeared 'out of place,' in other neighbourhoods or on urban transport vehicles, and when public space came to play an increasingly important symbolic and economic role in cities competing within a global marketplace for business, did massive anti-graffiti campaigns get under way. In times of fiscal austerity, city governments have also turned over key public spaces to private companies' management. The private sector has then used these opportunities to provide public amenities in exchange for more relaxed zoning and other regulations which contribute to corporate profits (Loukaitou-Sideris, 1993).

Whereas one might assume that the public space of the city is freely open to all inhabitants, the reality of privately provisioned public spaces such as corporate plazas and small parks is that a considerable amount of control is exercised over who may occupy those spaces and how they may be used by privately hired security forces. Trends towards the increased privatisation of public space as well as efforts to revitalise and gentrify key areas of the central city have thus generated numerous struggles over the definitions of, and public access to, urban space.

This is in stark contrast to the efforts of some urban youth to use public space to engender dialogue and/or implant new meanings of their own within the public landscape. As Tricia Rose and other cultural critics have pointed out, the constraints placed upon youth (whether in terms of gaining access to a public concert, a mall, or simply wall and street space) is very much about cultural politics, about 'not just what one says, . . . [but also] where one can say it, how others react to what one says, and whether one has the means with which to command public space' (1991: 276–7).

That urban youth are increasingly defined as 'undesirable' occupants of public space, and their access to it limited, is a function, in part, of privatisation trends, but also negative media images that portray young urban males as primarily dangerous, and young urban women as primarily vulnerable (with obvious racial and gender overtones) (Griffin, 1993; Breitbart, 1995c). Negative images of youth and the increased privatisation of public space both result in public policies that seek to remove young people from public places, delimit their geography and enforce their invisibility. Exemplary of these policies are increasingly popular curfews. Middle class suburbs with exceedingly low crime rates now join cities in the use of legal time curbs on the free access of citizens below the age of 18 to the out-of-doors. Indeed, President Clinton has come out publicly in support of curfews for *all* cities and towns in the US. This is in spite of ACLU (American Civil Liberties Union) challenges to curfews as 'house arrest for a chronological age', and other critical perspectives that point to the selective application of such laws to youth of colour (Breitbart, 1995c; Hanley, 1993: B1, 8).

These kinds of legal restrictions, combined with socially constructed popular images of urban youth as either victims or victimisers, have impacted upon all young people, but made it especially difficult for homeless youth who depend on access to safe public spaces to survive. These young people can only enter and occupy shopping malls, for example, by blending in, avoiding a certain type of dress and essentially becoming 'invisible' (Ruddick, 1996; Pfeffer, 1995).

Many middle class youths comply with restrictions on their free access to public space and seek refuge from what are perceived to be unsafe streets in increasingly commodified and expensive private indoor recreational settings (e.g. Discovery World, Kidsports[1]). In contrast, a number of less privileged youths are attempting to exert influence over their neighbourhood space and, along with concerned and involved adults, revision its appearance and purpose. Faced with numerous environmental and economic constraints, these young people are using street art, design and performance as mechanisms for reclaiming a space for themselves in urban life, or simply as outlets for creative expression and survival. They are attempting to revision and generate homelike qualities in otherwise unwelcoming and unsafe spaces. They are also using public space to contest preconceptions about themselves, and their lives, goals and aspirations.

This chapter presents three very brief case studies that document how such young people are currently involving themselves in the reassessment and modest alteration of urban space. Lessons derived from these case studies highlight the benefits as they are experienced and understood by the young people themselves.

YOUTH AND ART DESIGN INITIATIVES

Young people who live in declining parts of the city are profoundly aware of the influence that their local environments exert. They can literally see and feel the constraints that dangerous and/or inadequately provisioned neighbourhoods place upon them, and they can appreciate the opportunities that safe spaces, with ample resources, provide. Aside from supplying a physical backdrop for their activities and travel, these spaces send messages to young people about how an external world values or fails to value the quality of their lives. Indeed, it is often surprising to adults to discover how prominent observations about the built environment are in young people's expressions of hope or frustration.

DETROIT SUMMER

In Detroit, Michigan, adults and young people are currently collaborating to return small parts of the city to life sustaining pursuits. They are motivated, in part, by a desire to counter popular perceptions of the city as a violent and devastated former capital of automobile culture. With white flight to the suburbs (the city is now 75 per cent African American) and the shutdown of huge numbers of manufacturing facilities, formerly thriving working-class neighbourhoods are now devoid of jobs, businesses and services; they appear vacant (Darden *et al.*, 1987). Revitalisation efforts on the part of the city have directed what little money exists to corporate developments such as the Renaissance Center and platform people movers that elevate corporate executives high above the empty streets of the downtown. Detroit politicians have meanwhile embraced casino gambling as an economic development strategy and have taken to fencing off huge city blocks, removing them totally from public access or use. Inner city neighbourhoods surrounding downtown, once home to thriving retail shops, parks, businesses and homes, have taken on the appearance of a pre-urban landscape with huge numbers of vacant lots that are filled with burned out shells of buildings, tall grasses and debris.

As accurate as these visual images of devastation are, adult activists believe they obscure a rich history of grassroots organising around labour and civil rights issues, and ignore the current and deep commitments of residents to inventing a new local economy and new social relationships in Detroit. Contrary to the revitalisation efforts promoted by the city, these grassroots activities are encompassing in their breadth and accord an important role to the young.

Detroit Summer, 'a multi-cultural, inter-generational youth program to rebuild, redefine, and respirit Detroit from the ground up' is reflective of these activities (Dulzo, 1992). Its origins derive from the efforts of long time residents and activists to situate Detroit's problems within the larger context of changing global economic forces.

> Jimmy [Boggs] and I knew [we had] to establish a whole new social order. That's really when, for us, the idea of Detroit Summer was born. We have to establish new relationships – between old people and young, between workers and unemployed, between the educated and those who've never been to school, between the poor and the middle-class, between all of us and the earth.
>
> (*Grace Boggs, quoted in Dulzo, 1992*)

Begun in 1992, with the memory of Mississippi Freedom Summer 1964 in mind,[2] a collective of representatives from over fifty community organisations decided to invite local youth and college aged students from outside the city to come to participate with residents of all ages in a variety of projects and social programmes. Funded largely through donations, the programme provides volunteer youths with free housing, meals and transport to work sites.

For several summers, young people have worked side by side in collaborative teams with adults to construct local parks, design murals, paint houses for the elderly, build greenhouses and community gardens, rehab' housing, produce graphic art, and be trained in leadership and small business development (Plate 18.1). They have also sat together in discussion groups to identify pressing problems and trace their source. They have planned activities to address problems and have assessed the

Plate 18.1 Detroit Summer mural: part of a multi-cultural
youth project to improve Detroit.
Photograph: Gwyn Kirk.

meaning of their work in light of a larger movement for social and
economic change.

The emphasis placed on the change process as well as its tangible
products has enabled Detroit Summer to use relatively small and incre-
mental improvements to the built environment as a way of constructing a
foundation for a future and more encompassing transformation of urban
life. Through collective actions that involve artistic expression and concrete environmental intervention, youth are provided with models of
leadership development and opportunities for conversation and dialogue
that never existed before. Dialogues among Detroit neighbourhoods as
well as inter-generational exchanges further reinforce the importance of
crossing boundaries and learning from people of different age, class and
ethnic background.

While the participatory manner in which projects are brought to
completion helps to challenge stereotypical images of Detroit youth,
so too does the content of the creative work undertaken. Volunteers
come away with a much more encompassing definition of environment-
alism – one that incorporates and attempts to respond to the particular
environmental threats experienced by Detroit residents on a daily basis
(e.g. lead paint poisoning, violence, poor quality housing, etc.). While
these problems are attacked directly through physical labour, the murals

and other art and design work undertaken by young people appear to serve a more symbolic purpose. Through these artistic endeavours, youth have challenged stereotypes of themselves. They have demonstrated that they are less concerned with specific themes than they are with using their creations to introduce colour into a bleak landscape and establish a strong presence and sense of purpose for themselves in that environment. Both kinds of modifications of the physical landscape (practical/concrete and artistic/symbolic improvements) enable Detroit Summer youth to envision their own lives differently and to generate a new public vision of themselves and their neighbourhoods. This is then seen as a way of building a larger movement for social change, of 'being a part of some thing much bigger than oneself' (Boggs, 1993).

DUDLEY STREET NEIGHBOURHOOD INITIATIVE AND YOUTHWORKS/ARTWORKS

While Detroit Summer may be unique in its scope, similar efforts to address neighbourhood problems are currently underway in other US cities. One of the best documented is the Dudley Street Neighbourhood Initiative (DSNI), a community wide development project that was begun in 1981 and provides local youth with a central role in the revitalisation of one of Boston's most impoverished areas (Medoff and Sklar, 1994).

The 'guiding vision' of the DSNI is formulated in its 'Revitalisation Plan: A Comprehensive Community Controlled Strategy'. The plan seeks to create an 'urban village' with quality affordable housing and a much enhanced overall quality of life for its primarily low income residents (Medoff and Sklar, 1994: 108–10). What makes this organisation unique are its commitments to full resident control; the aggregation of enough land to affect the market[3] and an emphasis on both new construction and rehabilitation.

Children and teens make up one-third of the population of the ethnically diverse community served by the DSNI and from the earliest days of organising young people in the Dudley Street area have been actively recruited to participate in the DSNI's youth committee. They have helped identify critical problems and generate ideas for change. Children and youth are welcome in the DSNI offices, with evidence of their involvement pinned to the walls in the form of drawings and poems (Medoff and Sklar, 1994: 203).

The basic premise of the organisation is that low income communities have many resources of value to draw upon. The DSNI's Declaration of Community Rights begins 'We – the youth, adults, seniors of African,

Latin American, Caribbean, Native American, Asian and European ancestry – are the Dudley Community' (Medoff and Sklar, 1994: 202). Community organisers, architects and designers engage local youth in projects that are remarkably similar to those described in Detroit, ranging from neighbourhood clean-ups and park reclamation to mural painting and such activities as the architectural design of two proposed community centres and the open space in between (Plate 18.2).

In a foreword to a report on Dudley's Young Architects and Planners (Nagel and Sullivan, 1992), DSNI President Che Madyun explains how the design work undertaken by young people 'integrates DSNI's commitment to community control' by providing an important 'outlet for the transformative power of youth'.

> At a time when national concern about violence by and against young people produces curfews, increased jail sentences, and fear of the youngsters who are our future, Dudley's Young Architects and Planners Project . . . is based on the dreams and creativity of youngsters too often written off by others as worthless.
>
> *(Nagel and Sullivan, 1992: 3)*

Adult organisers believe that Dudley youth do not 'fit' the constructed images of inner city youth portrayed in the media (e.g. girls as potential welfare mothers, boys as 'savage gang members' and both as drug users or dealers). They challenge these stereotypes by involving local youth directly in physical revitalisation efforts and in collaborations with design professionals. In this way, they demonstrate how youth participation can become a critical part of the whole community development enterprise.

One example of this collaborative approach is YouthWorks/ArtWorks, a summer jobs programme connected to the DSNI, which was developed in 1992 by UrbanArts Inc., a non-profit agency that works with neighbourhoods to incorporate more art into the urban environment. The programme hires, educates and trains teens to investigate their communities using the vehicles of creative writing, photography, video and urban design. Youths also create artistic interventions that can have an impact on their immediate environment.

In the summer of 1993, UrbanArts Inc. began a collaboration with the DSNI, Action for Boston Community Development and Boston Urban Gardeners to involve youth in the design of green space between two proposed community centres. Working with an architect and paid an hourly wage, the youths learned and applied the techniques of surveying, interviewing, and design and model construction. They then planned the

Plate 18.2 YouthWorks/ArtWorks: re-designing Dudley's
open space in Massachusetts.
Photograph: Pamela Warden, Director of UrbanArts Inc.

'greenway' linking the two proposed community centre sites along the path of a former brook/stream.

In public presentations of their work that I attended, several groups of teenage participants revealed their plans and explained the three-dimensional models they had constructed. Several groups of girls talked about how they derived their design criteria with the likely users of the space in mind. For example, to uncover the needs of neighbourhood children, adult caretakers and the elderly, they interviewed extensively within the neighbourhood. This information was then reflected in their designs which evidenced a strong concern for safety and accessibility as well as aesthetics (Plate 18.3).

The boys' designs were often guided by different criteria. Several young men spoke of the boredom of their neighbourhood and of the 'lack of colour' and prevalence of small parks that 'all looked the same'. They said they wanted their open space designs to create something completely different, something that would announce one's presence in a new and revitalised place with 'special' elements that no other neighbourhood in Boston would have. Designs reflected these priorities containing such whimsical items as glass covered walkways with fluorescent coloured lights and highly decorative and monumental archways that could be seen from far away.

Plate 18.3 Girls' designs showed a strong concern for
safety, accessibility and aesthetics.
Photograph: Pamela Warden, Director of UrbanArts Inc.

In 1994, YouthWorks/ArtWorks sought to involve young people in similar artistic enterprises geared, this time, to the revitalisation of Blue Hill Avenue, a long thoroughfare that runs through the Boston neighbourhoods of Roxbury, Dorchester and Mattapan. The project also provided low income teenagers with summer jobs and paired them with professional artists from the community who ran a variety of workshops. The objective, according to director Pamela Warden, is

> not only to give kids a chance to learn new skills and to earn pay, but also to try to get the kids to look at their communities through the eyes of an artist, to see [them] . . . in a slightly different way and to understand that they have a relationship to their community that can be a productive one.
> *(Coleman, 1994: 1)*

The twelve youths involved in the design projects began by mapping everything along Blue Hill Avenue. Eventually they focused on the specific design of bus shelters. Though the architect in charge could detect few overall distinctions in approach between the boys and girls, the two most elaborate designs differed in ways similar to the greenway designs mentioned above. For example, one young woman, concerned about the safety of residents who might use a nearby African American community meeting house at night, took her design of a bus shelter below ground level in order to tie it physically to the centre. A young man, in contrast, used the exercise purely to have fun, emphasising its artistic and creative possibilities.

Similar to the experience of Detroit Summer youth, the time and effort spent surveying their neighbourhood and considering alternatives, reinforced for the young people of DSNI, the notion that all environments are *socially* constructed. This brought into question the inevitability of current conditions and provided an opening for positive intervention and change.

> 'What did I learn from the design exercise? I guess it [Dudley Street] won't be messed up forever.'
> *(DSNI youth participant commenting at the design presentation)*

Some DSNI youth also came away from their design initiatives with a sense of pleasure and new levels of personal confidence. Many were provided with an outlet for their creativity and seized the opportunity to express a desire to live in a place that was unlike other places, a place that contains elements of surprise and interest. Unlike Detroit Summer volunteers, they did not produce any immediate change in the environment. They did, however, use their art and design work to ask questions about the current state of the physical and social landscape and began to see a possible relationship between improvements in the built environment and the overall quality of people's lives. Design and art work thus provided an opportunity to re-image and, in at least some sense, reclaim a portion of the public space within their communities.

'BANNERS FOR THE STREET': PUBLIC ART WITH HOLYOKE YOUTH

Holyoke, Massachusetts, a nineteenth century industrial mill city, deemed a 'city in crisis' by the Massachusetts state legislature, was the site for a different form of youth-centred environmental intervention. This particular project focused on the southern portion of the city, near the canals. The area, once a vibrant immigrant enclave with an active industrial base, is now one of the most economically depressed parts of the city. The population includes many second and third generation Puerto Rican residents who have endured horrendous housing conditions and decades of arson destruction.

In 1982, several residents established a community non-profit development corporation, *Nueva Esperanza*, or 'New Hope'. This organisation continues to improve local housing conditions through rehab and new construction. Recently, the organisation broadened its scope to address wider concerns such as community health and the needs of teenagers. Along with two other Holyoke organisations, the CARE Centre for pregnant and parenting teens and the Holyoke Youth Alliance, *Nueva Esperanza* helped to establish an art centre for youth in the neighbourhood, *El Arco Iris* (The Rainbow).

In the summer of 1990, we began to explore the neighbourhood of South Holyoke with teenagers who were participating in a summer arts programme at *El Arco Iris*. The core group of approximately thirty ranged in age from 11 to 15 and were primarily of Latino descent. The goal was to encourage the youth to express their feelings about the neighbourhood and city, assess the built environment critically, and enhance it through the conception, design and production of public art (Plate 18.4).

The 'Banners' project began by asking the teens at *El Arco Iris* to draw their neighbourhood including anything they felt was interesting or important to them. Differing levels of personal happiness and despair as well as feelings about the environment emerged from these initial drawings and the individual commentaries that each young person made when presenting their work to the group. Images of stop signs loomed large in their pictures as did the rooftops of buildings, suggesting secret havens

Plate 18.4 Young people's art goes public, Holyoke, Massachusetts.
Photograph: Myrna Breitbart.

where the teens could be in control of a small space of their own. Several local 'hangouts' were also represented in the drawings with an emphasis on the social – i.e. *who* rather than *what* was to be found there. Teens depicted houses in need of 'fixing up' and complained of its being 'too quiet' and boring. On a positive note, they said they loved 'seeing people outside', hearing loud music and observing improvements in local housing.

Neighbourhood walks, planned by the youth, followed. While on these walks, teens photographed lots of desirable cars and took innumerable pictures of themselves. Familiar buildings and the homes of friends and family members were also popular targets in the viewfinder. Gradually, however, the angle widened to encompass tenements in disrepair, parks in various states of decay and a whole landscape of only *potential* delight. As we walked past many a broken fixture, the teens were quick to say that 'somebody should fix that up'. When we passed groups of men on a corner, they would comment that 'somebody ought to get rid of those drug dealers'. Young people had no shortage of ideas about improving their environment and were also very quick to comment on what they liked, e.g. *Nueva Esperanza's* new two-storey duplexes with a private yard and space for gardens. The teens made it clear that what they liked about these spaces was not the particular style, but their perception that the housing suited their needs and the needs of their families for increased privacy and freedom. If they could live there, they imagined that they could have their music turned up as loud as they enjoyed without having to worry about disturbing the sleeping tenement neighbour upstairs who works the night shift. Their attraction to the cars was also about the freedom to have *their* space and *their* music, as was the interest in gardens and having some room outdoors.

The drawing, mapping, route-marking, neighbourhood exploration and production of story boards, while generating opportunities for critique and the identification of neighbourhood strengths and weaknesses, became a source of ideas for themes for the design of street banners. Teens expressed a desire, however, to move beyond a documentation of the problems and causes to ways of addressing the symptoms and expressing their feelings and hopes for change. These latter themes include a desire for *water* as a source of beauty and recreation (e.g. such activities as paddle boating, swimming and fishing); having '*more fun things to do*' in the city as a whole including the opportunity to enjoy their *music*; providing *affordable housing* which would better support their families' needs and provide them with increased privacy and control; the importance of *warm and friendly people* in the neighbourhood and how valued they were; and finally, and perhaps most importantly, the desire to introduce more *peace, order and tranquillity* into their own lives. To give visual expression to these themes and draw public attention to the needs and desires of Holyoke youth, large street banners were produced. When asked whether outsiders might think their images reflect a desire to escape urban life, young people looked perplexed. To them, the image of two children holding hands in a boat symbolised survival and friendship, while more abstract designs with birds, sky and mountains expressed hopes for 'Peace. We want peace. Isn't that what everybody wants?' (Plate 18.5).

Plate 18.5 'Peace, We Want Peace. Isn't That What
Everybody Wants?' Holyoke Banners Project.
Photograph: Myrna Breitbart.

PLEASURABLE DISTRACTIONS, SURVIVAL
OR SOCIAL ACTION?

In the context of difficult life experiences and often hostile environments, public art, design work and other forms of local environmental intervention have enabled some young people to see possibilities in urban settings that otherwise form only neutral or constraining backdrops for their activity. The many benefits that derive from this involvement nevertheless raise the question of how such creative youth driven work fits within a larger movement for social change.

SKILLS ACQUISITION

The range of specific skills acquired by young people which can be applied in other settings, is certainly one significant outcome of all three projects described above. The development of youth leadership is reflected to varying degrees in the efforts made to invite and then sustain young people's active participation throughout the duration of each project and beyond. The amount of control given over to youth in identifying local issues and problems to focus upon for environmental intervention, implementation and follow-up was greatest for Detroit Summer and the DSNI. However, even the more limited aspects of decision making assumed by young people in Holyoke resulted in the acquisition of a generalisable set of skills, including techniques for fostering communication, making decisions, expressing judgements, organising events and collaborating with others to accomplish specific goals. Project work also provided many young people with an opportunity to acquire skills of a more technical nature such as interviewing, surveying, painting, writing, gardening and marketing.

PERSONAL DEVELOPMENT: SHARED FEELINGS; CHANGED ATTITUDES

Often, with public art, it is the *process* rather than product of art production that generates the most significant benefits and lasting outcomes (Raven, 1989; Breitbart and Worden, 1994). When young people are encouraged to re-examine the strengths and weaknesses of their surroundings and then act creatively to transform them, the experience can alter young people's attitudes towards each other and their future; it can also provide a much needed outlet for the expression of feelings. In the case of the 'Banners' project, discussions leading up to the production of visual images allowed Holyoke youth to share the common ways that negative messages about themselves, transmitted through the severely compromised quality of their neighbourhoods, affected them. The eventual movement of the banner designs from descriptions of what currently exists to what the city might (and should) become then provided an outlet for their more positive hopes, dreams and personal pride. As one child put it, 'When we first started we had no idea we could do this.'

These kinds of personal outcomes were also apparent in Detroit Summer, where inter-generational dialogues provided a setting within which differences in experience and perspective could be shared, and feelings of pain and happiness expressed. 'What dragged me in', said one young woman 'is that everyone listens . . . . Few people listen to teens and I was so happy when someone valued my opinion.' Young people in Detroit Summer were thus able to utilise the inter-generational dialogues that accompanied their design and environmental project work to explore differences among themselves and differences between their generation and prior generations of residents. Detroit Summer also enabled young people to examine how these differences affect their every day lives, their needs, their wants and their ability to effect change.

EXPANDED POLITICAL KNOWLEDGE

At times, the techniques used to produce the art and design work also become a medium for enhanced political understanding. Participants are not only able to make better sense of their own lives and the diverse perspectives of others, they are also able to gain insight into the under-lying economic, social and political forces that impinge on neighbour-hood life and space. For example, the expeditions and environmental surveys undertaken by young people within urban neighbourhoods involved in all three projects, provided the raw material for discussions that eventually helped youth to situate the local problems they saw within the much larger context of the history of urban de-industrialioa tion and ethnic politics. Confronting and developing a better under-standing of the inequalities which resulted from these larger economic and social trends, increased young people's dignity and provided some motivation to work for change.

CLAIMING SOCIAL SPACE: TELLING
STORIES/REPRESENTING COMPLEX LIVES

Youth involved with the DSNI also engaged in research in their commu-nity as part of the design process. Through this research, they were able to understand more how personal and political choices affect their neighbourhood and shape the built environment. This information then enabled young people to take back some control over image making. They were able to use their designs to address needs articulated by the community, represent their daily lives more accurately and challenge stereotypes imposed from outside. By observing the inordinate number of vacant lots and the scarcity of space for children, one young woman involved in the DSNI was prompted to fashion her greenway design 'so that people can stay in their own community and feel comfortable without having to take buses and cars longer distances' to find adequate and appealing leisure space. Another group of girls used their interviews with older people, children and the caretakers of children to design an outdoor community space that would jointly accommodate everyone's needs. This research taught them that, contrary to media portrayals, the elders and young people of their community cared a great deal about the Dudley Street area.

The research, writing, discussion and photography work that often accompanies the art and design projects also reminds youth of the specificity of their neighbourhoods and their unique, though often inter-secting, lives. Though it is common for the media to lump all low income communities together and present them as homogeneous and problematic, the exploratory techniques employed by each of the projects emphasised the richness and complexity of local life.

> 'I have learned a lot about Detroit – the city in which I was born. Not only have I been the recipient of many urban geography lessons, but also socio-economic political lessons from several dusty, rag-clad professors who hold all day office hours on street corners.'
>
> *(Carl McGowan, youth volunteer, Detroit Summer, 1992)*

The self-discovery of unique and sometimes unexpected materials, events and places provided youth with a rich source of ideas for their art projects and environmental work; it also presented them with an opportunity to insert themselves in a very visible way into the landscape and begin to claim a place in the public life of the neighbourhood.

In many urban neighbourhoods today, violence has generated a despair that has resulted in a number of different collective responses. In Detroit, there are marches on crack houses and public memorials for the young people who have died from gun violence. A revival of graffiti and the design of elaborate and artistically conceived wall murals to memorialise the dead or simply decry the influx of drugs and guns into neighbourhoods represent other types of responses (Cooper and Sciorra, 1994). Unlike earlier forms of graffiti and mural painting, however, contemporary young artists do not simply appropriate property; they generally seek permission to utilise public space in much the same way that Holyoke youth sought permission to hang their banners.

SURVIVAL MECHANISMS: ENLIVENING SPACE/ENVISIONING CHANGE

Memorialising murals and public art of other forms that seek to explore problems and solutions to a host of issues facing youth, provide one means for enabling young people to tell their stories. They also make otherwise difficult spaces more tolerable to live in. The public art generated by youth in the three projects described here, contributed as well to the creation of public spaces that provide either a vision of change or an occasion for celebration and enjoyment. While practical themes like safety, good housing, and the availability of educational, job and recreational opportunities appear in many of the designs that young people produced, these were by no means the only themes represented in their designs for public space. A strong desire for colour, vibrancy, sound and a high level of social activity were also key. These distinct elements of young people's environmental designs and productions indicate the importance of a high level of activity and visual stimulation to their daily survival and pleasure. Indeed, one can see young people's enthusiasm and investment in project work expand to the extent that they were able to use, or envision using, their art to enliven the public landscape and replace vacant and boring spaces with animated, personalised and decidedly youth-identified spaces.

In the YouthWorks/ArtWorks project, common structures like benches, greenhouses and tunnels appear in several designs of a public greenway between the two proposed community centres. Yet, when these designs were presented, youth were careful to distinguish *their* interpretation of such structures from the typical. One structure was described by a young man as 'almost like a greenhouse . . . but . . . different'. Another presented not an 'ordinary tunnel,' but 'Dana's mystical tunnel' with his name flashed in coloured lights. Three girls who named the space between their two community centres 'Club Paradise', included arches solely 'because people in Dudley don't usually have arches . . . and it's different'. Nearly all of the designs, even those that incorporate very practical components, thus include some sign of young people's unique presence in the area as well as elements of whimsy and colour.

'Banners for the Street' also involved a personal exploration by young people of their feelings and emotions. Though no lasting improvements were made to the built environment of South Holyoke, this project and the design exercises undertaken with youth in the DSNI, underscore how important even the temporary personalising of small spaces or the creation of symbolic new spaces can be for young people who live daily amidst stress and who, perhaps more than adults, notice and suffer the effects of the absence of colour and items of interest in their environments.

Robin D.G. Kelley has observed that in the context of economic crisis, permanent unemployment and the transformation and privatisation of public space, the meanings and practices of 'play' may have changed for inner city youth.

> While street performance is as old as cities themselves, new technologies and the peculiar circumstances of post-industrial decline have given rise to new cultural forms that are even more directly a product of grass-roots entrepreneurship and urban youth's struggle over public space.
>
> (Kelley, 1997)

While he was referring to the employment and investment opportunities presented to some urban youth through hip-hop and/or graffiti, this can also apply to the efforts made by the youths described here. Survival, the desire to say something about their lives, and the need to capture some moments of pleasure, led these participants to combine in their creative expression a desire for 'play' (e.g. a colourful and exciting versus dreary landscape) with a desire to meet some of the practical needs of the community, including the need for work and opportunities to make money. The public art and other forms of environmental intervention that resulted thus represent forms of direct action. They can provide urban youth, at least temporarily, with a means of escape, a voice in community affairs, and the reality or vision of a more pleasurable living environment.

CONCLUSION

Neighbourhood cultural and environmental projects such as those described here cannot eradicate or fill the gaps left by missing social and economic resources or a political environment that seeks to condemn young people for hardships beyond their control. They can, however, play a critical survival role by increasing the safe spaces within which urban youth can explore the sources of local problems, and envision, and sometimes create, alternatives.

The actual physical products of these young people's efforts – the greenhouses, community gardens, murals, designs and banners – were envisioned or placed in public spaces that have meaning for youth. Ideas were put out for others in the neighbourhood and beyond to draw strength from. As such, they provide a stark contrast between what *is* currently in place and what *could* be there instead. Youth who are attracted to such project work become involved in part because they recognise that in order to survive the absence of key resources and the negative stereotypes that make the future provision of those resources even harder to secure, they must utilise all opportunities that come their way to break silence, assert themselves into the life of the city and claim space. Creative forms of environmental intervention can at times contribute to this goal, producing visible albeit small changes while also becoming a potential resource for community building and mobilisation.

At the conclusion of one Detroit Summer and YouthWorks/ArtWorks project, I participated in group discussions with some of the adults and young people involved. Adults at these meetings, with the goal in mind of creating a social movement, searched for, and tried to uncover among youth an articulation of the larger meaning of the small projects they had just completed. This was done, in part, in reaction to the increased levels of responsibility, confidence and self-esteem that adults saw manifest in the young people involved – qualities that clearly contradicted media constructed images. Some of the adults present may also have harboured theories of personal and social change that they believed consonant with the values expressed by the young people through their project work – e.g. young people's resistance to permanent or hierarchical forms of self-leadership and their expansive definition of 'the environment' to include the familiar and close, the urban as well as the natural world.

The youth present at the Detroit Summer discussion shared many examples of how they have grown personally through their experiences and how, in some instances, the project work had enabled them to define future career goals. They also shared a desire with adults to build upon small accomplishments and consider how their combined experiences could help to generate a larger movement to respirit and revision urban life. In general, however, they resisted the imposition of abstract theoretical interpretations of their work. For them, involvement in transforming a small piece of their local scene through creative application was undertaken because it provided an opportunity to 'play', to do something slightly out of the ordinary and to gain access to new resources. These were all seen by the young people as valid and important goals for involvement. That some youths were then personally transformed by the experience and came to feel part of a larger mission or enterprise was an unanticipated additional outcome.

'we have not yet made this city what it could be or should be just by painting some houses and planting some gardens for free . . . more importantly, we have created hope that before didn't exist. We know what the future could be and we have the desire to make it come into being.'

(*Julia Pointer, youth volunteer, Detroit Summer, 1992*)

Those who have claimed, myself included, that there are substantial, largely unrecognised benefits to be derived from enabling youth to explore, critique, revision and creatively refashion their surroundings, will probably continue to look for and present examples of such activity as socially conscious resistance (Breitbart, 1995a, 1995c). In the process of imposing adult derived theories of social change, however, we must be careful not to obscure the important role that young people's creative environmental intervention can play as *a mode of survival* a survival that depends as much on youth seizing opportunities to 'play' and nurture hope as it does on generating new collective redefinitions of neighbourhood and lessons for adults about the necessary components of urban revitalisation.

NOTES

1 Discovery World and Kidsports are private, for-profit businesses that provide opportunities for young paying customers to use a variety of indoor recreational facilities.

2 During Mississippi Freedom Summer 1964, several hundred college-age volunteers (many from northern states) came to Mississippi and joined with black civil rights activists to establish 'freedom schools' for black children. They also engaged in other activities to oppose racist violence and promote civil and human rights.

3 DSNI is the first neighbourhood based development agency in the US to gain control over critical parcels of vacant land through the use of 'eminent domain' – the power of a governmental entity to take land or property with 'just compensation' for a compelling public interest.

REFERENCES

Boggs, G. (1993) 'Preparing people to work for social change', unpublished talk delivered at the Neighbourhood Academy, Detroit, Michigan, 7 June.

Breitbart, M.M. (1995a) '"Banners for the Street": reclaiming space and designing change with urban youth', *Journal of Education and Planning Research* 15: 101–14.

—— (1995b) 'Interview with architect Kathryn Firth', 25 January, Boston, Mass.

—— (1995c) '"It takes a child to inspire a village": revisioning and redesigning neighbourhoods with urban youth', unpublished paper delivered at the 'Building Identities: Gender Perspectives on Children and Urban Space' conference, Amsterdam, The Netherlands, 11–13 April.

—— and Caballer-Arce, G. (1993) 'Facing education: Holyoke children speak out', in *Facing Education: Portraits of Holyoke Children*, Exhibition Text Holyoke, Mass.: Michael Jacobson-Hardy.

—— and Worden, P. (1994) 'Creating a sense of purpose: public art and Boston's Orange Line', *Places* 9, 2: 80–6.

Coleman, S. (1994) 'Their aim is to create an "Avenue of the Arts"', *Boston Sunday Globe*, 17 April, 1: 1.

Cooper, M. and Sciorra, J. (1994) *R.I.P.: New York Spraycan Memorials*, London: Thames and Hudson.

Cresswell, T. (1992) 'The crucial "where" of graffiti: a geographical analysis of reactions to graffiti in New York', *Environment and Planning D: Society and Space* 10: 329–44.

Darden, J., Hill, R.C., Thomas, J. and Thomas, R. (1987) *Detroit: Race and Uneven Development*, Philadelphia: Temple University Press.

Dawsey, K.M. (1995) 'Surviving on instinct', *City Limits* 20, 9: 18–22.

Dulzo, J. (1992) 'Short cool summer', *Metro Times*, August.

Gooding-Williams, R. (1993) *Reading Rodney King, Reading Urban Uprising*, New York: Routledge.

Griffin, C. (1993) *Representations of Youth: The Study of Youth and Adolescence in Britain and America*, Cambridge: Polity Press.

Hanley, R. (1993) 'Night curfews growing to clear streets of teen-agers', *New York Times*, 8 November: B1, 8.

Jennings, K. (1992) 'Understanding the persisting crisis of black youth unemployment', in J. Jennings (ed.) *Race, Politics, and Economic Development: Community Perspectives*, London: Verso.

Kelley, D.G. (forthcoming) 'Playing for keeps: pleasure and profit in the post-industrial playground', in Wahneema Lubiano (ed.) *The House that Race Built and the U.S. Terrain*, New York: Random House.

Loukaitou-Sideris, A. (1993) 'Privatisation of public open space: the L.A. experience', *Town Planning Review* 64, 2: 139–67.

Males, M. (1993) 'Infantile arguments', *In These Times* 9 August: 18–20.

—— (forthcoming) *The Scapegoat Generation: America's War on Adolescents*, Monroe, Maine: Common Courage Press.

Medoff, P. and Sklar, H. (1994) *Streets of Hope: The Fall and Rise of an Urban Neighbourhood*, Boston, Mass.: South End Press.

Nagel, A. and Sullivan, G. (1992) *Youthful Visions: Building a Foundation for Community*, Roxbury, Mass.: DSNI.

Nauer, K. (1995) 'Chained reaction', *City Limits* 20, 9: 24–6.

Pfeffer, R. (1995) 'Researching across invisible borders: young punk women living on-their-own producing space and knowledge', unpublished paper delivered at the 'Building Identities: Gender Perspectives on Children and Urban Space' conference, Amsterdam, The Netherlands, 11–13 April.

Raven, A. (1989) *Art in the Public Interest*, Ann Arbor, Mich.: UMI Research Press.

Rose, T. (1991) '"Fear of a black planet": rap music and black cultural politics in the 1990s', *Journal of Negro Education* 60, 3: 276–90.

—— (1994) *Black Noise: Rap Music and Black Culture in Contemporary America*, Middletown, Conn.: Wesleyan University Press.

Ruddick, S. (1996) *Young and Homeless in Hollywood*, New York: Routledge.

19

VANLOADS OF UPROARIOUS HUMANITY

New Age Travellers and the utopics of the countryside

•

Kevin Hetherington

THE GEOGRAPHY OF YOUTH CULTURES

Stonehenge on a Victorian public holiday was a busy place: 'The pilgrim who goes there with his reverent mind full of Druids . . . undergoes a series of electric shocks. . . . He never bargained for vanloads of uproarious humanity, dressed in all the colours of the rainbow, and in many others of aniline origin. They come, they crack jokes, they feast, and they sing the latest sweet things from the music hall repertoire . . . while a fusillade of ginger beer adds to the general rudeness.'

(quoted in Chippindale, 1983: 173)

The geography of youth cultures provides us with an illustration of some of the issues associated with what Louis Marin (1984) has described as utopics. Utopics are a type of spatial play whereby a utopian outlook on society and the moral order that it wishes to project are translated into spatial practice through the attachment of ideas about the good society onto representations of particular places. Marin's best illustrations of this utopics can be found in his reading of Disneyland (1984) and America (1992). In this chapter I want to consider the geography of one particular youth culture, New Age Travellers, and their contested utopics of the British countryside.

The history of youth culture, whether that be spectacular sub-cultures or more ordinary and conformist practices, has always had an element of making space for oneself, of creating a turf and finding one's place, often on the margins of society (see Thrasher, 1927; Becker, 1946). Finding one's place has sometimes meant going elsewhere into a supposedly free space, a space perceived as more authentic (see E. Cohen, 1973, 1979) or more one's own, where issues of inclusion and exclusion can be determined by establishing categories of belonging and group identification. At other times, it has meant staying put and trying to change one's situation there. In either case, certain places come to be seen differently from the representation they have within society, they come to be invested with meanings that express values or beliefs just as important to those youth cultures as their identification with certain types of music or styles of dress.

This chapter looks at how, for New Age Travellers, being in a different space involves the utopic construction of the countryside as a different place from its main representation within British, notably English, culture as a picturesque arcadia of pastoral peace.[1] There is a small but growing literature on New Age Travellers (see Earle *et al.*, 1994; Halfacree, 1996; Hetherington, 1991, 1992, 1993, 1996b; Lowe and Shaw, 1993; McKay, 1996). Travellers are, however, more than just a youth culture; in some ways they are more like a new social movement or form of cultural politics (see Melucci, 1989) with an interest in issues such as those associated with the environment, communal and alternative living and lifestyles, anti-road protests, and access and rights to common land. New Age Travellers are a hybrid phenomenon. They remain a youth culture, in the sense that that was how their way of life originated and because most, though by no means all of those who travel and live a life on the road, have tended to be relatively young – but they are more besides. There is nothing singular about the Traveller identity; its elective tribal character permits many different types of identity that in some cases remain rather static while in others are highly adaptable and change over time. To some, to speak of New Age Travellers at all now might seem archaic, given all that has happened at the identity boundaries that Travellers have shared in recent years with ravers, anti-road protestors, environmentalists, crusties and poll tax protestors, since Travellers in the form of the so-called 'peace convoy' first came to public attention in 1982.[2]

There have been a number of similar types of youth cultures in other countries over the years. The early inspiration for the *Wandervogel*, the German Youth movement at the turn of the century was that of a nomadic, rural existence (see Becker, 1946). It was not uncommon for hippies during the 1960s to travel around North America in brightly decorated buses similar to those favoured by New Age Travellers. The phenomenon of drifter tourism, often associated with the hippy trail to India, also bears some resemblance (Cohen, 1973). In Australia, Ayers Rock has served as a site that has attracted 'new age pilgrims' (see Marcus, 1988), and one also finds so-called 'ferals', people living out in the bush, with a lifestyle free from the trappings of modern life, in an attempt to return to nature.

New Age Travellers are an important part of the contemporary terrain of British youth culture, even though they have formed and continue to form combinations that lead to new cultures that might be called something other than New Age Travellers (McKay, 1996).[3] They might also be seen as a good illustration of the sort of neo-tribal phenomenon that is said to characterise the contemporary forms of social identification (Maffesoli, 1988, 1996; see also Bauman, 1990, 1992).[4] In looking at Travellers in relation to representations of the countryside (see Halfacree, 1996), it is important to recognise that the countryside as a terrain is contested in terms of not only land use but also spatial representation, notably utopic representation. While for many the countryside is imagined as somewhere picturesque, tranquil and timeless, free from the social problems of the city, for New Age Travellers it represents something different: freedom, authenticity, mystery, spirituality and nomadism. The issue here is not about the 'truth' of either of these (or indeed other) representations of the countryside but that truth claims about how to live are expressed through a utopic practice that produces the countryside as a contested space of moral order.

NEW AGE TRAVELLER WAYS OF LIFE

The New Age Traveller lifestyle has its origins within the hippy counter-culture of the 1970s, notably the free festival scene that emerged at that time (see Mills, 1973; Clark, 1982; McKay, 1996). Free festivals grew up alongside large commercial festivals and were conceived as utopian models for an alternative society, often referring to an imagined ethos of freedom from constraints, carnivalesque transgressions of social norms associated with an ideal of the medieval fair. It was out of this free festival scene, especially that associated with the festival at Stonehenge held from 1974 to 1985 at the time of the summer solstice, that the New Age Traveller lifestyle emerged. The festival was the initial impetus for travelling. During the early 1980s Stonehenge became the focus of an annual pilgrimage, in the form of the convoy, at the time of the summer solstice. Initially, attending festivals was something people might do for just a weekend, but the idea of the festival and a nomadic way of life travelling between festivals soon caught on. The way of life between festivals, associated with the freedoms of the open road, became just as important as the festivals themselves. The vehicles, often old buses or vans, took on a significance of their own and became an important component of this lifestyle. Small groups of people would often meet at festivals and spend the summer travelling together in one or two large, highly decorated vehicles (Garrard, 1986a, 1986b; Garrard *et al.*, 1986). As another festival approached, small groups of Travellers in their vehicles would join up in convoys and travel to it *en masse*, in order to squat the site for the festival. A moral panic emerged around the so-called peace convoy in the media in Britain in the mid-1980s with supposedly dole-scrounging, dirty Travellers taking on the role of folk devils. The notoriety of a so-called 'peace convoy' began to appear in the national newspapers around 1983 and escalated until the convoy was smashed by the police in 1985 on its way to Stonehenge (Rojek, 1988; 1989) (Plates 19.1 and 19.2).

It is important to recognise, however, that New Age Travellers have not always moved around the country in convoys, nor are they always on the road. There is a distinctive seasonal pattern to their travelling. Generally festivals have been held between the months of May and September, partly for practical reasons associated with the desire to hold festivals in good weather and partly in association with key moments in the calendar, such as the dates of traditional fairs, and most notably because of the significance that the summer solstice has for many Travellers. During the winter it has been more common for Travellers to 'park up' in some remote or out of the way upland or rural area, or for them to return to live in towns and cities, with friends or relatives, or perhaps in squats. This way of life has been made increasingly more precarious by the introduction of the Criminal Justice and Public Order Act 1994, which gives the police increased powers to move or arrest people believed to be trespassing, and repeals aspects of earlier legislation requiring local authorities to provide sites for Gypsies. The resilience of Travellers remains. While it is believed increasing numbers are choosing to move from Britain to mainland Europe, the Traveller way of life continues to develop.

AUTHENTICITY, UTOPICS AND HOME

Utopics are associated with the translation of ideas about a good society into spatial practice, and can be described as a cultural performance of moral orderings through spatial practice. Utopics are an important

Plate 19.1 1986, a field in Hampshire. Travellers were attempting to reach Stonehenge but were under very heavy police surveillance. A meeting was held in the early morning before the police arrived to remove the vehicles. After the violence of the 1985 'Battle of the Beanfield' people voted against resistance and had to give up their vehicles and in some cases their children. They then had to pursue 'legal' channels to get their children and vehicles back.

Photograph: Alan Lodge.

Plate 19.2 Between 300 and 500 people from the Hampshire field decided to walk to Glastonbury to protest at their treatment. They encamped in the Greenfield, at the Glastonbury Festival site which is used to demonstrate alternative, ecological lifestyles.
Photograph: Alan Lodge.

aspect of the Travellers' way of life. This cultural performance takes place in a hiatus, a space of uncertainty, which Marin calls 'the neutral' and which emerges from the conditions of undecidability lying behind representations of space such as the countryside (1984). Drawing on Derrida's notion of différance (1976), whereby the meanings of texts are seen not to be fixed but endlessly deferred, Marin goes back to Thomas More's sixteenth century writing on Utopia to derive his concept of utopics (see More, 1985). More called his imaginary island Utopia but this was based on a pun, for u-topia means both *ou-topia*, no-place, and *eu-topia*, good place. For Marin, it is the tension between these two meanings that sets up the play of différance that constitutes the spatiality of utopics (see also Hetherington, 1997). The movement between a nowhere and an imagined perfect place and vice versa takes place across a space of uncertainty, ambivalence and undecidability, or what Foucault (1986) has described as a heterotopia (see also Genocchio, 1995; Soja, 1995; Hetherington, 1996a, 1997). This is the space in which representations of the good society and its moral order are articulated and contested. Utopias do not exist; what exist are the translations of ideas about the good life and about social and moral order into social reality, and that reality takes on a distinctly spatial character as it comes to be represented in particular representations of space (see Hetherington, 1997). Certain sites such as Stonehenge have always been important to the utopics of Travellers; they have had, in Shields's terms, a social centrality, a focal site for an identity something that is often important for youth cultures (Shields, 1992a; see also Hetherington, 1996a, 1996b). However, there have also been broader spatial issues around which this Traveller youth culture has developed and through which their identity has been performed. Utopic representations of the British countryside as a whole have been important to the politics that has surrounded New Age Travellers.

It is through this utopics that the values of a group are expressed in spatial terms. The way of life that New Age Travellers have adopted: nomadic, with an emphasis on free festivals and generally a rural way of life, involves a utopics that is based on the hiatus over the authenticity that is associated with the rural as home. For Travellers that authenticity is based on a nostalgia for an imaginary past in which rural rather than town life acts as a model for an ideal society in harmony with nature. Through the spatial practices of living in a small group or emotional community (*Bund*), emphasising the communion of a shared way of life, shunning much, though not all, modern technology and materialism, and identifying with ancient as opposed to modern freedoms associated with nomadism, the carnival atmosphere of the festival and ancient peasant common land rites, Travellers express an alternative moral ordering of home to that offered to them by modern societies (Hetherington, 1994).

The Travellers' utopic is one that looks to the past for its authenticity but selectively, often focusing on and identifying with the marginal and oppressed: the peasant, the vagrant, the circus figure, the mountebank, Gypsies, Native Americans and their ways of life. This utopic is based on a distinctly romantic outlook and aesthetic which places above all else a sense of the marginal at the centre of this notion of the authentic.

Some Travellers adopt a deep green attitude of living lightly on the land and see nature as a Being (Gaia) from whom we are drawn, and for whom we must care. The Tipi, modelled on the style used by Native Americans, is perhaps the favoured dwelling of such Travellers. For others, the old Digger idea of land for the people is more significant. In this sense, as with those who advocate rights of access to common land, the countryside is seen as belonging to everyone and such Travellers challenge the idea of land ownership and its concomitant laws of trespass.[5] Many Travellers also identify with nature religions and earth mysteries. Through these the countryside comes to be seen as a sacred and mysterious place, with a focus on ancient sacred sites, best understood through forms of rejected knowledge (Michell, 1982, 1986; Devereaux, 1990). 'Earth mysteries' generally refers to the study of ancient sites from a standpoint which is at the outset critical of modern science. Earth mysteries practitioners adopt a more holistic approach that refers back to ancient folk ways of understanding and interpreting the landscape: dowsing, ley line hunting, recovering folklore and customs associated with particular sites. The earth mysteries tradition challenges the modes of understanding offered by modern science and seeks to find in the landscape forgotten practices of knowing and understanding both natural and social. In its analysis of ancient landmarks and pathways, it treats ancient societies as technically accomplished and as having acquired knowledge about nature that has been lost to, or rejected and marginalised by, modern science. It is not the practices or epistemologies of earth mysteries researchers that have been adopted by New Age Travellers but they do share a similar utopic of authenticity that opposes the predominant ways of looking at the countryside as a site defined by agribusiness, commuters, tourists and scientists.

The contrast between urban and rural is also significant for the Travellers. Unlike many other youth cultures, they identify much more with the countryside than with the town. While it is true that a utopics of rural authenticity plays a significant part in shaping the identity of Travellers, this is not the picturesque authenticity of a Constable painting but an authenticity grounded in an identification with small scale communal solidarity, often expressed through the idea of tribes, through an identification with nomadism that is seen to be more authentic than the sociality of modern industrial societies. The countryside for most Travellers is not a rural arcadia expressed through images of pastoral peace, rather it is a place of mystery in which the sacred is reinvested in the landscape through a syncretist paganism and holism. Travel as a quest for meaning, community and self-discovery takes place in that landscape (Eliade, 1969; Turner, 1973). This utopic stands in contrast to the idea of authenticity that informs the representation of rural arcadia for many country dwellers, landowners and tourists. While they too may share aspects of the Travellers' romantic outlook on the countryside, the English representation of the rural is predominantly a pastoral one, based on culture having tamed and cultivated nature, turning it from a wilderness into a garden (Williams, 1985). In contrast, the utopics of Travellers wishes to turn the countryside, if not into a wilderness, into a place where wilderness can be found and nature restored to itself. Their authenticity is not the authenticity of the English country garden.

INVADERS, THIEVES AND GERMS: A GEOGRAPHY
OF DIRT

> As the nomads transgress all settled boundaries of 'home', they simultaneously map out the area which lies beyond cleanliness.
>
> *(Stallybrass and White, 1986: 129)*

Travellers have developed a utopics of the countryside based on an authenticity associated with mystery, festival, communion and freedom of access. For many local people the presence of Travellers in any place, especially if holding a festival, but even if only parked up in small numbers, usually creates uproar and horror, challenging their sense of rural authenticity. The three main ways in which the source of this horror is expressed is through concern with invasion, crime and dirt.

The source of anxiety and the strength of the (hostile) reaction by local people are due not to any physical danger, although they may be expressed as such, but are related to uncertainties surrounding the visible persona of the stranger that are identified, by default, as the source of the stranger's capacity to disrupt their familiar utopic of rural peace and quiet. These anxieties are expressed as a horror of pollution by dirt, disease and drug taking; an attempt is made to make visible unseen anxieties by defining those with a different lifestyle as a source of threat (see Young, 1990 for a similar study of the reaction to the women of Greenham Common). In Douglas's (1984) terms the authentic space of the rural home or village is polluted by the nomadism of the Travellers whose otherness transgresses the boundary of the familiar. It is notable that the dirty appearance of the Travellers, which is often a part of their style, is singled out and taken as a condition of moral disorder. The condition of the stranger has always been seen as bringing dirt. The space which they inhabit is viewed as having been contaminated, made filthy, obscene and contagious, as the following series of quotes from letters in a local Salisbury newspaper at times when Travellers have visited the area reveal:

> We lost customers who did not enjoy stepping over these dirty scruffy people sitting on the doorstep outside. . . . There was no organization then, and the condition they left the place in was appalling. And as for the hygiene – I was there, I watched these people. It makes one wonder what breed of beings these people are.
>
> *(Salisbury Journal, 10 June 1976)*

Those of us who live on the west side of Amesbury have had to endure an invasion of private property by this rabble who, for the past week, have taken over the River Avon, from Woodford to Countess Road, doing exactly as they please, despoiling the countryside and cutting down trees for firewood.

(Salisbury Journal, 1 July 1976)

Commenting on the festival held at Stonehenge,

I have a special interest in Stonehenge and deplore such disrespect to the nation and its monuments. Surely some bog or moor or other unwanted spot could be found for these alien lifestyles without costing the taxpayers and ratepayers.

(Salisbury Journal, 1 July 1976)

Why are we then invaded annually by these drop-outs, who come into our stores and handle fresh fruit and veg, and other unwrapped commodities, picking and choosing and leaving the regular customers very little choice.

(Salisbury Journal, 23 June 1983)

Similarly:

When you read of 30,000 trespassers camping illegally on farmland you might imagine the excretion, filth, plastic garbage and wrecked vehicles they have left behind them.

(Salisbury Journal, 12 June 1986)

Another commented on how the Travellers violated the locale by their presence:

This short lived event, despite nationwide warnings to hippies that no assembly would be tolerated, must be compared with what, in the past, can only be described as a two-week rape of the entire area by many thousands.

(Salisbury Journal, 29 June 1989)

The local MP, Robert Key, was reported in the local newspapers to be investigating whether the Travellers were a possible source of hepatitis and HIV (*Salisbury Journal*, 22 May 1986).

Collectively, these responses identify the Travellers as other. This Other is an invading, polluting stranger who, in Simmel's words, '[C]omes today and stays tomorrow' (1971: 143). Just as has been the case with Jews, Gypsies and vagrants down the centuries (see Biere, 1985; Mayall, 1988), the New Age Travellers are hated not because they are always on the move, but because they might stay and the remoteness they introduce into the familiarity of the near might contaminate it and bring down all manner of horrors upon the local people (see Bauman, 1990; Shields, 1992b). They are out of place not because they belong somewhere else but because they belong nowhere. They are not simply unplaceable, outside of time and space, they inhabit the disjuncture between experiences of place and the moral order by which it is represented; 'unclean', 'slimy' (Douglas, 1984), Travellers have a status that is as uncertain as their origins. This youth culture strikes directly at an opposing utopics of the countryside as something picturesque and peaceful (see Halfacree, 1996).

CONCLUSION

In this chapter I have tried to locate New Age Travellers within Marin's realm of the neutral and associated that neutral with the countryside. This is a space that in the political sense has been far from neutral. The anxieties that Travellers bring and the legislation that has been brought down on them in recent years have all been based on a contested spatiality that informs our understanding of nature and the countryside. While New Age Travellers and those who live in the countryside may share what Urry (1995) has described as a romantic gaze that emphasises the authentic as a source of moral order, that is the basis of an idea of home, their senses of what constitutes authenticity are quite different. For Travellers, the authenticity of the rural lies not in the picturesque and the pastoral as it does for most British people, but in an authenticity of nature as something mysterious and spiritual and of society as something expressive and communitarian that exists in harmony with this view of nature. The cultural politics that Travellers engage in is a distinctly spatial politics. It is one where ideas of freedom, nomadism, tribalism and harmony with nature are expressed through a utopics in which this outlook on society is translated into spatial practice. That spatial practice comes into conflict with the spatial practice of others, local people, farmers, landowners, local authorities, police and guardians of the countryside like the National Trust and English Heritage. Their utopics are challenged by those of Travellers and this opposition produces the countryside as a contested space. While it may be manifested in issues of trespass, rights of access, land use, rights of assembly, respect for property and so on, it is principally a contest of representation. The politics that Travellers engage in with their opponents is a politics of representation where one utopic comes up against another. Travellers have come into existence and developed in the space constituted by this politics of representation, they exist in a hiatus between the no-place and the good-place, in an ambivalent ou/eu-topic space in which contested representations of rural authenticity exist. The space of youth cultures, like that of New Age Travellers, is a space of différance as well as difference.

NOTES

1 Shields (1991) uses the term 'social spatialisation' to account for the social construction of spatial representations through spatial practice. Useful as this is to our overall understanding of how space is constituted through practice, it tends to generalise over the social construction of spaces and loses sight of some of the different practices involved. Utopies are a form of social spatialisation but are specifically concerned with issues of value; translating ideas about goodness and order into space. It is this specificity that concerns me here.

2 Such archaism is one of the inevitable consequences of studying youth culture; one works in a different social time to the groups amongst which one is carrying out research. My research was carried out, principally, through semi-structured interviews with key informants, primary documentary research and a small amount of participant observation at free festivals for my Ph.D., between April 1990 and September 1992 (see Hetherington, 1993).

3 Both 'counter-culture' and 'sub-culture' are problematic terms, ones that I generally prefer not to use. Counter- and sub- what? In both cases it implies a unified and hegemonic mainstream culture. Recent arguments both from within the study of youth cultures (Thornton, 1995) and in more general arguments about postmodernism would suggest that this so-called mainstream does not exist as a unity and that there is a symbiosis between the culture of consumption and commercialisation and that which has always opposed it (see Martin, 1981). Sub-cultures construct the mainstream just as much as they are constructed in relation to it.

4 I have preferred to use the German concept of a *Bund* (1992, 1994) to characterise the type of sociation often connected with the emotional communities of so-called neo-tribes. For me this term is both conceptually clearer and more limited in its scope than 'neo-tribe'; focusing on a type of elective and affectual *grouping* in which a strong identification and solidarity with others emerge. Neo-tribalism is rather too sweeping in the way it generalises, on the basis of a select few groups of people who play with identities and identifications with one another, the defining condition of the whole of Western society.

5 The Diggers were a group of egalitarians led by William Everard and Gerrard Winstanley and influenced by the Anabaptist movement in Europe. The Diggers, during the English Civil War, promoted the view that all property

should be held in common. In 1649 they took up camp on St George's Hill near Walton on Thames, a piece of land they did not own, and began to start cultivating it as if it were common land.

REFERENCES

Bauman, Z. (1990) 'Effacing the face: on the social management of moral proximity' *Theory, Culture and Society* 7, 1: 5–38.

—— (1992) *Intimations of Postmodernity*, London: Routledge.

Becker, H. (1946) *German Youth: Bond or Free?* London: Keegan, Paul, Trench, Trubner.

Biere, A. (1985) *Masterless Men: the Vagrancy Problem in England 1560–1640*, London: Methuen.

Chippindale, C. (1983) 'What future for stonehenge?', *Antiquity* 57: 172–80.

Clark, M. (1982) *The Politics of Pop Festivals*, London: Junction Books.

Cohen, E. (1973) 'Nomads from affluence: notes on the phenomenon of drifter tourism', *International Journal of Comparative Sociology* 14, 1–2: 89–103.

—— (1979) 'A phenomenology of tourist experiences', *Sociology* 13, 2: 179–201.

Derrida, J. (1976) *Of Grammatology*, Baltimore: Johns Hopkins University Press.

Devereaux, P. (1990) 'Stonehenge as an earth mystery', in C. Chippindale, P. Devereux, R. Jones and T. Sebastian (eds), *Who Owns Stonehenge?* London: Batsford, pp. 35–61.

Douglas, M. (1984) *Purity and Danger: An Analysis of the Origins of Pollution and Taboo*, London: Ark/Routledge.

Earle, F., Dearling, A., Whittle, H., Glasse, R. and Gubby (1994) *A Time to Travel? An Introduction to Britain's Newer Travellers*, Lyme Regis: Enabler Publication.

Eliade, M. (1969) *The Quest: History and Meaning in Religion*, Chicago: University of Chicago Press.

Foucault, M. (1986) 'Of other spaces', *Diacritics* 16, 1: 22–7.

Garrard, B. (1986a) *The Last Night of Rainbow Fields Village at Molesworth*, rev. edn, Glastonbury: Unique Publications.

—— (ed.) (1986b) *Greenlands Farm*, Glastonbury: Unique Publications.

—— Rainbow, Jo and McKay, A. (eds) (1986) *Rainbow Village On the Road*, Glastonbury: Unique Publications.

Genocchio, B. (1995) 'Discourse, discontinuity, difference: the question of "other" spaces', in S. Watson and K. Gibson (eds), *Postmodern Cities and Spaces*, Oxford: Basil Blackwell, pp. 35–46.

Halfacree, K. (1996) 'Out of place in the country: Travellers and the "rural idyll"', *Antipode* 28, 1: 42–72.

Hetherington, K. (1991) 'The geography of the Other: Stonehenge, Greenham and the politics of trust', *Lancaster Regionalism Group Working Paper* No. 41, Lancaster: Department of Sociology.

—— (1992) 'Stonehenge and its festival: spaces of consumption', in R. Shields (ed.) *Lifestyle Shopping: The Subject of Consumption*, London: Routledge, pp. 83–98.

—— (1993) 'The geography of the Other: lifestyle, performance and identity', unpublished Ph.D. thesis, Department of Sociology, Lancaster University.

—— (1994) 'The contemporary significance of Schmalenbach's concept of the Bund', *Sociological Review* 42, 1: 1–25.

—— (1996a) 'The utopics of Social Ordering: Stonehenge as a museum without walls', in S. Macdonald and G. Fyfe (eds) *Theorising Museums*, Oxford: Basil Blackwell, pp. 153–76.

(1996b) 'Identity formation, space and social centrality', *Theory, Culture and Society* 13, 4: 33–51.

—— (1997) *The Badlands of Modernity: Heterotopia and Social Ordering*, London: Routledge.

Lowe, R. and Shaw, W. (1993) *Travellers: Voices of the New Age Nomads*, London: Fourth Estate.

McKay, G. (1996) *Senseless Acts of Beauty*, London: Verso.

Maffesoli, M. (1988) 'Jeux de masques: postmodern tribalism', *Design Issues* 4, 1–2: 141–51.

—— (1996) *The Time of the Tribes*, London: Sage.

Marcus, J. (1988) 'The journey out to the centre: the cultural appropriation of Ayers Rock', in A. Rutherford (ed.), *Aboriginal Culture Today*, Kunapipi: Dangeroo Press.

Marin, L. (1984) *Utopics: Spatial Play*, London: Macmillan.

—— (1992) 'Frontiers of Utopia: past and present,' *Critical Inquiry* 19, 3: 397–420.

Martin, B. (1981) *Sociology of Contemporary Cultural Change*, Oxford: Basil Blackwell.

Mayall, D. (1988) *Gypsy-Travellers in Nineteenth Century Society*, Cambridge: Cambridge University Press.

Melucci, A. (1989) *Nomads of the Present*, London: Radius/Hutchinson.

Michell, J. (1982) *Megalithomania: Artists, Antiquarians and Archaeologists at the Old Stone Monuments*, London: Thames and Hudson.

—— (1986) *Stonehenge: Its History, Meaning, Festival, Unlawful Management, Police Riot '85 and Future Prospects*, London: Radical Traditionalist Papers.

Mills, R. (1973) *Young Outsiders*, London: Routledge and Kegan Paul.

More, T. (1985) *Utopia*, London: J.M. Dent [Everyman edn, 1910].

Rojek, C. (1988) 'The convoy of pollution', *Leisure Studies* 7: 21–31.

—— (1989) *Leisure for Leisure: Critical Essays*, Basingstoke: Macmillan.

Shields, R. (1991) *Places on the Margin: Alternative Geographies of Modernity*, London: Routledge.

—— (1992a) 'Individuals, consumption cultures and the fate of community', in R. Shields (ed.), *Lifestyle Shopping: The Subject of Consumption*, London: Routledge, pp. 99–113.

—— (1992b) 'A truant proximity: presence and absence in the space of modernity', *Environment and Planning D: Society and Space* 10, 2: 181–98.

Simmel, G. (1971) 'The stranger', in D. Levine (ed.), *On Individuality and Social Forms*, Chicago: University of Chicago Press, pp. 143–9.

Soja, E. (1995) 'Heterotopologies: a remembrance of other spaces in the Citadel-LA', in S. Watson and K. Gibson (eds) *Postmodern Cities and Spaces*, Oxford: Blackwell, pp. 13–34.

Stallybrass, P. and White, A. (1986) *The Politics and Poetics of Transgression*, London: Methuen.

Thornton, S. (1995) *Club Cultures*, Cambridge: Polity Press.

Thrasher, F. (1927) *The Gang*, Chicago: University of Chicago Press.

Turner, V. (1973) 'The center out there: pilgrim's goal', *History of Religions* 12, 3: 191–230.

Urry, J. (1995) *Consuming Places*, London: Routledge.

Williams, R. (1985) *The Country and the City*, London: Hogarth Press.

Young, A. (1990) *Femininity in Dissent*, London: Routledge.

....

20

MODERNISM AND RESISTANCE

how 'homeless' youth sub-cultures make a difference

•

Susan Ruddick

YOUTH/SPACE/MODERNISM

For many decades scholars have debated about the limits and possibilities of oppositional sub-cultures of youth and adolescents. Early works focused on the extent to which these sub-cultures constituted 'real' acts of resistance or the extent to which, conversely, they were simply symbolic acts serving to reproduce the very structures of inequality that they challenged. The conclusions of various analysts depended largely on whether they framed 'successful' resistance solely in term of class struggle and the manifestations of a working class consciousness, or whether they accepted acts of resistance by youth on their own terms – in their ability to 'win space for the young: cultural space in the neighborhood and institutions, real time for leisure and recreation, actual room on the street or street-corner' (Clarke *et al.*, 1976: 45).

Even within this more extended approach, the role of space has been treated as almost incidental by cultural theorists – a by-product of acts of resistance. More recently feminist scholars and those writing in the area of identity politics have begun to recognise the importance of spaces, and in particular safe space or third space – where marginality can be affirmed or even celebrated (Soja, 1996). But the crucial role of space in the production of sub cultures of resistance, the interdependency of the production of 'space' and 'self', have not been fully explored. Nor has the relative availability of such spaces in cities been considered. Yet if we look at cities in successive phases of modernisation it becomes clear that certain spaces and neighbourhoods by virtue of their transitional, indeterminate nature, their lack of fixity of meaning, become more fruitful ground for the production of new identities than others.

In a larger study on the emergence of a service network for homeless youth in California, I seek to understand the emergence of oppositional sub-cultures as embedded within and contributing to larger processes of place production: neither operating wholly 'within' specific places nor dominating the meaning of those places but as part of a complex contingent process in which the new identities of social groups and the meanings of specific neighbourhoods become stabilised together. In that study I demonstrate the ways in which these acts of opposition indeed 'make a difference', both in the extent to which they have won space and resources for marginalised youth on their own terms and in the extent to which they have contributed to changes in the location, philosophies and organisation of services intended to 'get them off the streets' (Ruddick, 1996.) This essay reflects on one aspect of that process: the emergence of a punk squatting sub-culture in Hollywood in the late 1970s and early 1980s.

I would argue these acts of resistance can be thought of as occurring within a larger crisis of modernisation, as described by Berman:

> To be modern is to experience personal and social life as a maelstrom, to find one's world in perpetual disintegration and renewal, trouble and anguish, ambiguity and contradiction: to be part of a universe in which all that is solid melts into air. To be modernist is to make oneself somehow at home in this maelstrom . . . to grasp and confront the world that modernization makes, and to strive to make it our own.
>
> *(Berman, 1982: 15)*

One scarcely needs to restate, in the 1990s, that we are living through a period of fundamental dissolutions, the breakdown of an old order and transition to a new one. What has interested me most about this process, however, is not the substantive form of the new order, be it new modernism, postmodernism, neo-fordism, post-fordism, or whatever else one might choose to call it, but rather the dynamics through which the new is constituted. How does meaning become fixed (even if only partially)? How do we leap from the 'raw material of process' to 'new universals'?

For youth this transition has involved the dissolution of one social imaginary, the juvenile delinquent, and its replacement with others, including homeless youth. Juvenile delinquency is a modern concept, and the array of institutions that 'treat' juvenile delinquency constitute the modern system of juvenile care. Since the mid-1970s, a number of new and dominant images of youth and youthful misbehaviour have emerged on the horizon. Two of the most immediately recognisable are the 'youth gang' and 'homeless youth'. During the modern period, the treatment of both was fused within the concept of juvenile delinquency, and addressed within an all encompassing (if internally differentiated) system of juvenile care. Since the mid-1970s, however, these two images have come to represent very different understandings of youthful mis-behaviour and treatment, which is part and parcel of a new modernism. What role, if any, have youth played in the emergence of this new modernism?

To address this question one must consider three premises that have become fashionable within critical theory, but are all too often agreed upon by fiat. The first premise suggests that *agency matters*. The object here is, fittingly, those people who are denied agency in all but the most banal forms: adolescents. Adolescents are generally considered too old to be ascribed the power of 'nature', yet too young to be reasoned actors in the sense one might consider adults to be.

The second premise is the belief that *new social subjects are created and create themselves in and through the social space of the city.* Here one must investigate whether and how the control of public space and resources plays a role in privileging one story over another, in creating and sustaining a social imaginary.

The third premise is that *symbolism embodies a material force* that is crucial in the creation and maintenance of social identity in and through space. Literature on oppositional youth sub-cultures says much about identity but little about space itself as a medium of sub-cultural and sym-bolic production. Space is a container where sub-culture 'takes place' as a by-product of acts of resistance. Literature on the homeless is more attentive to the role space plays in the constitution of identities, but it demonstrates the way that status is predetermined and confirmed by marginal space, never transcended. Here homeless youth presented an interesting challenge, sus-taining an image which negotiated between the Scylla of 'the homeless' and the Charybdis of youth as 'the dangerous classes'. Unlike other margin-alised youth (e.g. youth gangs) and other homeless, they have developed a social identity which confronted their stigma.

PUNK HOLLYWOOD

A culture includes the 'maps of meaning' which make things intelligible to its members. These 'maps of meaning' are not simply carried around in the head: they are objectified in the patterns of social organization and relationship through which the individual becomes a 'social individual'.

(*Clarke* et al., 1976: 10)

When runaway youth were de-institutionalised in the mid-1970s, they soon joined the ranks of 'the homeless', in the absence of alternative services. But they defined themselves first and foremost as *youth*. This struggle for self-definition was expressed in a tactical inhabitation of space resulting in a shifting regional concentration of services for homeless youth from places specifically for 'the homeless' (e.g. Skid Row), or appropriately suburban areas to Hollywood, where the youth themselves preferred to congregate. For a limited time in the late 1970s these youths enjoyed a strategic control of space within Hollywood, defining themselves as punk squatters. By the 1990s, however, the loss of alternative supports within a counter-culture and the gradual closing off and control of the social space of Hollywood by planners and providers of social services to homeless youth reduced the degree to which youth could sustain a self-defined social and public identity.

The de-institutionalisation of runaway youth in the mid-1970s was part of a massive expulsion of young people from the juvenile correctional system onto the streets of American cities, and in California, once on the streets, they turned primarily to Hollywood.

Although Hollywood has always had a seamy underside, since the mid-1960s this seamy side has expanded. With the liberalisation of obscenity laws and relaxed restrictions on what could be shown in movie houses, avant-garde theatres in Hollywood, popular in the 1950s art scene, switched to pornographic film. The burgeoning adult entertainment industry coincided and clashed with the growth of youth sub-cultures. It was in this volatile geography that juvenile prostitution first consolidated in the Hollywood area.

The picture emerging over the past three decades in Hollywood is one of increasingly restricted options for homeless and runaway youth. It was in the bleakest period, the mid-1970s, that the punk squatting scene emerged. The dynamics of its formation have much more to do with broad development in youth sub-cultures, than with the particularities of survival options for runaway youth in Hollywood. But in an atmosphere of rising exploitation of juveniles, and absence of services it represented a crucial alternative for survival on the streets.

Punk squatters, living in the area between 1976 and 1982, numbered well over two hundred at their peak. They resisted (sometimes with violence) any attempts to incorporate them into fledgling services for homeless youth that were being introduced into the area at the time. During this period, punk squatters defined the dominant spectacular youth sub-culture in the area. Like contemporary homeless youth punk squatters were often adolescents running from family abuse.

The larger study of the history of service provision to youth in the area is based on investigation of newspaper archives, archives of various

services in the area and extensive interviews with over 40 service providers, planners and public officials as well as accounts from 15 youths. Of
the youth I interviewed, one, Frank, was able to give detailed accounts of
the punk squatting scene. It was difficult to find other youth who had
participated in this sub-culture who were (1) still alive, (2) willing to talk
about it and (3) able to remember with any semblance of accuracy what
had occurred there. The several hours of interview material that Frank
gave me I have supplemented with a more general account provided in
Lee and 'Shreader''s contribution in *Hard Core California* (1983).

As Frank recalls:

'There was a whole bunch of us [the tail end of the 'first wave'] who were
obviously spoiled little upper middle class kids from families with too
much power, who decided they were going to go off and try something
else . . . .

  '[But] some kids were really running from being beaten or sexually
molested or having alcoholic parents, and then there were some who were
running because their parents were being too restrictive, because their
parents wanted them to dress this way or go to that school . . . . Actually,
Tony was a good case of a lot of these kids. His Dad had left and his Mom
was alcoholic and really abusive . . . . So I mean he'd go home sometimes,
and it wasn't really as much of a problem any more, 'cos he was big
enough that he could beat the crap out of her if she went on a bender with
him. But he generally didn't want to be there. But he'd go there sometimes
to eat. It's hard to sit there and feel bad for yourself when you're in that
kind of situation, and there's someone else who's got scars up and down
their back from their Mom burning them with a cigarette when she was
drunk.'

(Frank, 1990)

In spite of these similarities, punks did not see themselves as homeless
youth. As Frank (1990) noted early on in our interview:

'When you told me you were doing your project on homeless youth, I
thought of Covenant House, and these kids who come here from the mid
west . . . . I definitely thought 14 year old girls on the street. *But I never
thought me, 'cos we weren't "homeless youth".*'

The sustenance of a distinct identity by punks, their self identification
as *punks* rather than as runaways or homeless youth, was intimately
bound up with the perpetuation of their squatting sub-culture, itself
dependent on access to and control over particular material and symbolic spaces within the Hollywood area. Their sub-culture serves to

remind us that these youth are not solely victims, but, like other city dwellers, are also *creative subjects* in the environments within which they live.

Of the 3,900 runaways that were estimated to come through Hollywood each year in the late 1970s, punk squatters (be they stable or 'commuters') at the outside were not more than 400 strong. In the early years punks congregated in two large, abandoned residential estates located in the Hollywood Hills, known locally as Doheny Manor and Errol Flynn Manor. There were about 50–70 punks squatting in each of the two manors – Doheny and Errol Flynn – and 15 to 20 more in each of three or four condemned apartments around the Hollywood area. After a gig numbers hanging out could run up to two hundred (Lee and 'Shreader', 1983: 30).

There were three critical types of spaces supporting the punk sub-culture in Hollywood which, according to Frank's ironic inference, could be likened to the cultural spaces of the new middle class (Frank, 1990):

'In terms of the space you really had three types, you had where you were crashing essentially, you had where you were going to gigs, and you had where you were hanging out after the gigs. So it's easy to think of it in terms of – ooh – before you go to the show, you enter the "master bedroom" and dress appropriately and then you "sit in the theater and enjoy the show" and afterwards you'd go out, "perhaps promenade on the town and have an expresso".'

There were places for gigs (clubs and bars); places to crash (abandoned manors and condemned buildings); and places to hang out after a gig (marginal social spaces like Hollywood Cemetery, or certain accepted joints that were quasi-marginal, such as Danny's Oki Dogs and the Astro burger – which tolerated punk clientele but were also the target of frequent police raids). These were not simply used as marginal *spaces* but at marginal *times* as well – a significant contrast to the use of Hollywood time–space by homeless youth by the mid- to late 1980s. For the squatting punks, a typical day would begin at 2 or 2.30 in the afternoon. The day revolved around hanging out and getting ready for a gig – which typically began at 10 or 11 p.m. The gig was usually followed by a major social congregation – in Hollywood Cemetery, or Wattles Park – which might begin as late as 2 the next morning.

Clubs and bars were initially concentrated in Hollywood, but beginning in 1979 suburban locations began to dominate. This process of decentring the sub-culture outside of Hollywood – which in turn began to redefine the social space of the sub-culture in the centre – has curious parallels to other 'decentrings' affecting Hollywood's decline, such as the rise to dominance of the suburban shopping mall and the regionalised theme park. Moreover, it had very immediate implications for the squatting punks.

> 'The Hollywood area was really sort of a central hub for the scene. It's a lot like the relationship between downtown and Los Angeles. Downtown isn't of itself that interesting but it winds up being a place where people go anyway. Hollywood was the same way. There was a big concentration of clubs nearby. A lot of people would come in and eventually they would end up hanging out in Hollywood.'
>
> *(Frank, 1990)*

That the squatting punk sub culture was able to sustain itself for so long in Hollywood was partly due to the ready availability of condemned buildings for use and, initially, clubs to play in. But the erosion of the quality of marginal space and the transformation of the sub-culture went hand in hand.

Finding condemned apartment buildings was often preferable to staying in the manors, because the manors were illegal: these places weren't technically condemned, and if they were too obvious the squatters would draw the police. But in the early stages nobody cared if punks squatted a condemned building – they would hang out as long as possible until somebody came to knock the thing down. These places were trashed both as an expression of the sub-culture and also at times to make them (strategically) habitable:

> 'You never want one door to a room unless the wall is weak enough that you could easily kick it through. The best buildings were the ones that were so heavily decomposed that you could make a few holes in places with a hammer and you could pound your way through . . . . Posters and newspapers over all the windows, and usually the way to do it was you would break the window, and then you would poster or paper it over, so it would look like it was completely derelict, and no one was in it, but if someone would come in you could easily just dive right through it . . . . There was a lot of pure vandalism too, like "let's spray paint the walls", "let's burn part of the carpet" . . . but a lot of it is trying to accommodate a building to illegal occupation . . . we had carpenters, we had construction people in the punk scene, these people with backgrounds in that to rehabilitate buildings. The issue was: Why the hell do that? Why the hell fix it? You're only gonna be in it until they throw you out. So you had no investment in the place, except "What kind of fun can we have with it and how can we get out of it once they push us out?"'
>
> *(Frank, 1990)*

In fact, the punk squatters were a fairly stable group compared to the current population of homeless street youth, who in the words of one service provider, 'spend a fair amount of time looking each day for a place to stay' (Weaver, 1991). Contemporary homeless youth use many of the same strategies to 'ready' a building for squatting.

THE ELIMINATION OF MARGINAL SPACE AND DECLINE OF PUNK SQUATTING

'We're talking about the erosion of free space. You know, not just open space in the sense of, "oh yeah, nice parks". We're talking about space with any kind of latitude for independent action.'

(Frank, 1990)

The dissolution of this fairly self-sustained and self-contained squatting culture began in 1979, and was attributable to three interrelated factors. The first factor was a generational shift: as the sub-culture became popular it went through a series of transformations, different generations, which might be seen as the 'natural' stages of formation, dissolution and death. The second factor, emanating from the first was an aspect of this generational shift: a changing attitude towards the use and abuse of marginal space. Increased destruction of marginal space led to enclosure of key marginal spaces which denied squatting punks a valuable resource for their sub-culture. The third was the consequence of the loss of space. As the quality and availability of marginal space deteriorated, punks increasingly had to fight for use of marginal space on streets and in alleys, both to squat and to hang out, and formed gangs to protect themselves against other street sub-cultures.

The sub-culture went through three generations according to local lore – the first, more outspokenly political, the second oriented to the musical and symbolic aspects of the sub-culture and the third – dominated largely by lower class kids – more demonstratively violent (Lee and 'Shreader', 1983: 38). The 'first wave' of Hollywood punks, beginning in 1977, could be seen as part of an intellectual avant-garde – disenchanted youth of the upper and middle classes, who spent a lot of time patrolling the boundaries of the culture.

This first wave was concerned about the authenticity of those who adhered to the scene, which drew later criticisms of their 'cliquishness' (Lee and 'Shreader', 1983). But compared to the later concerns in the Hollywood scene, which focused increasingly on street survival, this cliquishness, and the energy spent patrolling sub-cultural boundaries, spoke to a considerable amount of time the first wave had at their disposal for careful recruiting to the sub-culture and life in the manors.

Frank was a member of the second wave: 'too late to be avant-garde, too early to be trendy'. As he described it, the second wave shared some of the political leanings of the first, and spent some time in the manors but also began to move into squats that they named, as the manors began to be closed off and patrolled by police. The second wave had some awareness of the anarchist underpinnings of the punk sub-culture, and were for the most part 'mentored in' by the first wave, but were generally more concerned with music and the style of the sub-culture than any larger political objective. They were not involved in publishing political tracts.

The third wave, beginning in 1979, tended to be lower class suburban youth, who were generally more bellicose and more influenced by the media image of the sub-culture than its early roots.

> 19/9: there was a separation between the South Bay working class kids and the post-glitter Hollywood punks. The Hollywood kids didn't stand a chance. The new breed of suburban Punk was physically tougher, angrier and more immediately REAL about their intention than the original party people.
>
> *(Lee and 'Shreader', 1983: 38)*

Frank described the shift this way:

> 'A lot of the first wave people were pretty educated – in a lot of cases upper middle class or even upper class people who'd rejected that for one reason or another. We did have skills. We had people who could translate – Spanish, Dutch, German . . . . The "first wavers" were really politically active – they used to publish these tracts . . . . This was one of the big tricks you did if you were politically active groups, and the first wavers were. A lot of them would be anarchist type tracts or Zero Work tracts, and then you'd sneak down to a, like, copy shop, and they don't really monitor those machines, so, when they weren't looking you'd reach over punch the authorise buttons, and then just go Xerox up all these copies and run out with a stack full of paper . . . . The original punks had a distinctly anarchist bent and in many cases a formal one – they were anarchists, and they understand the difference between Proudhon and Bakunin, and the Nazi punks generally came in after the movement had spread its way out to classes other than the . . . dilettante educated intelligentsia, and they didn't have a lot of interest in the niceties of communal organisation or whether throwing bombs is politically a good thing to do . . . they were more interested in the symbolism and

the formalism that was inherent in the Nazi movement and "oh wow, isn't it cool to goose step around and be part of this hierarchy". Whereas we had worn swastikas as such as a means of saying fuck you, they wore them 'cos they thought it was cool and when they came in they really forced a change in the scene. You'd see black kids or Chicano kids on the scene wearing an SS overcoat, and suddenly, when these guys came in who were taking it seriously, it was no fun anymore.'

The third wave were more demonstratively destructive of property. This feature was criticised by the first two generations of punks on the scene, because it resulted in the elimination of important marginal spaces for punks in Hollywood.

'Hollywood Cemetery. Really important, that was like our living room. That was where we would go to party. What killed it was they put up this barbed wire fence all around it, because apparently some people did go in and vandalise it. *We actually did discover some of it and tried to put it back to rights, just because we didn't want to lose the space* [Frank's emphasised intonation].

'There was a problem that the first and second wave of punks had as the third wave came in, 'cos the third wave tended to be more media oriented, and what they'd seen on TV, where we would kick the hell out of some-body because they gave us shit, they would just kick the hell out of somebody to kick the hell out of somebody, 'cos that's what the media said they were supposed to do . . . .

'This younger generation thought "yeah that's what it's about, you trash things and you beat people up". These places that we would go as places that we could hang out and not get hassled – they were part of us, they would come too – and inevitably they would start doing stuff like – "oh it's cool, let's kick over the grave stones, let's break into one of these mausoleums" – it must have been end of 80/81 that somebody really fucked up some of the grave sites, and then you got these Nazi punks coming in out of nowhere going into the Jewish section and kicking over the gravestones – and after that, that was the end of that space.'

(Frank, 1990)

In 1980, Hollywood Cemetery was barbed-wired off after it had been seriously vandalised. Punks then shifted their 'social centre' to Wattles Park, a place which did not have the same homological resonance as the cemetery. There is some suggestion, moreover, that the third generation did not depend as heavily on the marginal spaces in Hollywood for the reproduction of their sub-culture, or if they did they made different use of them. While a few of them squatted 'Skinhead Manor', around Argyle

Avenue north of Hollywood Boulevard, others lived in their vans (Frank, 1990). The map of marginal spaces used by punks in Hollywood raises some fascinating questions about the way sub-cultures both orient themselves to and draw upon space as part of the homological construction of the sub-culture: 'the symbolic fit between values and lifestyle of a group' (Hebdige, 1979: 113–15). First generation punks recognised and drew upon the potent symbolism of marginal spaces in Hollywood, and the importance of this symbolism for the sub-culture offers us as powerful a reason for the initial concentration of punks in Hollywood as a simple availability of space. These spaces were not simply the place where sub-culture 'takes place'. They were themselves, the object of bricolage. (On bricolage as it applies to oppositional sub-cultures, see Clarke, 1976: 177.) First generation punk sub-culture had distinctive signifying practices, concentrating on the 'act of transformation performed on an object – the act of subverting meaning rather than solely on what is meant'. (Hebdige, 1979: 124). The dissonance of punk sub-culture was expressed in the symbolism of the spaces they chose as much as their music or other practices. The manors and apartments often had a prior history as sites of spectacular crimes of violence – together with the Cemetery, which functioned as a living-room, these spaces echoed the dissonance of Hollywood's punk culture.

In the early years in Hollywood, punks were not usually chased out of condemned buildings (Doheny and Errol Flynn manors excepted) but only vacated them as they were knocked down. But they were chased from the public spaces outside clubs and bars where they hung around before and after the gigs. As one punk noted 'there was no consistent pressure to make us disappear, there was only pressure whenever we did appear'.

But in 1979, this changed. Police began to enter clubs to disrupt shows, beginning with a police induced riot in the Elks Lodge in 1979 (Lee and 'Shreader', 1983: 34). By the beginning of the 1980s, increased demolition of condemned buildings and policing of the manors forced many punks to spend more time literally 'on the street' and in parks and alleys when they couldn't find places to hang out or squat. Before this time, towards the very end of the 1970s, some loosely structured gangs had begun to form within the punk scene, but they tended to be focused on defining internal philosophical and stylistic differences. Somewhat more formalised gangs of the period included LA Death Squad, the largest, and its self ironic Doppelgänger LA Wimp Squad; and Mikes Mercenaries – which went on to form the band Suicidal Tendencies.

In 1980 a new gang formed, the Hollywood Street Survivors (HSS), which concentrated more on the practical matters of self-preservation; it included somewhere between 100 and 200 members. Its members did not engage in turf wars like more 'traditional' gangs, but operated more 'like mutual protection associations' (Frank, 1990). The transformation from squatting sub-culture to street protection league had two critical implications. Firstly, it forced members to focus on physical survival. Secondly, it severed their contact with 'commuters'. Even three years after the formation of Hollywood Street Survivors, and when punk music was definitely in decline, the protective benefits of punk street sub-culture remained. Lois Lee, founder of Children of the Night, an organisation to help adolescent prostitutes noted:

> 'Many of the kids hanging out here are not prostitutes. They are runaways or kids alone. They seek refuge among the punkers. The pimps don't like that. They can't sell kids with Mohawks, pink or green hair.'
>
> *(Lee, 1983)*

Nevertheless, the loss of 'prime' marginal space, in the form of squats, forced some changes in punk survival strategies, with a rise in panhandling and a tendency among punks to frequent Hollywood Boulevard towards the end of the third wave. The type of survival strategies that homeless youth have adopted today were generally shunned by punks, especially in the earlier years. Survival sex – that is the exchange of sex for food, money or shelter by young people who do not consider themselves as prostitutes – was practised by at least half of homeless youth in Hollywood by the early 1990s. For punks, however, survival sex and prostitution were definitely frowned upon.

> 'We weren't hustling for a living. In the manor houses, particularly, you didn't want to bring in an outsider, it was a dumb thing to do. You wouldn't want someone you didn't know. Prostitution was *really* frowned on – regardless of whether you're talking gay or straight prostitution. There were maybe people I knew, one or two people, who were doing that, but no, you'd live off whatever you could get your hands on. And maybe once, only once, I actually heard of someone even doing something like rolling a drunk to get some money. We just generally ignored money. It was more a matter of taking material goods that we needed as opposed to taking money to get material goods.'
>
> *(Frank, 1990)*

In these instances the heightened visibility and distinction of a 'punk style', which might arguably make shoplifters easier to identify, was in fact used as a mask. But to change the mask required access to private spaces such as manors or squats. A typical shoplift went like this:

'You'd go in with whatever haircut you had – you know "the purple Mohawk" – and you'd grab all the food, macaroni, cheese and noodles, . . . Top Raman is the best, they're easy to steal and easy to cook . . . and the first thing you'd do is go into one of the manor houses and you'd shave off your hair or you'd change the cut, and you'd bleach it and dye it a different color.'

(Frank, 1990)

Panhandling was not part of the early punk scene, in the mid- to late 1970s – at least in the form that is currently visible in Hollywood:

'Generally we'd panhandle each other, in front of a gig . . . it was "wherever you're getting it why don't you share it with me?" . . . Usually you wouldn't pay to get in, unless you could bum the money to do it. You'd look for punks who'd driven up to a gig in cars, or, the best, punks whose Mom had dropped them off . . . [laughter] . . . and try to hit them up for money just to get into a gig.'

(Frank, 1990)

First and second generation punks never panhandled and rarely frequented Hollywood Boulevard, a prime panhandling area for contemporary homeless youth: it expressed too readily the values of the dominant culture that they shunned. The anarchist underpinnings of punk sub-culture expressed themselves in the music scene; pretensions to stardom were frowned upon, everyone and no-one was a star. Desires for stardom represented a wish to become successful within and on the terms of the dominant culture.

In the late 1970s, the maintenance of the boundaries which defined punk squatters as an 'outsider culture' included shunning money, and pilfering goods directly and as needed rather than begging for handouts. The appearance of the store 'Poseurs', which sold punk paraphernalia, was a second generation phenomenon. The name 'Poseurs' was, itself, a play on the fact that only 'pretenders' to the punk culture would actually *buy* paraphernalia associated with it: 'real' punks made their own hair gel and dye, ripped their own T-shirts and so on.

By 1980 both the social and spatial orientation of the sub-culture began to shift. More punks frequented Hollywood Boulevard and Melrose and panhandled tourists for money, and they were more likely to accept the offerings of social services. Just as first and second generation punks frowned on panhandling, so they shunned services. As a 'second waver' Frank noted:

'We didn't care. We wanted nothing to do with services. We really didn't. I mean, any kind of outreach or outreach programme that was moved in our direction we would meet with absolute derision and tell people to get the fuck out . . . any kind of social service workers who would try to come in and take somebody back to their parents . . . there was one social service worker who got the shit beaten out of him for coming in and trying to do something like that . . . we didn't want anything to do with the whole system of, you know, smiling, smirking, good for you control. We wanted nothing to do with it. So it had better not come around looking for us.'

According to service providers in the area, it is definitely different now – at least for youth under 18. Outreach workers still talk about a certain initial resistance on behalf of street kids to accept services (Levins, 1991; Cruks, 1990; de Paul, 1990).

Whatever effect the loss of marginal space had on punk sub-culture as a whole, it is clear that it made it increasingly difficult for punks in the Hollywood area to sustain themselves as an independent group of squatters. Loss of space contributed firstly to the isolation of squatting punks in Hollywood from other punks in Los Angeles as the area became less and less of a node and punks from other areas stopped in effect 'migrating to Hollywood'. LA Death Squad was a regional 'gang' centred in Hollywood but it included commuter punks and valley punks, and numbered several hundreds. Its successor, Hollywood Street Survivors was definitely a local affair, restricted to Hollywood punks. As squatting life in Hollywood became increasingly difficult, it also became less attractive to punks from other areas. Parallel with this erosion of free space was an increased need to focus on strategies for more elemental survival – as exemplified in the development of the Hollywood Street Survivors – which changed the orientation of the punk street subculture substantially.

CONCLUSION

The devolution of spaces available to punks can be thought of within the larger process of loss of space to homeless and runaway youth in Hollywood, which transpired in three phases from the early 1970s on. In the first phase, youth had some limited access to and control over strategic spaces in Hollywood – a tacit support by counter-cultural elements which gave access to places to crash such as counter-cultural newspaper offices. This control over strategic space continued in a limited fashion with the dominance of punk in Hollywood. But tactical use of space began to predominate with the rise of manor houses as preferred squats. Then, beginning in the 1980s, the quality and availability of tactical space began to deteriorate as condemned buildings were demolished and punks were driven from open public spaces, and forced to compete with other groups for marginal spaces in back alleys and parks. This deterioration was accompanied by a shift in the social and public definition of these youths. They were no longer viewed primarily as members of a particular sub-culture, but first and foremost as runaway and homeless youth who participated in various sub-cultures. The loss of strategic space controlled by the youth themselves was countered to some extent by the rise in shelters and services which youth had access to. Increasingly it was the shelters and services and not the youth who began to define their public identity.

The crisis of the social imaginary 'runaway' occurred with the onset of de-institutionalisation. With the lack of prepared space for runaway youth, there was no fixed centre to organise their meaning, no single or dominant image around which various ideas about runaways could be ordered, nothing to 'stop the flow of difference'.

For the homeless, and homeless youth among them, the capacities to challenge the pre-given meaning of space are expressed in the tactical appropriation of space and the use of subversion. Subversion has little transformative potential when it represents the routine subterfuges of agents whose actions are *already inscribed and submerged within particular structures.* Here the homeless *subvert the meaning of structures that were not intended for them, where they would otherwise 'have no legitimate business',* except by virtue of their subversive activity (e.g. using movie theatres to sleep in).

Tactics, in the sense that de Certeau introduces them, are the first clue to understanding how new meanings might become fixed, or at the very least temporarily subverted, in and through the use of space (de Certeau, 1984). Tactics are, in this sense, the glowing coal that keeps a particular meaning alive until it can find a place to sustain light. If there is a common characteristic running through the sites that homeless youth in Los Angeles County occupied by tactical appropriation, it was their quality as sites of leisure, the beaches, Hollywood, Sunset Strip, or Disneyland for example. Their return to these sites in spite of attempts to relocate them elsewhere was the manifest insistence that they be considered first and foremost as youth, rather than as 'homeless' (to be served by the Skid Row missions of downtown Los Angeles) or as 'delinquents from normal families' (to be relocated in more appropriate suburban family environments). Similarly if we look at the larger process of polynucleation of services for the homeless throughout Los Angeles, this suggests that this new pattern of service locations (outside of Skid Row) was a response to the tactical inhabitation of space by the homeless. But is this enough? What does it matter to the meaning of homelessness if shelters and services are scattered through the city rather than concentrated in Skid Row? The concept of tactics alone cannot help us at this point. And if we do not move beyond it, it presents us with a danger far greater than that of simply describing how power is exerted over marginalised peoples. To speak exclusively of tactical forms of resistance is to risk normalising, even romanticising, the condition of the marginalised people, humanising the face of poverty in a way that demands no further action.

Here I think the concept of tactics bears further refinement in order to explain the various ways that space (and meaning) can be seized. The tactics described thus far are the *tactics of invisibility*. In one sense homeless youth 'stockpiled their winnings' through a tactics of invisibility, simply by returning repeatedly to Hollywood simply as 'youth' and holding out long enough to force services to move to Hollywood. But this was only a limited victory. The second kind of tactics might be called a *tactics of rupture*. Here, once again, the meaning of the space is manipulated or diverted, but with an important difference. The space is manipulated in such a way as not to simply *conceal* the identity of the marginalised but to seize or *affirm* that identity. In its most extreme form, this tactics of rupture manifests itself as 'Punk Hollywood' – extreme because in and through the control of a material and symbolic space within Hollywood, youth, who would otherwise be considered runaways or homeless, were able to create and sustain a different image of themselves. The concept of rupture is already well developed in literature on youth sub-cultures, but it overlooks the crucial role of space and control over space in affirming a new identity. Here we must draw on other concepts of space and self.

Punk Hollywood expresses a social space where the ordering of front and back region were clearly inverted. Punks made a tactical appropriation of this space, these back regions within Hollywood. But they did so in a way which changed the value of the space. These spaces did not confirm for them their own marginality, but became their privileged front region. Punk squatters thus claimed the 'essence' of marginality in social and spatial terms. Through the tactical inhabitation of particular spaces, punks created a rupture of meaning. Rupture is the process by which an order of equivalencies is reversed, the very attributes valued by 'society at large' are shunned by those who are denied those attributes. Rupture is the act of choosing the position of outlaw. Rupture does not so much attempt to 'magically resolve experience of contradictions' as represent the experience of contradiction itself (Hebdige, 1979: 121). Here, moreover, it is more than the simple material availability of space; the space must have a symbolic meaning that resonates, is homologous, with the particular identity of the group. The meaning of punk squatting sub-culture was destroyed as much by denying punks the potent symbolism of particular spaces, the sealing off of these 'spaces of rupture' – Hollywood Cemetery, Errol Flynn and Doheny manors and finally Oki Dogs – as by denying them any space at all within Hollywood. But even this was not enough. One cannot simply 'enter the space of difference', one has to make difference overflow the space it occupies, allow it to insinuate itself into other spaces.

The destruction of punk squatting space, and the decline and transformation of the sub-culture as it was pushed literally into the streets, precluded this possibility. But the tactical inhabitation of Hollywood by homeless youth ultimately forced a *suturing* of meaning, between metaphorical street and family, between a sacred and profane existence (Ruddick, 1996). The new understanding of homeless youth that emerged from this tension is not just a simple co-optation of oppositional sub-cultures, which would imply that the sub-culture was simply incorporated into a pre-existing understanding of homeless youth. Service providers were forced to change their understanding and mode of treatment of youth in this act of suturing the positive identities that the youth chose for themselves and the images they had of runaways in a new space within Hollywood.

Oppositional sub-cultures and stylised acts of resistance of youth and adolescents have often been thought of primarily in terms of their subjective expressive component – as representations of particular philosophies and lifestyles, symbolic solutions to particular problems which over the long term may or may not be resolved. Often, following the studies of high culture, the styles and symbols of sub-cultures (including artefacts of clothing and music) have been examined in a self-referential and self-contained history – each considered in terms of its reaction to or embellishment upon the messages of an earlier or sometimes opposing culture (e.g. Mods versus Rockers, Disco versus Punk and so on). To fully appreciate these activities as acts of resistance which contain within them a potential for societal transformation, we need to focus more upon their relationship to a wider social and spatial setting for youth and adolescents, including systems of juvenile care.

REFERENCES

Alleman, R. (1985) *Movie Lovers Guide to Hollywood*, New York: Harper and Row.

Berman, M. (1982) *All that is Solid Melts into Air: The Experience of Modernity*, New York: Penguin.

Carlson, G. (1990) Director, Angel's Flight, personal interview, 9 April.

Clarke, J. (1976) 'Style', in S. Hall and T. Jefferson (eds) *Resistance Through Rituals: Youth Subcultures in Post-war Britain*, London: Hutchinson.

——, Hall, S., Jefferson, T. and Roberts, B. (1976) 'Sub-cultures, cultures and class: a theoretical overview', in S. Hall and T. Jefferson (eds) *Resistance Through Rituals: Youth Subcultures in Post-war Britain*, London: Hutchinson.

Cruks, G. (1990) Director of Gay and Lesbian Center, West Hollywood, personal interview, 30 May.

de Certeau, M. (1984) *The Practice of Everyday Life*, trans. S.F. Randall, Los Angeles: University of California Press.

de Paul, M. (1990) Caseworker, Covenant House, personal interview, 16 April.

Frank (1990) Personal interview, 22 September.

Hebdige, D. (1979) *Sub-culture: The Meaning of Style*, London: Methuen.

Lee, C. and 'Shreader' (1983) 'Los Angeles', in P. Belsito and B. Davis (eds) *Hard Core California: A History of Punk and New Wave*, Berkeley: The Last Gasp of San Francisco.

Lee, Lois (1982) Personal interview by Mitchell Fink, *Herald Examiner*, August.

Levins, M. (1991) Caseworker, Outreach Assistant, Los Angeles Youth Network; Former Caseworker, Teen Canteen, personal interview, 10 April.

McRobbie, A. and Garber, J. (1976) 'Girls in sub-cultures: an exploration', in S. Hall and T. Jefferson (eds) *Resistance Through Rituals: Youth Subcultures in Post-war Britain*, London: Hutchinson.

Ruddick, S. (1996) *Young and Homeless in Hollywood: Mapping Social Identities*, London and New York: Routledge.

Soja, E. (1996) *Thirdspace. A Journey to Los Angeles and Other Real and Imagined Places*, Oxford: Blackwell.

Weaver, D. (1991) Director, Teen Canteen, personal interview, 15 April.

....

LIST OF CONTRIBUTORS

Shane Blackman is a Senior Lecturer in the Department of Applied Social Sciences, Canterbury Christ Church College, UK. Shane has formerly taught at the University of Surrey and at the Institute of Education, University of London. He has undertaken qualitative research on social and cultural processes, including schools, training, equal opportunities, workplaces, youth cultures, drugs, the police and currently youth homelessness. He is the author of *Youth: Positions and Opposition: Style, Sexuality and Schooling* (Avebury, 1995).

Sophie Bowlby is a Senior Lecturer in the Department of Geography at the University of Reading. Her research interests are in feminist analysis in geography with particular reference to shopping and more generally to the gendered social organisation of activities in urban spaces. Current research topics include the presentation of self in public space, the experiences of young Muslim women entering the labour market and the links between the home and work-based social networks of men and women.

Myrna Margulies Breitbart is a Professor of Geography and Urban Studies at Hampshire College in Amherst, Massachusetts, USA, where she has taught for nearly twenty years. Her teaching and research interests focus on the gender, race and class dimensions of urban development, environmental design and social struggle; the role of the built environment, environmental conflicts and participatory planning in social change; youth culture, environmental activism and the use of neighbourhood environments as resources for critical learning. She has collaborated with women's housing development organisations, public art agencies and urban youth programmes.

Ruth Butler is completing a Ph.D. thesis on social constructions of disability and their effects on disabled people's access to public space at the Department of Geography, Reading University, UK. Her current research interests include disabled people's social identities, disabled people's sexualities and disabled women's rights to have and raise children.

Deborah Chambers is a Reader in the Sociology of Communications and Media at Nottingham Trent University. She has research interests in feminist cultural politics and is currently working on a cultural study of the 'The Family Album'.

Luke Desforges is completing a Ph.D. on tourism, representation and identity at the Department of Geography, University College London, UK. He is a lecturer in human geography at the University of Wales, Lampeter.

Claire Dwyer is a Lecturer in the Department of Geography, University College London, UK. Her research interests are gender, ethnicity and multi-culturalism. She has completed her Ph.D. which focuses on questions of identity for young British Muslim women.

Kevin Hetherington is a Lecturer in Sociology in the Department of Sociology and Social Anthropology at Keele University, UK. His main research interest is in the sociology of space. He has recently co-edited *Consumption Matters* (with Stephen Edgell and Alan Warde; Blackwell, 1996). Forthcoming books include *The Badlands of Modernity: Heterotopia and Social Ordering* (Routledge, 1997), *Vanloads of Uproarious Humanity: Britain's New Age Travellers* (Cassell, 1998) and *Ideas of Difference: Social Space and the Labour of Division* (co-edited with Rolland Munro; Sage, 1997).

Cindi Katz teaches at the Graduate School of the City University of New York, where she is Chair of the Environmental Psychology Program. She is co-editor (with Janice Monk) of *Full Circles: Geographies of Women over the Life Course*, published by Routledge, and author of the forthcoming *Disintegrating Developments: Global Economic Restructuring and the Struggle over Social Reproduction*. Her work over the past two decades has focused on the geographies of children's everyday lives and the questions of social reproduction.

Heinz Hermann Kruger holds a chair in education at the University of Halle-Wittenberg in Germany and has published widely in the fields of educational theory, qualitative research methods, and youth history and ethnography. His current interests include the critical analysis of post modernism and childhood/youth in the east–west German comparison.

Marion Leonard is a doctoral student and part-time lecturer at the Institute of Popular Music, University of Liverpool, UK. She is author of '"Rebel Girl" You are the Queen of My World: feminism, "subculture" and grrrl power', in *Sexing the Groove*, edited by Sheila Whitely and Sean Hawkins (Routledge)

Sally Lloyd Evans is a Lecturer in the Department of Geography at the University of Reading, UK. She completed both her B.A and Ph.D. in the Department of Geography, Royal Holloway, University of London. Her main research interests are in work and employment, with particular reference to racialised gendering in the informal sector. Following the completion of her Ph.D. on 'Ethnicity, gender and the informal sector in Trinidad', she has continued to work in the Caribbean on informality, social capital and small business development. In the UK context, her main research topics include the experiences of young Muslim women in the labour market, the employment and work identites of Afro-Caribbean youth, and the importance of social capital and local networks in the labour market.

Tim Lucas is a Research Assistant at the University of East London, UK, working on the 'Finding The Way Home' project. He has recently completed a Ph.D. thesis at the University of Sheffield titled 'Youth gangs in Santa Cruz, California: constructing and contesting a moral panic'.

Sara McNamee is a postgraduate student/graduate teaching assistant in the Department of Sociology and Social Anthropology at the University of Hull, UK, where she teaches methodology and data analysis. Her thesis is a sociological analysis of the extent of ownership, and an examination of the use of computer and video games among children and young people aged 5–16. Her research interests include the sociology of childhood and youth; gender issues and leisure sociology.

Ben Malbon leads a predominantly nocturnal life in London and is, amongst other projects, completing his thesis on clubbing at the Department of Geography, University College London, UK.

Doreen Massey is Professor of Geography at the Open University, UK.

Her most recent books include *Space, Place and Gender* (Polity, 1994), *High-Tech Fantasies* (with P. Quintas and D. Wield) (Routledge, 1992), *Geographical Worlds* (edited with J. Allen) (Oxford University Press and Open University Press, 1995), *A Place in the World? Places, Cultures and Globalization* (edited with P. Jess) (Oxford University Press and Open University Press, 1995).

Robina Mohammad did a degree in Geography at the University of Reading as a mature student. She then went on to do the Society and Space M.Sc. at Bristol University and is now doing a Ph.D. at Kings College, London University on 'Spanish regionalism and the politics of water'. Her research interests are the construction of gendered and racialised identities and the politics of representation.

David Oswell is Director of the Ph.D. Programme at the Centre for Research into Innovation, Culture and Technology and lectures in Media and Communication Studies in the Department of Human Sciences, Brunel University, London, UK. He writes and researches in the area of youth and children's television, history and policy and co-ordinates the Children and Media Network, which is a UK based network of academics, regulators and those working in the media industries.

David Parker is a Lecturer in the Department of Cultural Studies, University of Birmingham, UK. His research interests are the British Chinese and overseas Chinese communities; social identities and the development of social theory of diaspora. Publications include *Through Different Eyes: The Cultural Identities of Young Chinese People in Britain* (Avebury, 1995).

Birgit Richard is an assistant at the University of Essen, art and design department, Germany and has published in the fields of art and contemporary youth cultures and media. Her special interests are interactive media especially current developments concerning the Internet and the question of the death in real and artificial realities.

Susan Ruddick is a Professor in the Geography Department at the University of Toronto, Canada. Her research focuses on the relationship between space and social identity in the development of policy for marginalised groups in global cities, including youth and minorities. Recent publications include: *Young and Homeless in Hollywood: Mapping Social Identities* (Routledge, 1996).

Tracey Skelton is a Lecturer in the Department of International Studies at Nottingham Trent University, UK, where she teaches cultural geography, gender and development and Caribbean studies. She is the co-editor (with Tim Allen) of *Culture and Global Change* (Routledge, 1998) and is an author in the Women and Geography Study Group collective which has written *Feminist Geographies: Explorations in Diversity and Difference* (Longman, 1997).

Fiona Smith is a Lecturer in Human Geography at the University of Dundee, UK. Her research interests focus on eastern Germany and the transformation affecting central and eastern Europe. She has recently completed a doctoral thesis on neighbourhood activism in eastern Germany during reunification and is currently engaged in work on women and space in central Europe.

Kevin Stenson is a Lecturer in Criminology at Buckinghamshire College, a College of Brunel University, UK. He obtained his Ph.D. from Brunel University and his publications include (with David Cowell) *The Politics of Crime Control* (Sage, 1991). His current interests include policing, crime prevention, youth and crime.

Gill Valentine is a Lecturer in the Department of Geography at the University of Sheffield, UK, where she teaches courses in Social Geography and Qualitative Methods. She is the co-editor (with David Bell) of *Mapping Desire: Geographies of Sexualities* (Routledge, 1995), co-author (with David Bell) of *Consuming Geographies: Cultural Politics of Food* (Routledge, 1997) and author of *Stranger Danger: Children, Parenting and the Production of Public Space* (Cassell, forthcoming).

Paul Watt is a Lecturer in Sociology at Buckinghamshire College, a College of Brunel University, UK. His current research interests include the socio-economic position of council tenants, housing and social inequality in London, and youth in urban settings.

ILLUSTRATORS

Linda Dawes is the Cartographer/Resource Manager in the Department of International Studies at Nottingham Trent University. Her interests include the broad area of cartographic representations, particularly mapping for tourists, which was the title of her M.Phil. research. Linda has had cartographic commissions from the Guinness World of Records and currently, Fens Tourism and South Holland District Council.

Alan Lodge is a photographer with a special interest in documenting the lives of travelling people and those who attend festivals. His work focuses on representing the varied and diverse lifestyles of people engaged in 'alternative' and youth sub-cultures. His photographs have been used in a variety of publications including the *Guardian*, the *Independent*, *i-D Magazine*, *Select*, *Sounds*, the *Radio Times* and *New Statesman and Society.*

Alan Tien provided the self-portrait lino-cut for the Part One title page. He writes: I was born in Saigon in 1978. I first moved to England in 1981 and I have lived in Nottingham ever since. I am 18 years old and am currently studying Art and Design at South Nottingham college. After I have finished this two year course I would like to go on to Higher Education and study for a degree in Interior Design. My hobbies include playing football and going out clubbing. I like House music, I also like bands like Shed Seven and Mansun.

Tracy Smith provided the self-portrait lino-cut for the Part Two title page. She is in her early 20s and is of Caribbean heritage. She is a student at the Art and Design department of South Nottingham College.

Christina Poxon provided the self-portrait lino-cut for the Part Three title page. She writes: I am 17 years old. I live at home with my mum, dad and sister. I work eight hours a week at a bakers near where I live. I am currently studying Art and Design at South Nottingham College. I am really enjoying the course, especially the fashion and textiles areas. After my two years at South Nottingham I am hoping to go on to higher education to study fashion and textile more thoroughly.

Nicholas Wright provided the self-portrait lino-cut for the Part Four title page. He writes: I'm 16 years old, I was born in Nottinghamshire and I live in Cotgrave. I have one younger brother, Jonathan. My dad is a window cleaner and my mum is a hairdresser. I went to the local primary and junior schools and then to Dayncourt Comprehensive. I am now studying at South Nottingham College in the Art and Design department and enjoy my work. My outside interests include playing the guitar, judo, playing on the computer, fishing, going to the movies and reading. I hope you like the work I've done for this book.

College Statement: South Nottingham College has a large and successful Art and Design Department and has been a major provider in this field since 1970. Courses range from general diagnostic at all levels together with specialist courses in a variety of disciplines. The students participating in this book are studying on a GNVQ Advanced programme.

INDEX

•

Page numbers in bold refer to illustrations. An *n* following a page reference refers to information in a note.

Weston, Simon 88, 89, 97
Wheelock, J. 199
White, A. 336
Wiking Jugend 298
Williams, Rachel 45
Willis, Paul 14–15, 208
Wilson, Amelia 113
women: and cultural
   transformation 139–40; *see also*
   femininity; feminism; girls/young
   women; lesbian youth; sexism
Women's Study Group, Centre for
   Contemporary Cultural Studies,
   University of Birmingham 16, 18
Wong, Faye 70
woodland 7–8, 8
Worden, Pamela 315
working-class youth 4, 305–6; and
   Ecstasy 168; and education 15,
   251; employment and
   unemployment 14, 15, 134, 251,

252; 'Kevs' 259–60; research on 10,
   11, 13–15, 24, 195–6;
   territorialism 253, 262
World Council of Churches 157

Yablonski, L. 12
Yip, David 69
Young, I.M. 85
'Young Team' (Glasgow gang) 12
young women *see* girls/young women
youth: defining 2–6, 3,5–6, 38, 4,5–6,
   127
Youth Law of the GDR 289, 290, 302*n*
Youth Works/Art Works project
   312–16, 323, 324
Yucatán, Mexico 121–2, 123, 124, 129
Yuen Fong Ling 79–80, 81

zines 25, 101, 103–11, 113–16; on the
   Internet (e-zines) 111–13